T0266277

Fuel and Combustion Systems Safety

Fuel and Combustion Systems Safety

What You Don't Know Can Kill You!

John R. Puskar
Puskar Publishing LLC

Published by John Wiley & Sons, Inc., Hoboken, New Jersey.
Published simultaneously in Canada.

***Library of Congress Cataloging-in-Publication Data*:**

Puskar, John R., 1958-
Fuel and combustion systems safety: what you don't know can kill you! / John R. Puskar. – First edition.
pages cm
Includes index.
ISBN 978-0-470-53360-4 (hardback)
1. Furnaces–Safety measures. 2. Boilers–Safety measures. 3. Furnaces–Combustion. 4. Combustion engineering. I. Title.
TJ320.P87 2014
621.402′5–dc23

2013020444

10 9 8 7 6 5 4 3 2 1

Contents

Foreword

During my tenure as a board member and chairman at the United States Chemical Safety Board (CSB) from 2002 to 2012, the CSB investigated several incidents involving releases and explosions of flammable gases. Earlier, while I was an executive at Honeywell, for several years I had responsibility for the operation of three natural-gas-fired boilers. Both of these professional experiences have reinforced to me how critically important it is to understand the hazards of flammable gases and to be certain that all appropriate safety precautions are taken whenever anyone is working with and around flammable gases.

Here are three examples of what can happen if such safeguards are not taken:

In Connecticut, a major gas-fired electric power generating facility was under construction and within a few weeks of starting up. To clean their newly installed gas inlet lines, workers blew natural gas through the lines at high pressure, venting the gas to the atmosphere. Unfortunately, the gas came in contact with an ignition source and exploded. Six workers were killed and the explosion caused catastrophic damage to the almost-completed facility. Following its investigation, the CSB recommended that natural gas be banned as a cleaning agent for utility piping.

In West Virginia, a propane explosion occurred at a convenience store when a 500-gallon propane tank was being taken out of service so that it could be replaced with a new tank. During the tank switchover, propane leaked from a valve on the older tank. Emergency crews were called but the convenience store was not evacuated. The propane seeped into the store, found an ignition source, and exploded. Two emergency responders and two propane technicians were killed, the convenience store was demolished, and workers in the convenience store were injured. All of this could have been avoided if the area had been evacuated when the propane leak was first detected.

In Garner, North Carolina, a new industrial water heater was being installed in a meat factory. It was to be heated with natural gas. A newly installed natural gas line was being purged to remove air from the line. The purging agent was natural gas and it was directed into an enclosed room in the factory. The concentration of natural gas built up in the room, contacted an ignition source, and exploded. Four workers died and dozens of other were injured in the collapse of sections of the factory. Upon completion of its investigation, the CSB recommended that when facilities are purging gas piping, the purged gas should be directed to a safe location outdoors.

Each of these terrible accidents could have been avoided by applying common sense prevention techniques: don't clean process lines with high pressure natural gas, use a safer alternative; know the appropriate safety precautions when you are working with propane tanks and evacuate to a safe distance at the first sign of a leak; don't vent a natural gas line into an enclosed space.

More generally, how do companies prevent catastrophic accidents that kill and injure employees, destroy facilities, and cause untold damage to finances and reputations? There are no simple answers but here are some practical suggestions: provide active leadership on process safety at all levels of the company, hire and educate the right people, understand and control your hazards, focus on both personal safety and process safety, develop and analyze process-safety metrics.

In short, companies need to understand the hazards in their operations and take the necessary steps to control those hazards.

If your industry's hazards include the use of combustible gases, John Puskar's book will point you in the correct direction. John has spent many years in the business of helping clients and companies understand and deal with flammable gases in their facilities. He has 30 years of experience dealing with combustion systems and he is a well-known and respected expert in the investigation of combustion incidents like the ones described above. We are very fortunate that he has taken the time to put his knowledge and expertise into writing. This book will be a very valuable asset for industry professionals.

In particular, John's book should be read by anyone involved in the natural gas business, whether as a producer or as a consumer. There is no room for complacency in dealing with flammable gases. It is especially important that company executives read some of the fifty case histories in the text. They need to understand what the hazards and risks are in dealing with this very important energy source. Quoting Carolynn Merrit, the late chair of the Chemical Safety Board: "If you think safety is expensive, try having an accident."

John Bresland

Shepherdstown, West Virginia
May 2013

Preface

I wrote this book with the intention of passing along what I have learned in 30 years of working in the field of industrial fuels and combustion equipment. I had never thought of myself as an author or an academic – I was always focused on just trying to be a good engineer. I am and have always been very passionate about being an engineer. I believe with my heart and soul that it is truly a profession where one person can change the entire world. I still tell this enthusiastically to young people today. I thank God that my mother encouraged me to follow in the footsteps of my brother in law Michael Novak who got his electrical engineering degree at Youngstown State University in the 1960's. I too would end up there after graduating from high school in Youngstown, Ohio in 1976 to earn my degree in Mechanical Engineering in 1981.

In 1985, at the tender age of 27, I started my own business as a one-man part-time operation working from my home doing mechanical engineering work for industrial clients. I did this while I pursued my MBA from the Weatherhead School of Management at Case Western Reserve University in Cleveland, Ohio. My first client was my previous employer SOHIO where I worked as a corporate Energy Systems Engineer doing boiler- and steam-related projects at refineries and chemical plants. You might know the company better as John D. Rockefellers place, Standard Oil of Ohio. What a fantastic place to start a career! I appreciate my early mentors William "Bill" Frink and Ralph Ender who shaped my early views of how to conduct oneself as a young engineer. I will never forget Bill interviewing me and asking how I picked YSU as a college. I was really taken back by the question because in my world, being the son of first generation immigrants with my father being disabled for my entire life, mom working full time, and me paying for my own school I had never actually known there was a choice.

I was excited about a career where the things that I worked with were on a surreal scale. The boilers and furnaces were multiple stories and the energy release was on a massive scale. It was awe inspiring to just be near such equipment let alone to actually be part of improving it. The dark side of the size and scale of this equipment was what I also witnessed when things went wrong. I was moved by the size and scale of disasters where I witnessed the aftermath. When I saw the pain and devastation of these disasters my way to personally change the world as an engineer was to focus my little business on creating systems to make sure these kinds of disasters would never happen again.

The company I founded and nurtured for over 28 years, CEC Combustion Safety, inspected and tested thousands of fuel trains at over 500 plant sites in dozens of countries. We trained tens of thousands of people in how to recognize fuels and combustion systems hazards and how to operate and maintain this equipment safely. This was done with a very unique and rigorous protocol that embodied applicable codes and lessons learned. This work has been and is today being done for the world's most widely known and successful companies, including U.S. Steel, Ford Motor Co., General Motors, ConAgra, Tyson Foods, Pfizer, Owens Corning, Alcoa and dozens of others throughout the world. Although I am no longer associated with the firm, (selling it in 2011 to Eclipse Inc.), Lach and Doug Perks, owners of Eclipse, a worldwide leader in combustion systems and burner technologies, are continuing to pursue the company's core safety message.

I am proud of what was accomplished from an idea and the difference the dedicated staff has made in the lives of many throughout the years. After witnessing deaths, horrible injuries, personal tragedies, lawsuits, plant closings, and the blanket destruction of lives, my former staff and I had dedicated ourselves to doing our part to prevent these things from ever happening again. This one desire, to help people with our knowledge and skills, drove us all every day. There's nothing else that can motivate the dozens of staff to have given up time with their families to be on the road 5 days a week for years in hot boiler rooms and industrial plants all over the world. I cannot name them all but my most sincere thanks go out to every CEC employee through the years. I know they got their motivation daily in the eyes of everyone they helped when they left every plant and gave operating staff peace of mind.

My boys, John and Zack once asked me, "Dad, so what is life all about? Is it just that we work every day and have a family and that's it?" I told them that at the end of the day you need to leave the world a better place than you found it. Find something to do that will be your legacy that improves society as a whole. Leave your mark. And so I say to you reading this book, if you can apply even some of the most basic knowledge you find here and prevent even one minor incident, you will have helped to make my life even more worth living. I hope that this book in some small way contributes to the greater good, and I thank God for the opportunity that He has given me to put it together.

I have dedicated this book to the thousands of people who have been injured or have lost their lives in tragic fuel or combustion equipment-related incidents. This book contains over 50 stories and "real-life incidents" to provide practical learning experiences. These incidents resulted in 46 deaths, hundreds of serious injuries, and billions of dollars in economic losses. It is not an attempt to scare you, nor is it an attempt to point fingers. The stories you will read are meant to help you understand that bad things can happen to anyone operating fuel systems and combustion equipment if the proper (PPE) people, policies, and equipment countermeasures are not in place.

The stories that are presented in this book are not extreme or rare cases. For every incident that results in injury or death, there are dozens of incidents you do not hear about. With this type of work, if just a few things do not go your way, you too can be part of another tragedy. Your best defense is knowledge: knowing about fuels, piping, combustion, controls, and risks. This book is just a start. As you read on, I hope you will see that ensuring fuel systems and combustion equipment safety is an ongoing process and journey, one that takes real time, effort and teamwork in order to be successful.

I owe a great debt to many friends and colleagues in industry and elsewhere who have supplied the information and ideas that are discussed in this book. I especially want to thank the people at Ford Motor Company and the United Auto Workers and the members of their fuel-fired combustion equipment safety program. When I look back on my life, being able to work with these fine people in the aftermath of an incident there I realize it changed me forever. Some of these people included Scott Bell, Danny Berry, Dwayne Atkins, Hank Budesky, Bob Konicki, Don Schmid, Chris Petersen, Dr. Greg Stone, and many others. We worked together for more than 10 years almost weekly to develop programs that I am confident are the best in the world.

Others who have contributed greatly to this book include Bryan Baesel of CEC Combustion Safety Inc., who worked with me proofreading manuscripts and helping to suggest information that would improve this book. Other who have been influential in my journey include Joel Amato State of Minnesota Boiler Chief, David Douin Executive Director National Board of Boiler and Pressure Vessel Inspectors, Dave Peterson, Loss Engineer Cincinnati Insurance, Al Sobol Combustion Engineer, Walt Luker former President of Matrix Risk Consultants, and Steve Mezsaros of Pfizer were all friends and colleagues that contributed greatly throughout my career. I also want to thank my friend Sean George whose life was changed by a personal tragedy related to this topic. Instead of leaving the world he spent his life advocating for things that would help avoid others from experiencing what he went through.

Other organizations that helped with the book and that continue to contribute daily to the body of work related to fuels and combustion equipment safety are the U.S. Chemical Safety Board, the National Fire Protection Association, the American Society of Mechanical Engineers, and the National Board of Boiler and Pressure Vessel Inspectors. I interfaced with Dan Tillema from the U.S. Chemical safety board on a number of occasions. He has helped me to locate several sources of information and has been unwavering in his support for anything that can improve safety. John Bresland former Chairman of the U.S. Chemical Safety Board provided a foreword for the book and has also been an inspiration. NFPA, ASME, and NBBI have responded enthusiastically for the many requests I have made for materials and extracts to be used in this book. They have also published many of my articles through the years and have accepted me as a speaker at dozens of conferences.

Richard Hoffmann, a metallurgist provided assistance in the materials area, and his contributions are appreciated. Ted Lemoff, recently retired from NFPA has been a longtime friend and contributor to everything about fuels and combustion safety. He assisted me in finalizing the book and made contributions in the fuel gas and purging areas, and with the artwork. Ted is truly a competent, caring and very special person I am lucky to be able to call a friend. The following members of the NFPA Technical Committee on Ovens and Furnaces reviewed and improved parts of the book related to their expertise; Franklin Switzer of S-Safe, Tom George of Tokio Marine Insurance, and Ted Jablkowski of Fives North American Combustion Inc. Also, Tom Witte contributed to the information on nitrogen for purging.

Thank you also to the following who provided photos and drawings that make sometimes difficult concepts easier to understand to many readers; Tony McErlean of Air Products; Jim Griffin of Emerson Process Management; Stephanie Taylor of Dwyer Instruments; Chad W. Bryan, Chief Boiler Inspector, Tennessee Department of Labor & Workforce Development; George Nichols Technical Illustrator, John Finn, FEBA Technologies; Robert Ratkus, DCP Midstream; and Paul Wehnert, Heath Consultants.

I hope that this book inspires you to truly change the culture at your organization and introduce these important topics as a means to enhance the overall safety. I hope you are not picking this up after a tragedy. If you are, then it is not too late to avoid another one. If your site or sites have not had issues, consider yourself lucky. Do not wait until it is too late, start now and start the journey to address People, Policies, and Equipment in this fuels and combustion systems world.

Last but certainly not least, my special thanks also go to the love of my life, Lisa Roe. I remember turning to her in 2009 and saying from the basement of our home, "Hey, today I am starting this book thing." That launched literally a couple thousand hours of work and her being patient with me and supportive through all the toil.

JOHN R. PUSKAR

1

What You Don't Know Can Kill You

How the Book Is Structured and What This Chapter Is About

Most chapters in this book begin with a small introduction and then a story that provides the flavor of an experience related to the topics in that chapter. These stories introduce real concepts that under certain conditions can lead to devastation.

The book also features a number of emphasis boxes that give key concepts special attention. In many cases the boxes describe seemingly obvious and simple, but nonetheless vital, issues. In this chapter we give you a flavor of some high-level concepts involved in the fuels and combustion systems safety world. We form a basis for an understanding of the more technical issues presented in later chapters.

There's something in this book for everyone: from operators of equipment and hands-on maintenance personnel to corporate risk managers and global safety directors. The book's perspective switches frequently and addresses issues of concern for all these groups. This chapter will mean more to corporate staff, managers, and those in charge of fuel and combustion equipment safety and risk management programs. It is my hope that many of you will read the chapter and realize the importance of this topic. This will give you the perspective to be supportive of, and wanting to implement, the concepts and strategies presented in subsequent chapters.

Real-Life Story 1: Innocent Lives Lost from a Hot Water Heater Explosion

A lot can be learned from reviewing a terrible disaster that left five children and one teacher dead at an elementary school in Spencer, Oklahoma, in 1982.[1] A hot water heater explosion changed the town, and many families, forever. Shedding light on the

Fuel and Combustion Systems Safety: What You Don't Know Can Kill You!, First Edition. John R. Puskar.

underlying causes behind the explosion at Star Spencer Elementary School provides an opportunity to review your own combustion equipment testing, repair, and preventive maintenance schedules and to reframe your thinking about which of your combustion systems may be important.

It was shortly after noon in a busy school cafeteria on an average day. Children were seated at tables enjoying lunch when their secure little world was torn from them. Suddenly, a concrete wall that separated the lunchroom from the kitchen blew in as an 80-gallon water heater exploded and launched itself skyward. The children seated nearest the wall were crushed and killed as concrete and steel were propelled from the epicenter of the blast. In all, seven people died and 36 were injured.

Tragic warning signs at Star Spencer screamed out loudly to those trained to listen. Sadly, most building managers and facility staff would never have heard them. For example, would you know that when people complain that the water is too hot in the sinks, it could be a sign that you are about to have an explosion in your building? What about that safety relief valve that keeps dripping, or the little gas leak? It was a combination of issues as subtle as these that contributed to the deaths and injuries that day.

The first employees arrived at the school at 7:00 A.M. The cafeteria workers noticed that the domestic hot water was much hotter than normal. The custodian was called, and the gas water heater was shut down to await the arrival of a maintenance technician. The technician's fix was to replace the gas valve and relight the water heater. He replaced the valve with the only valve immediately available, a used valve that had been sitting on the shelf in the maintenance shop. The technician returned within the hour and noted that the water heater seemed to be working normally.

The cafeteria workers soon noticed that the water temperature was again much too hot, and getting hotter. They placed another call for service, which tragically went unanswered. At 12:13 P.M., the explosion ripped through the school.

Investigators found that the hot water heater's burner would not shut off. The used replacement gas valve was defective. These have been the subject of safety recalls. It is recommended that you check the Consumer Products Safety Commission website (www.cpsc.gov) to see if you might have one of these at your facility or even at home.

Hot water heaters are also required to have safety relief valves, which are supposed to relieve water if an excess temperature or pressure condition occurs. The safety relief valve on the water heater was found to be altered and could not function as designed. Someone had cut off the temperature probe part of the device, rendering this part of the protection ineffective.

Because the water had no place to go, it continued to increase in temperature and pressure as the heater continued to fire and store the energy input. At some point the limits of the strength of the steel and its connections were reached and the tank tore itself apart. At failure, when the tank gave way, all the water inside expanded at 1600 times its volume in an instant. This created a pressure pulse that blew out walls and moved anything in its path, sending debris flying in all directions.

Had the proper procedures and inspections been put in place, this accident could have been prevented. If there was a preventive maintenance schedule in this case, it was ineffective. The safety relief valve was installed incorrectly, the high-temperature-limit shutoff did not function, and the gas valve was defective. These problems were all avoidable.

Oklahoma's boiler inspection law at the time covered high-pressure steam boilers but not smaller equipment such as water heaters. This situation was not unique to Oklahoma. Most states do not provide much in the way of inspections for certain classes of combustion equipment, even in educational facilities or places of public assembly. The school system, then, had to determine what would constitute adequate inspection, maintenance, and repair of the water heater. This is an often-overlooked responsibility. Most facility managers do not understand that they are responsible for proper inspections and maintenance when it comes to fuel systems and combustion equipment. They often think that an insurance or state boiler inspector is providing some overall safety evaluation, but generally this is not the case. In most states, boiler inspection laws call for inspecting only the pressure vessel part of each boiler system, and not for looking at combustion issues.

Boiler inspectors (who are often hired by insurance companies or are employees of the state) have their hands tied when it comes to what they can ask someone to do. They have little enforcement authority. What they are inspecting for is often limited to exactly what is called for by the letter of the law. For example, in many cases they can only evaluate equipment for code compliance based on when it was installed even though important code changes occur almost annually because of advances in knowledge and experience.

Typically, there is no screening for how far away the "grandfathered" technology is from the most recent codes. This type of inspection sometimes means that archaic and antiquated equipment that has little in the way of modern safety features could "technically" be in compliance.

Lessons Learned You as a safety, risk, or maintenance professional and your facilities staff are responsible for all fuel lines, boilers, and hot water heaters, regardless of size and fuel source (even electric). You are responsible even if they are inspected annually by an outside entity and even if there are no local or state requirements to do so. Your guide must be that of standards and codes that exist relative to this equipment. This book gives you a chance to discover these requirements and apply them to your facilities and equipment before an incident occurs and lawyers tell you that you should have been fulfilling these requirements. Ignorance is not a defense.

Equipment does not have to be massive in size or Btu capacity to cause death and destruction. I began the chapter with this story to bring the reality of this type of "small" often overlooked equipment directly into view. It is my hope that at some point someone will go down to their basement and change out a safety relief valve or locate a problem on a hot water heater that will save a life.

The energy that can be stored in water is incredible. Check out the popular television show *Mythbusters's* video of blowing up a water heater[2] to see what can happen. This video shows a water heater launching through a simulated house roof hundreds of feet into the air when the tank ruptures. You must, on a regular basis, review these often ignored pieces of equipment and replace the safety relief valves. This is an inexpensive task that is simple to do. If you review the manufacturers' instructions for most small relief valves, you will see that they call for regular inspections. In most cases this inspection is supposed to include their removal to look inside. Rather, given their cost, you might as well replace them. Be careful lift-testing these; I have often found that they do not reseat effectively. Again, replacement on a schedule is your best option, whether or not they need to be serviced. Also make sure that the burners on hot water heaters are evaluated periodically. Their flames need to be reviewed to see that they are burning cleanly, and drafts need to be checked. It is also important to ensure that over-temperature shutoff safety devices are functional.

1.1 KNOWLEDGE GAPS IN OPERATING FUEL SYSTEMS AND COMBUSTION EQUIPMENT

Even though humankind has used fire for millions of years, a continuing pattern of accidents, deaths, and injuries tells us that there is still a long way to go when it comes to fuel systems and combustion equipment safety. You just read a story about how a simple hot water heater in a school cost six lives because of seemingly simple issues. And this equipment was not nearly as large or complex as that at most industrial sites.

My 30 years in the business has shown me that many incidents happen because personnel at industrial sites are often not trained adequately in the safe startup and shutdown of combustion equipment, daily operations, or proper testing of safety devices and maintenance of critical systems. Very few formal classes exist regarding burners, gas piping, safety systems, or fuel/air ratio controls. The fuel and combustion equipment industry itself is very fragmented. Although there are numerous technical books and articles on the subject and codes and standards provide safety information, much of the knowledge in use seems to be tribal, passed on from person to person from word of mouth among those who operate and maintain combustion equipment. Much of this information is wrong and misinterpreted.

To be sure, most companies attempt to maintain their equipment. However, the maintenance is usually done by someone with little formal training, and equipment is often far removed from its optimum configuration after years of being cobbled together so as to "just run." Over the years, while inspecting thousands of pieces of equipment, I've experienced my share of horror stories—and not just the kind that end in explosions. I've witnessed alarms ignored and safety devices turned off. I've even seen wooden sticks and cardboard shoved into relays and safety devices to keep the safety interlocks from shutting equipment down—a potentially deadly fix.

There seem to be two worlds out there: one where large organizations have enough of this equipment to have very good staff members and practices and another where there are just a few pieces of this equipment and no one with much knowledge of it. The more common situation is that fuel systems and combustion equipment are some small ancillary part of the operation and no one is really fully trained or completely understands the equipment or the hazards. Many of the tragedies related to fuels and combustion equipment throughout history could have been prevented with the right PPE. In this case I'm not referring to personal protective equipment but to *people, policies, and equipment*. In my opinion, every fuel or combustion equipment incident

TABLE 1.1 Fires That Caused $10 Million or More in Property Damage, 2000–2009

			Direct Property Damage (Millions)	
Year	Number of Fires	Fires More Than $10 Million Damage	As Reported	In 2000 Dollars
2000	31	31	$1814	$1814
2001[a]	19	15	762	702
2002	25	22	562	509
2003	21	17	2623	2417
2004	16	9	337	242
2005	16	6	217	101
2006	16	13	380	305
2007	45	33	3393	2709
2008	35	23	2372	1794
2009	24	17	940	693

Source: Stephen G. Badger, *Large-Loss Fires in the United States—2009*, Fire Analysis and Research Division, Copyright © 2010, National Fire Protection Association; reprinted with permission.
[a]Excluding the 9/11/01 World Trade Center attack from the loss totals but not from the fire incident totals.

begins as a people, policy, or equipment issue. My goal is to provide you with knowledge of these PPE tools that will dramatically reduce the risks of an incident.

From my contacts in industries that use fuels and combustion equipment, I have found that most owners of fuel-fired equipment do not understand the obligations that they have for the safety of their workers and plants. Many professionals responsible for facilities with fuel-fired equipment work within a culture of ignorance, misunderstandings, or denial about the impact of an explosion or fire caused by the operation of this equipment. I am aware of numerous disasters from gas piping or equipment that failed or had not been installed correctly. Many of the real-life stories presented in this book are from my personal experience and witness.

According to the National Fire Protection Association (NFPA), large-loss fires between 2000 and 2009[3] (the most recent statistics available at the time of publication) made for losses that averaged $1.1 billion per year during this 10 year period (Table 1.1). Most of these incidents occurred in manufacturing and industrial settings.

Furthermore, NFPA reports that U.S. fire departments responded to an estimated annual average of 10,500 structural fires in industrial and manufacturing properties in 2005–2009. These fires caused annual averages of 11 civilian deaths, 254 civilian fire injuries, and $726 million in direct property damage. These types of losses have been experienced in the United States for many years, often resulting from fuel systems and combustion equipment issues that could easily have been prevented. These losses don't count the hundreds of other significant business interruptions, facilities damage, lawsuits, fines, litigation, and lost-market-share issues that have also occurred. Smaller but more frequent production outages also cost millions in business interruptions, supply chain delays, lost orders, and decreased competitiveness. These losses are often deemed to be culturally accepted as a general cost of doing business. It doesn't have to be this way.

Fuel systems and combustion equipment safety is critical to the daily operation of many facilities and their employees. Unfortunately, many companies act only when a very large and tragic event occurs. Many companies believe that explosions, fires, or outages from fuel-fired equipment only happen to others—that their company is immune. Only loss of life seems to make the 11 o'clock news. Headlines soon fade or rarely get the follow-up attention required to highlight the pitfalls of equipment that has been poorly maintained and operated. Today's corporate public relations departments are also very good at shutting down the flow of information that may leak to the media. My experience has been that little "poofs," "pops," bulging furnace walls, and "pregnant boilers" are more prevalent than not and imply that incident headlines are only the tip of the iceberg. For each incident reported there are undoubtedly many left unreported because they did not result in death, injury, or significant loss of production. Hence, they are never clearly researched and the lessons are not adequately learned.

1.2 MANAGING FUEL SYSTEMS AND COMBUSTION EQUIPMENT RISKS

OK, so by now you accept that there is risk and possibly danger associated with fuel systems and combustion equipment. If you wanted to better understand how that applies specifically to your facilities and equipment, how would you know what *safe* is? In the fuel and combustion systems world, *safe* is not necessarily a destination but a journey. The state of the art is constantly evolving in the codes and standards world. These are documents and sources of information that are out there to help you better understand what level of safety your organization is at and the type of journey you should be considering. Many proactive corporations manage their fuel and combustion system risks successfully by creating programs that address people, policy, and equipment issues. They spend millions of dollars to develop, implement, and update ongoing programs.

The journey will also depend somewhat on your organization's culture. *Culture* can be described as the subconscious knowledge that is embedded in people without them even knowing. Culture is slowly absorbed through habits and regular reinforcement. It becomes the emotional guiding light in the presence of confusion, stress, and dangerous situations. It's what's there when subconscious thought must take over because things are happening too fast. Combustion equipment culture is no different.

In some organizations the culture is such that there are willing and eager minds open to learning. Information is shared openly and there is little fear of retribution for making honest mistakes. These can be characterized as *rapid acceptance cultures*. In other cases, where someone will be harmed if even a slight mistake is perceived, acceptance and new thinking will be difficult. The following story demonstrates how culture might influence your experience in managing these risks.

Real-Life Story 2: The Bulge in the Boiler Firebox

I walked into the boiler room, took one look down the side of the boiler, and said to the boiler operator on duty: "Excuse me, can you tell me anything about what happened here in the past to make that bulge?" When looking down the side of what used to be a flat surface, it was clear that the side of the boiler bulged outward. Obviously, sometime in the past there was an explosion. It's not clear that anyone reported it. It's also not clear that there was damage beyond the outer boiler skin being deformed, but then how would anyone know? Remember, sometimes staff, even those who might be managing the boiler house facilities, might not want to report something like this and attract a lot of scrutiny. The culture may be that it's not a good career move for anyone to do so. When you review facilities, make sure to ask about anything you see, even if it's old or very obvious, as it may be the first time that it has really been brought to anyone's attention and been acted on.

The operator responded: "Oh, that, that was a small poof we had. It just happened once at light-off but it went away." It was clear that this investigation now needed to focus on a few key issues, such as low-fire gas flow control valve positioning for light-off, malfunctioning gas pressure regulators, the possibility of leaking valves, faulty purge timing, and incorrect fuel/air ratios.

The investigation was begun using a combustion analyzer to check the flue products. This identified an excess of 1500 ppm carbon monoxide (CO) at a low fire level. The maximum CO expected for a well-tuned burner of this type for this application should have been less than 100 ppm. It got worse as the boiler moved off low fire and pegged the meter at over 2000 ppm CO. Significant levels of CO are an indicator of poor fuel/air mixing and incomplete combustion. No one could remember reviewing or adjusting the fuel/air ratio since the boiler was installed and commissioned over 10 years ago.

Many combustion analyzers can only read CO to about 2000 ppm. The generally accepted lower flammable limit of carbon monoxide is 12,500 ppm,[4] or 12.5%. It would be rare to have an explosion due to accumulating this much CO. However, because CO can be read by an analyzer, it serves as a surrogate indicator of other additional combustibles present, including unburned fuels. Partially burned fuels represent things that come from the cracking of some of the hydrocarbon molecules when they are partially burned. The elevated temperatures and mixtures of fuels and fuel derivatives can have a wide range of ignition temperatures. It's these other combinations of things, not the CO, that usually end up being what ignites and makes for explosions.

Lessons Learned Cultures must be such that personnel are not afraid to report issues. An unreported minor issue can easily degrade quickly and cost someone his or her life. Always question any damage you see to a firebox or fuel train. Never assume that since it looks old, everyone is aware of what happened and that the actual cause has been identified and the problem abated properly. Best-practice organizations conduct daily flame observations and logging of findings. They also conduct fuel/air ratio adjustments (burner tuning) at least annually.

Incomplete combustion can make for flammable mixtures in the firebox and flue passages of combustion equipment. If flue gas oxygen metering and reporting are part of the equipment instrumentation, be aware of the limitations of meters being used, exactly what they measure, and all of the implications of the meter readings. Train the staff and drill them about what levels of oxygen should be a cause for concern and what actions should be taken.

1.3 THE CREATION OF FUEL SYSTEMS AND COMBUSTION EQUIPMENT CODES AND STANDARDS

Because of numerous tragedies and the continuing human desire to use heat processes to advance society, many smart people have gotten together over the years to learn from past mistakes and to create definitions and examples of what safe fuel and combustion systems should look like. Some of the most important of these documents are called codes and standards. Today's many codes and standards organizations consist primarily of volunteer experts working under well-defined protocols to assemble informative documents that include much hard-learned wisdom. It's your challenge to read and incorporate the knowledge and experiences contained in the applicable standards and codes and apply this information to enhance the level of fuel and combustion equipment safety within your organization.

To learn why these documents and organizations came into existence let's look at the history of combustion equipment in industry. During the nineteenth century, boilers and steam engines became the heart and soul of the industrial revolution. At the same time, accidents related to boilers and pressure vessels became commonplace. From 1870 to 1910 there were more than 10,000 recorded boiler explosions in North America[5] (an average of 250 per year). By 1901, the rate had climbed to between 1300 and 1400 recorded boiler explosions per year. When these incidents occurred, they were often horrific and involved many people. There were public outcries for remedial action. It soon became clear that this technology needed to be made safer if it was to proliferate. The American Society of Mechanical Engineers (ASME), answered the call with groups of volunteer mechanical engineers coming together to create the first boiler code committee in 1911.[6] The first ASME boiler code was published in 1914–1915. The code documents produced by ASME identified safe practices for the construction of boilers and pressure vessels and for pressure piping systems. These documents provided specifications for steels required, their thicknesses, welding practices, and many other fabrication and installation issues that enhance safety. ASME codes, standards, and more information about the group may be found at www.asme.org.

Once these documents were developed, industry experts realized that there needed to be another group that actually enforced the rules. This group would need to consist of paid professionals who could be on the job every day acting as code enforcers or inspectors. They would need to visit fabrication shops, review welds, and measure and verify thicknesses of pipes when boilers were being fabricated. Identifying this need gave birth to the National Board of Boiler and Pressure Vessel Inspectors (NBBI) in 1919.[7] Headquartered in Columbus, Ohio, this group was created to promote greater safety to life and property through uniformity in the construction, installation, repair, maintenance, and inspection of pressure equipment, most of it boilers.

The National Board membership oversees adherence to laws, rules, and regulations relating to boilers and pressure vessels. The National Board members are chief boiler inspectors, representing most states and all provinces in North America, as well as many major cities in the United States. More information about the NBBI is available at www.nationalboard.org.

NBBI functions include the following:

- Promoting safety and educating the public and government officials on the need for manufacturing, maintenance, and repair standards.
- Offering comprehensive training programs in the form of continuing education for both inspectors and pressure equipment professionals.
- Enabling a qualified inspection process by commissioning inspectors through a comprehensive examination administered by the National Board.
- Setting worldwide industry standards for pressure relief devices and other appurtenances through operation of an international pressure relief testing laboratory.
- Providing a repository of manufacturers' data reports through a registration process.
- Accrediting qualified repair and alteration companies, in-service authorized inspection agencies, and owner–user inspection organizations.
- Investigating pressure equipment accidents and issues involving code compliance.
- Developing installation, inspection, repair, and alteration standards (the National Board Inspection Code).

As additional emphasis was put on having safe standards for the use of fuels such as natural gas, which led to a group called the National Fire Protection Association (NFPA) being created.[8] The organization was formed by a group of sprinkler manufacturers, installers, insurance, and enforcement officials, who developed the first code for the installation of fire sprinklers, issued in 1896. NFPA's mission is to reduce the worldwide burden of fire and other hazards on the quality of life by providing and advocating consensus codes and standards, research, training, and education. NFPA publishes more than 300 codes and standards. Among them are codes and standards regarding the safe design and installation of fuel train controls and combustion systems. These include specific codes and standards for boilers and for devices other than boilers. The NFPA's website is www.nfpa.org.

Before describing some of the codes and standards developed by these organizations and others, it's important that you understand the difference between a standard and a code. A *standard* is prepared and presented by a recognized national organization that collaborates on technical issues and identifies state-of-the-art best practices for safety. A *code*, on the other hand, is intended to be adopted as a law. Standards usually say how to do something for safety, whereas codes require when and where to do something for safety.

Each code and standard is managed by a combination of staff and dedicated volunteer committees with members from among end users, insurance companies, manufacturers, testing laboratories, special experts, and trade associations. These groups usually meet several times a year and are responsible for maintaining, updating, and eventually gaining consensus for the final published standard. Codes and standards are typically updated regularly, usually every three to five years. Some are reaffirmed where the technology has not changed. Codes and standards are often adopted into law by federal government departments, states, provinces, and other jurisdictions to become legally enforceable.

1.3.1 How Codes and Standards Are Structured

There are several parts to a code or standard. Much of what is described here is from the perspective of NFPA documents. These documents normally include a table of contents, definitions, the body of the document, and appendixes (often called an annex). In most cases, only the body of the code is enforceable—not the appendices, which contain explanatory materials. When NFPA codes are revised, there are vertical markings on the sides of pages to indicate what has changed. Asterisks indicate that there is appendix material.

Remember, codes and standards are typically not prepared only by scientists and testing agencies. Consensus code and standard developers have rules that ensure adequate representation and balance of the participants in the process: people on committees from a wide variety of backgrounds and perspectives. Recommendations for revisions can be submitted by anyone (even the general public) and are to be considered, debated, and voted on. Most standards developers have forms and guidelines available on their websites to assist those wanting to submit proposals for revisions.

Following the letter of the code does not guarantee safety in all cases. Each document has many pages covering the requirements for safe design, installation, operations, and maintenance of the respective equipment, but sound engineering judgment in applying this information is still required. It must also be remembered that these codes and standards are minimum requirements. Best-practice organizations understand this and often try to do more.

In Sections 1.4 and 1.5 we provide brief overviews of the most applicable codes for common combustion equipment, highlighting NFPA and ASME codes. (All NFPA codes can be purchased at www.nfpa.org and ASME codes are available at www.asme.org.)

1.3.2 Applying Codes and Standards

In most cases, codes are not retroactive. Most are meant to be applied when the equipment is installed. Most plants have equipment that is "grandfathered" in: that is, it is exempt from certain current requirements as long as the equipment was installed in compliance with the codes and standards in effect at the time of installation, continues to be used for the same purposes, and is not changed significantly. Almost no one has equipment for very long that meets all current codes. Some proactive organizations conduct gap analyses to learn where they do not meet current codes and then do something about it. Keeping up with codes and standards after equipment is installed is recommended. It allows an organization to remain current with newly discovered risks and changes to technologies that can make for reduced risks.

When codes and standards are changed, or new code and standards documents are added, it is usually for very good reasons. An example of a code change is the requirement in NFPA 86, the Standard for Ovens and Furnaces, that there be two automatic pilot valves in series. In many older ovens and furnaces, only one such valve exists. The code committee learned after a number of incidents that it's safer to have two of these to prevent gas leakage when a system is in a closed or off state. These two valves in series minimize the chances of fuel leaking past a defective valve and accumulating in the combustion chamber, thus posing an explosion risk. Adding another valve in series to minimize this risk usually costs less than $500. Even though the equipment may be grandfathered in with one valve, why would you not want to add this additional protection when considering the low cost and reduced risk?

1.4 FUEL SYSTEM CODES AND STANDARDS

NFPA publishes codes and standards in many areas of fire protection. The following are publications are related directly to fuels.

- *NFPA 54, the National Fuel Gas Code* NFPA 54 is a safety code that applies to the installation of fuel gas piping systems, fuel gas appliances, and related accessories. It covers pipe materials, pipe joining methods, pressure testing, purging, and certain other gas piping installation issues. Don't be fooled by the word *appliances* here if you are an industrial user. This code is somewhat general, and elements of it can be applied to many systems. It covers both natural gas piping systems from the point of the utility's delivery to the appliance shutoff valve on individual appliances and to some propane piping systems from the final-stage pressure regulator to the appliance shutoff valve. In the case of most utility connections for natural gas, the point of service to a customer starts with the discharge flange of the natural gas meter or at the service shutoff valve where a meter is not installed.
- *NFPA 31, the Standard for the Installation of Oil-Burning Equipment* NFPA 31 applies to the installation of stationary liquid fuel–burning appliances. It also covers the storage and supply piping for liquid fuels. As many appliances (i.e., boilers and furnaces) can use liquid fuel as well as other fuels, the language for this standard is similar to that of NFPA 54, although NFPA 31 deals specifically with fuel oil.
- *NFPA 58, the Liquefied Petroleum Gas Code* NFPA 58 covers the storage and use of liquefied petroleum (LP) gases. The LP gases included in NFPA 58 include propane and butane. Propane is a gas at normal temperature and pressure (72°F and atmospheric pressure) but is often compressed to be liquefied for ease

of storage and distribution. NFPA 58 applies to the storage of liquid LP's, their piping, and use in a facility of liquid and vapor at over 20 psig. Because the fuel is often transported by truck and rail, the U.S. Department of Transportation (DOT) regulations are also to be consulted for truck unloading operations at a receiving facility.

1.5 COMBUSTION EQUIPMENT CODES AND STANDARDS

In this section we discuss popular combustion equipment–related codes and standards. The ones discussed below are published by ASME and NFPA. ASME codes cover primarily boilers, while NFPA codes cover many other types of combustion equipment as well as boiler combustion systems.

Most codes and standards related to combustion equipment call for the testing of safety devices, training, and the existence of startup and shutdown procedures. A lot of ASME codes and standards deal with such subjects as the type of materials to be used, their thickness, and information on installations and repairs.

ASME Boiler and Pressure Vessel Codes

ASME distributes over 600 codes and standards all over the world. The Boiler and Pressure Vessel Code originated in 1914. It has been adopted in whole or in part by all 50 states, many municipalities, and in all Canadian provinces. The Boiler and Pressure Vessel Code is organized in 12 sections:

Section I: Power boilers
Section II: Materials
Section III: Rules for the construction of nuclear power plant components
Section IV: Heating boilers
Section V: Nondestructive examination
Section VI: Recommended rules for the care and operation of boilers
Section VII: Recommended guidelines for the care of power boilers
Section VIII: Pressure vessels
Section IX: Welding and brazing qualifications
Section X: Fiber-reinforced plastic pressure vessels
Section XI: Rules for in-service inspection of nuclear power plant components
Section XII: Rules for the construction and continued service of transport tanks

ASMECSD-1: Boilers Up to 12.5 Million Btu/h

ASME CSD-1 (CSD, Controls and Safety Devices) is a code that applies to boilers that have a fuel input rating of less than 12.5 million Btu/h. This code is applied and enforced in at least 26 states and some major municipalities. It is unique in that it is the only code that actually covers the fire or combustion equipment safety side of smaller boilers. In most cases where it is applicable, jurisdictional inspectors will ask to see evidence that annual testing of safety interlocks is taking place correctly. This code's jurisdiction includes requirements for the assembly, installation, maintenance, and operation of controls and safety devices on boilers operated automatically and fired directly with gas, oil, gas–oil, or electricity.

NFPA 85, the Boiler and Combustion Systems Hazards Code

NFPA 85 applies to single- and multiple-burner boilers, waste heat or heat recovery steam generators (HRSGs), stokers, and atmospheric fluidized-bed boilers with a fuel input rating of greater than 12.5 million Btu/h. It also applies to unfired steam generators used to recover heat from combustion turbines. NFPA 85 covers fabrication issues, operation and maintenance procedures, combustion and draft control equipment, safety interlocks, alarms, trips, and related controls that are essential to safe equipment operation.

NFPA 86, the Standard for Ovens and Furnaces

NFPA 86 includes extensive information about categories of ovens and furnaces, their installation, design issues, required safety devices, testing of safety devices, and issues related to training of operators and maintenance staff. There is also information about the safe operation of the various types of ovens and furnaces and any other heated enclosure used for processing of materials and it's related equipment.

NFPA 87, the Recommended Practice for Fluid Heaters

NFPA 87 is a 2011 edition document that identifies safety issues related to fluid heaters. These fluids would include heat transfer fluids but not petrochemical process-related fluids. This document excludes certain petrochemical process heaters and refers the reader to American Petroleum Institute materials.

NFPA 70, the National Electrical Code

The National Electrical Code is the bible of the electrical installation industry. It is a very comprehensive document and covers all issues related to residential, commercial, and industrial wiring and electrical device installation. This code is not specific to combustion equipment. It covers electrical panels, devices, switches, conduits, grounding, arc flash, and other installation issues that interface with combustion equipment.

1.6 OTHER WIDELY RECOGNIZED CODE- AND STANDARDS-RELATED ORGANIZATIONS

There are a number of other organizations that play an important role in fuel systems and combustion equipment safety. Those most common in North America are described in this section. In Chapter 11 we discuss those most popular in Europe and the developing world.

Factory Mutual (www.fmglobal.com)

Factory Mutual (FM) is an insurer well known in the industry for its very high standards, well-trained staff, and extensive risk management guidelines for just about every issue and type of occupancy or process in existence. Many of Factory Mutual's data sheets and other materials are available on the Web. FM's loss prevention data sheets are well written, based on a great deal of experience, and full of very practical information.

FM also has a testing laboratory that "approves" components and devices. Their Approval Guide is widely distributed and used throughout the industry. FM's approval logo is a key to look for when codes require that a component in a fuel train must be approved by a nationally recognized testing agency for the service for which it is intended.

Underwriters' Laboratories (www.UL.com)

Underwriters' Laboratories (UL) is a nationally recognized testing agency that has been in existence for over 50 years. One of the most respected names in the history of electrical and fire protection safety, UL reviews and tests hundreds of components to certify them for safe use and applicability.

When working with fuel trains, devices, or equipment that is "listed" and "approved," be very careful when making modifications other than direct replacement of components. If you change something that is listed and labeled and there's a problem later, the insurer may have a reason to fight a claim. Make sure that your insurer is involved in approving changes made to fuel trains or control systems and that you receive written authorizations from them before proceeding.

1.6.1 Other Standards Developers and Related Industry Organizations

In addition to the agencies listed above, there are other organizations that publish or enforce codes and standards for the safe operation of combustion devices. You may see tags on equipment that reference these groups:

1. American Gas Association (AGA), www.aga.org.
2. CSA International (Canadian Standards Association), www.csa-international .org.
3. Industrial Risk Insurers (IRI); no longer exists, purchased by GE Gaps.
4. American National Standards Institute (ANSI), www.ansi.org. ANSI is the "umbrella" organization for all American national standards and is the U.S. representative to the International Standards Organization (ISO). ANSI does not develop standards, but oversees the creation, promulgation, and use of thousands of American national codes and standards.
5. Technical Safety Standards Authority (TSSA), www.tssa.org. TSSA is a not-for-profit, self-funded delegated administrative authority that administers and enforces public safety laws for pressure vessels, fuels, and other areas in the

province of Ontario, Canada. TSSA, established under Ontario's Technical Standards and Safety Act, does not develop standards but provides guidance on complying with the government standards established in Ontario, Canada.

6. American Petroleum Institute (API), www.api.org. The API is the only national trade association that represents all aspects of America's oil and natural gas industry. The API has extensive standards available related to refining and petrochemical process equipment and combustion systems.
7. PHMSA (Pipelines and Hazardous Materials Administration) is the section of the U.S. Department of Transportation that deals with issues related to hazardous pipelines, www.phmsa.dot.gov. DOT is a cabinet-level branch of the U.S. government that was established by an act of Congress in October 1966. It ensures a fast, safe, efficient, accessible, and convenient transportation system that meets the nation's interests. DOT regulates air, rail, pipeline, and road transportation in the United States.
8. International Society of Automation (ISA), www.isa.org. Founded in 1945, ISA is a leading global nonprofit organization that is setting the standard for automation by helping over 30,000 worldwide members and other professionals solve difficult technical problems while enhancing their leadership and personal career capabilities. Based in Research Triangle Park, North Carolina, ISA develops standards, certifies industry professionals, provides education and training, publishes books and technical articles, and hosts conferences and exhibitions for automation professionals. Many ISA standards address combustion equipment-related issues.

1.7 SAFETY INSTRUMENTED SYSTEMS AND SAFETY INTEGRITY LEVELS

In addition to complying with the most current fuel and combustion system codes and standards, risks can also be minimized by applying safety instrumented systems and safety integrity levels. Safety instrumented systems (SISs) and safety integrity levels (SILs) are becoming ever more important methods of describing the overall safety level and risk associated with combustion systems. These terms are used regularly in the process combustion equipment industry. There seem to be two different worlds when it comes to fuel systems and combustion equipment safety. One world consists of those involved with industrial manufacturing and large commercial and institutional venues; the other is the entire process and petrochemical refining world. The interesting thing is that these two worlds don't seem to talk much and don't seem to share many requirements.

The process and refining worlds have valuable information, including many standards and papers that have been created and maintained by organizations such as ISA, API, and the American Institute of Chemical Engineers (AIChE; www.AIChE.org). The chemical process and petrochemical industries use an approach to safety that differs from that of much of the commercial and manufacturing industrial combustion industry.

In the process industry, many systems, including combustion systems, are designed according to the SIS/SIF/SIL approach. A great explanation of this topic exists at the General Monitors website.[9] An SIS is a combination of inputs (sensors) with a logic solver (hardwire or computer-based, PLC device), outputs, and final elements (e.g., valve actuators).

In the process and petrochemical-fired equipment world, small dedicated computers called programmable logic controllers (PLCs) have replaced many hardwired relay-based logic solvers. PLCs, in combination with sensors, instruments, and components such as control valves, can make up an SIS. Each SIS may have a number of SIFs, safety instrumented functions. Every SIF within a SIS will have a SIL, a measure of a SIS's probability of failure on demand. SIL 1 is the highest level (most risk of failure on demand). The scale generally runs from 1 to 4, with 4 being the highest achievable level of risk reduction, the lowest probability of failure on demand. As stated at the website:

> If something needs to be a SIL 4 system it is likely very complex and costly and probably not usually economical to implement. For the process industries, if a process has so much risk that a SIL 4 system is required to bring it to a safe state, there is a fundamental problem in the process design that needs to be addressed by a process change or other method.

The hierarchy is this: SIS is the overall system; it may contain several SIFs; each SIF has an SIL. The SIL describes the SIF's probability of failure.

> It is a very common misconception that individual products or components have SIL ratings. SIL levels apply to the entire instrument safety function and safety instrumented system (SIF and SIS). The logic solvers, sensors, and final elements are only defined as suitable for use in a specific SIL environment, but only the end user can ensure that the safety system is implemented correctly. The equipment or system must be used in the manner for which it was intended to successfully obtain the desired risk reduction level. Just buying SIL 2 or SIL 3 suitable components does not ensure a SIL 2 or SIL 3 system in service.

Although ratings can be identified, depending on where the boundaries of the system are defined, ratings do not usually take into account an actual "in service" probability of failure on demand that takes into account human factors such as the maintenance of final elements or the possibility that someone can leave a safety system in a bypassed mode.[10]

The identification of risk tolerance is subjective and site-specific. The owner or operator must determine the acceptable level of risk to personnel and capital assets based on company philosophy, insurance requirements, budgets, and a variety of other factors. A risk level that one owner determines is tolerable may be unacceptable to another owner. When determining whether a SIL 1, SIL 2, or SIL 3 system is needed, the first step may be to conduct a process hazard analysis to determine the functional safety need and identify the tolerable risk level.

Mike Scott and Bud Adler of AESolutions posted an article[11] on reviewing SIL levels for boiler burner management systems in light of NFPA 85 requirements. In their evaluation of the requirements of both prescriptive and performance approaches, they indicated that the NFPA 85 requirements are very prescriptive;

that is, specific types of components and approaches were required. The SIS approach is more performance based. Note that NFPA 85 allows alternative approaches to safety with the approval of local authorities.

A SIS approach means that concepts are given and the designer of a performance-based system is freer to address the specific hazards and situations than in the prescriptive approach. The performance-based approach requires that process hazards be evaluated and risk analysis be performed. The risk analysis effort should consider many factors, including hazards associated with the process, the sequence of events that could lead to a hazardous event, human factors, opportunities to mitigate risks through design and layers of protection, and the organization's overall risk tolerance level. Following the risk analysis, a SIL level can be assigned. The ANSI/ISA 84.00.01-2004 standard provides extensive information on this approach.

1.8 THE WORLD OF INSURANCE AND COMBUSTION EQUIPMENT

Every day, whether or not we think about it consciously, we all use a kind of "risk radar" to evaluate life's challenges. This means that we need to accept that something is a risk and that we have to plan for how much of this risk we may want to try to control and how much of it we may want to transfer to an outside vendor or an insurer. For example, tremendous risks exist when we drive a car. Not only do we have to guide thousands of pounds of stamped steel and molded plastic from point A to point B without incident but we must make sure that we stay out of the way of others. There are many driving risks that we choose to control and manage on our own, such as driving at the speed limit, wearing seat belts, and perhaps taking a defensive driving course. However, since so many other things about driving, like the weather, are not under our direct control, most of us transfer some of the unknown risks to an insurance company.

Fuels and combustion system risks can be thought of and managed in the same way. For example, an owner or plant may transfer some liability to a boiler insurer, which will try to manage some of the risk with boiler inspectors and property risk engineers. The remaining facility risks might be managed by an organization's internal engineering, maintenance, safety, and training staffs through specifications for equipment, maintenance programs, and training to address the culture related to operating fuel systems and combustion equipment.

Before we take on this topic more fully, let's cover a few basic definitions related to this world. There are many different types of insurance, four of which might be related to fuel systems and combustion equipment coverage:

1. *Property:* protects structures and equipment and is usually related to fires. This would also typically cover fire-side explosions in a boiler firebox.
2. *Boiler and machinery:* also often called *breakdown insurance*. In this case it offers coverage, usually, for issues related to the water side of the boiler (i.e., the pressure vessel itself). This would typically not be for failed maintenance issues such as poor water treatment. This would not typically cover fire-side explosions related to fuels in a boiler firebox.

3. *Business interruption:* covers the loss of profits if an incident were to occur and the facility were to be out of service.
4. *Casualty:* can have many meanings but usually covers disability and/or accidents to people.

The four types of insurance identified above represent four different types of risk that organizations may want to manage. Managing these risks means accepting some of them and taking steps to mitigate them or to transfer them to someone else in exchange for payments. Let's look at how the management and transfer of this risk works on an enterprise level. As a first step, many large organizations have to decide how much risk to self-insure or accept and how much risk they want to pay to transfer. They can also transfer some to a "captive," their own insurance entity set up to fund some part of a possible loss.

The insurance industry is a unique world. It has differing amounts of capacity each year. Unlike a normal purchase such as buying a car, where the consumer wants a product and picks from several offerings, when you purchase insurance, you have to put together a package of what you want in terms of boiler insurance, property insurance, casualty insurance, business interruption insurance, and so on. This package is then presented to a number of insurance organizations to see if they have the financial capacity for the risk and want to assume it. Financial capacity depends on such things as how many disasters occurred during a year for the insurance industry and the specific potential provider.

If an insurer wants to provide a price quote, it passes the information on to one of its underwriters. The underwriter works at an insurance company and has the job of reviewing statistics and using formulas to come up with a price and with qualifying conditions. There are detailed databases on each potential customer or related industry, with historical data on losses and previous coverage providers, which they use for reference. The underwriter considers the limits of liability sought, deductibles, and loss histories and presents an offer for your consideration.

The process can be complex; that's why most large companies work through a broker. A company seeking insurance coverage can have a dozen different entities somehow involved in providing the overall levels of coverage. For example, one entity might handle the first $100 million in losses, another the next $300 million, and so on, a process called *layering*. In some cases the insurers sell off some of these risks to reinsurers, which would reinsure all or part of a package that a particular insurer might have just signed up to provide.

Once a company decides on an amount for property insurance and a vendor, it may engage a risk engineering service. This could be to help mitigate self-insured risks or it could be required by the underwriter as part of the coverage. These types of services may be furnished by the property insurance provider or an outside organization such as AXA Matrix Risk Consultants[12] or Factory Mutual.[13] One of the qualifying conditions for the coverage may be that there are loss prevention inspections at the customer's highest-risk and or most valuable sites. These are usually identified by the maximum foreseeable loss value at a site. *Maximum foreseeable loss value* (MFLV) is an insurance term that quantifies the overall impact of a site being completely destroyed. It is unlikely that every site will be visited for inspections, so MFLV provides a means to prioritize sites.

When it comes to focus on fuel systems and combustion equipment, most industrial organizations experience periodic visits by a property insurance inspector for at least their major facilities. These reviews are often completed within a single annual visit and can be rather broad in scope, covering a range of insurance concerns, such as fire, wind, earthquake, flood, fleet, and business interruption. Considering the property loss inspectors' short duration at a site and broad scope, these visits cannot be relied on to verify that fuel systems and combustion equipment are properly designed, maintained, and operated. In fact, it is not their intent to give you general overall approval of the way you operate, or maintain what you have. Instead, it is at best usually an overview of some key risk reduction points related to the equipment.

When industrial facilities have some type of boiler or pressure vessel, these may require periodic inspections by a state or provincial certified boiler or pressure vessel inspector. For these objects (called *jurisdictional objects* by the inspectors), the intent of the state and provincial inspection is often limited to the pressure vessel portion of the object. Considering the specific focus of the inspection (i.e. pressure vessels and not fuel and combustion systems) the state and provincial certified boiler and machinery inspectors cannot be relied on to verify that the combustion equipment is properly designed, maintained, and operated. Even in those jurisdictions where the inspector is required to inspect the fuel train, more often than not, the inspection will consist of verifying that the installation meets originally installed code requirements from what can readily be seen without touching, opening, or testing anything. In most cases the inspector will not be verifying that the equipment is properly designed and maintained. In the end, in almost all cases, state, provincial, and local codes will stipulate that the overall responsibility for the safe operation and maintenance of the equipment always lies with its owner or user.

Another way that insurers try to control risk is by requiring customers to submit plans for new or substantially modified combustion equipment for review and approval before they are installed or retrofit work starts. Always remember to make sure to ask whether your insurer needs this information if you are involved with a new project or a retrofit. You don't want to create situations in which there is reason to fight or deny a claim. This could happen if an unapproved system later has a problem.

Real-Life Story 3: But It Was Inspected Recently!

Relying on jurisdictional boiler inspectors to manage your risks is not enough. You must do your own inspections and preventive maintenance safety testing as well. You must also address gas piping systems in your inspection and testing program. The following is an account from a local newspaper about a tragic incident in the U.S. Midwest that occurred at a nursing home, killing five people in 1999.

A blast centered in the nursing home's boiler room blew out the walls and caused the ceiling to collapse. More than 90 residents and about 20 other people were in the building at the time of the incident. Even as emergency crews were searching for survivors and evacuating the elderly residents, the incident was being attributed to a boiler failure. Press reports[14] indicated that the facility manager had recently had the boilers inspected and that the inspection found them to be in good working condition. Most people would have thought the equipment to be safe. Litigation followed, indicating that the incident may have been due to a gas piping leak. Gas piping leaks or corrosion of gas piping systems are not typically within the jurisdiction of state boiler inspectors.

The facility manager probably thought that everything was covered, including the gas piping, when the state issued an annual boiler inspection certificate. Like many others, he probably did not completely understand what the inspection did and did not cover. This is an issue within the boiler inspection world that no one seems to want to address. There is an ongoing culture of letting owners think that the state certificate is a completely protective all-encompassing shield against any and all concerns.

Lessons Learned Most official "boiler inspections" are jurisdictional and cover only the "water side" or the pressure-retaining components, water-level controls, and boiler internals. This incident occurred in a state that had adopted ASME CSD-1 as law, meaning that this code was enforced by jurisdictional inspectors. The code requires fuel train safety components to be tested but does not have jurisdiction for gas piping feeding the boiler.

As often happens, once an incident occurs, there is an important legal scramble by every insurer to try to define the perceived liability. Because of this, owners need to understand very clearly that jurisdictional and state-mandated inspections are *very* limited. They absolutely do not guarantee that fuel piping and every boiler safety system have been inspected or tested and that everything is safe. If you don't understand this and you manage a facility, you can lull yourself into a false sense of security and end up very disappointed! It is recommended that in addition to a satisfactory inspection by a state or jurisdictional boiler inspector, facility managers should contract with a capable service provider for an inspection of gas piping and controls and boiler fuel train safety device testing on at least an annual basis.

IMPORTANT

Insurers, including their inspectors and equipment review team, need to be part of your overall efforts toward fuel and combustion system safety. However, you need to understand the limitations of their approvals and what their scope really is. It is often much more limited than you think!

1.9 PERSONAL CRIMINAL LIABILITY

Ever been to jail? No? Neither have I, so far. However, I can cite numerous examples of situations where criminal prosecution for workplace accidents has occurred, and it's a growing trend. Most people think that if something happens, perhaps the company will be involved in a worker's compensation claim or even a civil action.

Rarely is thought given to being criminally liable for one's actions at work. The following should give you food for thought. Whenever there is a serious injury or death, it's always a prosecutor's prerogative to conduct an investigation and submit information to a grand jury for an indictment. It could be a supervisor telling employees to ignore obvious safety issues with a boiler or furnace, to bypass safety controls, or even not to respond to an obvious gas leak. In our world today, your actions may be second-guessed by a grand jury that lacks a technical background but, instead, is simply considering the probability of a successful conviction by a jury of your peers, who also lack technical know-how. If this does not cause you to err on the side of safety, nothing will.

Real-Life Story 4: Personal Criminal Liability at Work?

Take, for example, the case of a construction equipment company owner on the U.S. east coast who was charged with manslaughter, assault, criminally negligent homicide, and reckless endangerment for his perceived role in a 2008 accident that killed a crane operator and caused major financial losses to a project. According to press reports,[15–17] the criminal action happened even though the owner was never cited by the Occupational Safety and Health Administration (OSHA) for anything at the job site. The owner of the company plead not guilty to the charges.

This case and others like it could have serious implications for high-level executives who are never involved in the day-to-day details of running a business. Just because your plant is compliant with OSHA standards does not mean that you and the company are off the hook if something bad happens.

This incident involved a crane that was made in the early 1980s. This equipment was somewhat high tech and needed very specialized maintenance, periodic testing, and training—kind of like a boiler, an industrial furnace, or an oven. Press reports indicated that the crane manufacturer was out of business and that many replacement parts were no longer available. This can happen with older burner management systems and combustion controls on boilers and furnaces. My guess is that at least 25% of all burner or furnace controls in use today throughout industry are obsolete and no longer supported by manufacturers. When a major gear needed to be replaced, the crane owner tried to buy one from alternate suppliers and found three bidders. One Chinese company offered to provide a gear at less cost than that proposed by the other two bidders and offered a four-month faster delivery time. The company owner did what most others would do and chose the lowest bidder with the best delivery schedule.

Forensic evidence showed that a structural weld made by the successful bidder failed catastrophically. Prosecutors stated that the owner did not conduct due diligence on this company and that the vendor was supplied with grossly inadequate welding specifications. They went on to state that the owner did not follow generally accepted engineering and workplace standards to ensure that the weld was safe.

The owner said that he had the weld inspected properly and that city officials even signed off on it. The owner was eventually found not guilty. There is no doubt that he suffered greatly through the entire legal process.

Lessons Learned This case is an important lesson to upper management everywhere. The message is that if negligent actions are found, or even perceived, an individual can be prosecuted and held criminally liable. Prosecutors are under tremendous pressure, especially in very public incidents, to hold someone accountable. They can be very aggressive about this. It all starts with an OSHA report. Lots of times, if there's an OSHA citation, there's presumed guilt and presumed civil negligence. The best defense is not to have an incident in the first place. If you take advantage of what is presented in subsequent chapters, your chances of an incident will be greatly diminished. Much of what is presented was learned the hard way. The more than 50 real life stories in this book reflect what can happen if things go wrong.

NOTES AND REFERENCES

1. Star Spencer School, www.waterheatersafety.com/files/Spencer_Oklahoma_Hot_water_heater_in_school_kills_seven.pdf.
2. *Mythbusters'* hot water heater explosion: http://dsc.discovery.com/tv-shows/mythbusters/videos/exploding-water-heater.htm.
3. Stephen G. Badger, Large-Loss Fires in the United States—2009, Fire Analysis and Research Division, National Fire Protection Association.
4. Praxair, MSDS for carbon monoxide, www.praxair.com/~/media/Files/SDS/p4576j.ashx.
5. American Society of Mechanical Engineers website, history, www.asme.org.
6. Ibid.
7. National Board of Boiler and Pressure Vessel Inspectors. www.nationalboard.org.
8. National Fire Protection Association website, history, www.nfpa.org.
9. General Monitors website, SIL 101, How safe do I need to be? www.gmsystemsgroup.com/sil/sil_info_101.html.
10. William G. Bridges, Process Improvement Institute, Inc., and Harold W. Thomas, Exida.com, Accounting for Human Error Probability in SIL Verification Calculations, Global Congress on Process Safety, 2012.
11. Mike Scott and Bud Adler, article on NFPA 85 SIL levels and BMS systems, www.tappi.org/Downloads/unsorted/UNTITLED—05ppm66pdf.aspx.
12. AXA Matrix Risk Consultants, www.axa-matrixrc.com.
13. Factory Mutual, www.fmglobal.com.
14. Flint *Herald*, November 12, 1999, and State of Michigan Fire Marshal's report, www.wsws.org/en/articles/1999/11/nurs-n12.html.
15. *New York Times*, Owner of Crane Company Indicted in 2008 Collapse, www.nytimes.com/2010/03/06/nyregion/06crane.html?_r-0.
16. Crane Owner James Lomma Acquitted of Manslaughter Charges, www.metro.us/newyork/local/article/1141479—crane-owner-james-lomma-acquitted-....
17. James Lomma Acquitted of All Charges in Crane Collapse, www.nytimes.com/2012/04/27/nyregion/james-lomma-acquitted-of-all-charges-in-crane-collapse.

2

Combustion Basics

Laying a Foundation

In this chapter we lay a foundation for more complete technical understanding of fuel systems and combustion equipment. If you've been associated with this world, there may be little here that is new. If not, this is a chapter you may refer to over and over again in your career. The information in this chapter is out there in many forms and places. I have never seen it assembled in one place. Make sure that you read this chapter very carefully and understand the concepts. These fundamental concepts will be very important to understanding many of the risk reduction fixes that are provided in later chapters.

Real-Life Story 5: The Power of Propane

"Imagine having an explosion in your home that, when you saw it, the only thing left was toothpicks." That's how a fire marshal described the scene at a convenience store after an explosion tore the building apart.[1] What occurred that day was a tragedy, taking the lives of four men. According to the U.S. Chemical Safety Board report, it could have been avoided with proper training and safety precautions.

In early 2007, a junior propane service technician who had only been on the job for about a month and a half was transferring roughly 350 gallons of propane from an old 500-gallon tank to a new tank at a popular convenience store. When the technician removed what eventually proved to be a defective safety plug from a valve, propane began to pour out of the valve opening. He called his off-site supervisor, and eventually called 911. But he didn't realize that the leaking propane had a direct

Fuel and Combustion Systems Safety: What You Don't Know Can Kill You!, First Edition. John R. Puskar.

path into the store. Shortly after the supervisor and emergency crews arrived on the scene, the propane that had accumulated inside the store ignited.

The resulting explosion covered 100 square yards and knocked over several vehicles, including an ambulance and a fire truck. The shock waves were felt as far as a mile away. Seven nearby homes were damaged, and the windows of a local school were blown out. The two propane service technicians and two emergency responders on site at that time were killed. Five others were seriously injured. Nothing was left of the building but some twisted metal and debris. Fortunately, the two propane tanks were left intact: The explosion might have been even worse had they ruptured.

Investigations into the explosion uncovered the fact that the junior technician, who lacked formal propane-related training, "likely did not observe a telltale sign that the valve he was servicing was defective: The valve's safety plug was designed with a small hole through which propane may be seen leaking if the valve is stuck open, before the plug is fully removed." Before the incident, West Virginia and 35 other states had no requirements for the training or qualification of propane technicians.

Also troubling was the fact that the 10-year-old propane tank was located directly next to the store's exterior wall, in violation of state and OSHA regulations. This location allowed the leaking propane to pass directly into the store through building openings during the leak. The investigation concluded that if the tank had been placed at least 10 feet from the building, propane probably would not have entered the store in large quantities and an explosion would likely have been prevented.

Lesson Learned It's very important to know some of the basic properties of fuels with which you are working. Although propane is heavier than air, it, like other fluids, goes first to where its pressure or kinetic energy takes it before gravity forces take over. The leaked propane shot up into vents and building openings and then settled inside.

There are codes and standards (NFPA 58) for locating propane storage tanks and vents. The investigation indicated that if the tank had been even 10 feet away from the building, the incident might not have happened. There were many people involved with this incident that were not adequately trained. This included the service technician, who was not aware of the release that could occur as a result of his actions, and others, who did not evacuate more completely once the release started. An excellent video and detailed report describing this incident is available from the U. S. Chemical Safety Board web site, www.csb.gov.

2.1 COMBUSTION DEFINED

Let's start off with something very basic: a definition of combustion. *Combustion* is a rapid chemical oxidation reaction that takes place between fuel and oxygen to produce carbon dioxide, heat, water, and visible light.[2] Another common oxidation reaction occurs when iron or steel rusts. However, this is a slow oxidation reaction. You can think of combustion as very rapid rusting of fuel since the oxygen and fuel molecules combine very quickly when combustion occurs.

EXHIBIT 2.1 The fire triangle.

Explosions occur when fuel and oxygen accumulate, combine, or mix to the right proportions, and then are ignited all at once. The expansion and subsequent pressure pulse of the ignited fuel–air mixture happens in milliseconds. The pressure pulse depends on many things, but it's usually measured in relatively small units such as a couple of pounds per square inch, and it usually lasts for a very short time, such as milliseconds. You may recall that *force* is defined as pressure multiplied by area. Small forces and large surface areas result in large forces. This means that explosions can destroy large structures (buildings, furnaces, ovens) with seemingly small forces even if they are applied for just milliseconds. The challenge for everyone involved with combustion equipment is to control all of this potential energy and use it in a way that is safe.

The most basic way to think about combustion is in the context of the fire triangle (Exhibit 2.1). The fire triangle concept illustrates that when you combine fuel, heat, and oxygen, you produce combustion. There is also a concept called the *fire tetrahedron*, which adds the element of a chain reaction to the fire triangle. The chain reaction element represents the release of heat from the exothermic fire reaction that keeps the reaction sustainable and ongoing. We discuss this later in the chapter when we review burner designs and how the design of burners can support flame stability. When you remove any one of the three legs of the fire triangle, you stop combustion. Let's look at each leg of the triangle in more detail so that we can better understand how combustion systems work.

2.2 FUELS

Let's start our discussion of the fire triangle by focusing on fuels. Fuels have many different properties that have to be considered carefully when designing a system to burn them safely and efficiently. When fuels are burned thoroughly, all of the fuel molecules react completely with oxygen and liberate all of their useful heat.

Fuels are defined as any materials that can undergo combustion (i.e., be oxidized). Most fuels are organic, containing carbon. The most common fuels are some combination of carbon and hydrogen. However, there are also inorganic materials that will burn which contain no carbon; these include metals such as magnesium and sodium.

One of the most common fuels is methane (CH_4), which is the largest component of natural gas. It's composed of one carbon atom and four hydrogen atoms. Natural gas can vary widely both in composition by source and seasonally. Percentages of methane in the commercially available natural gas range are generally between 72 and 85.[3] A common mistake is thinking that natural gas is all methane. If more information is needed on the specific typical composition of a site's natural gas or its variability, contact the natural gas supplier.

Another common commercially available fuel is propane (C_3H_8). Propane is composed of three carbon atoms and eight hydrogen atoms. Propane occurs naturally like methane in gas wells but is produced primarily in petroleum refineries for commercial purposes. Commercial propane is usually specified with a corresponding grade or classification and heating value. Other hydrocarbons are present in commercial propane, including propylene (up to 5%, depending on the grade), ethane, and butane. These vary depending on the source of the propane. Normally, these variations are not a problem where the propane is used as a heating fuel.

2.2.1 Fuel Properties

As you can imagine, many different fuels and hydrocarbon-based derivatives can be made with carbon and hydrogen. The number of molecules of each in a given combination determines the fuel's characteristics and the amount of air or oxygen that's needed to consume or burn it completely.

Matter can exist in one of four phases: solid, liquid, gas, or plasma. Combustion of solids or liquids actually takes place above the fuel surface in a region of vapors created by heating the fuel. The form of a fuel, solid or liquid, has a lot to do with how well it ignites and burns. Wood, for example, does not light or burn well as a solid chunk, but ignites easily and burns vigorously as sawdust. It's the same when we burn liquids. They are usually atomized or sprayed to make them into fine droplets to increase their surface area. The increased surface area makes them easier to vaporize and burn.

Fuels have many properties that depend largely on their chemical composition, including specific gravity, flame temperature, heating value, lower and upper flammable limits, flame speed, ignition temperature, and cost. This discussion will focus on gaseous fuels. The *specific gravity* of a gas is a comparison of its density to that of air. As a point of reference, air has a specific gravity of 1. Natural gas is generally considered to be mostly methane but also to include other hydrocarbons. It is lighter than air. Methane has a specific gravity of about 0.56,[4] meaning that it is only about 56% of the weight of air. Depending on the other hydrocarbons, natural gas has a specific gravity of 0.6 to 0.7.[5] The specific gravity of propane is about 1.5.[6] This means that it's one-and-a-half times heavier than air and about twice as heavy as natural gas.

A release of natural gas will tend to rise and dissipate in air after kinetic energy forces have dissipated. A release of gaseous propane in air will tend to fall under the same circumstances. In both cases a release would have both gases mixing with air and the resulting mixture having some intermediate specific gravity of something between that of the fuel and the air.

Pure propane vapor can sometimes accumulate in a low spot, so the venting of fuel train components such as regulators on these systems must be done carefully. A practical word of caution here regards your home propane barbecue grill. Explosions have often occurred when burners have flamed out inside the grilling area with the cover down. If this were to happen, the propane fuel could accumulate in the burner area under the cover. When the cover is lifted, air could mix with the fuel that is already at ignition temperatures (completing the fire triangle), and an explosion could occur. This has happened many times, causing horrible injuries to those involved. Make sure that you use extreme caution if you think burners have flamed out. Keep people away, try to shut off the fuel source, and let the grill cool down on its own for a long while before attempting to open the lid.

Flame temperature depends on the type of fuel being burned, although both natural gas and propane can have a flame temperature of around 3500°F. The amount of energy released when fuel is burned is measured in British thermal units (Btu) in the United States and in calories in many other countries (metric system). One Btu raises the temperature of 1 pound of water 1°F, and 1 calorie raises the temperature of 1 gram of water 1°C. A chemical reaction that releases heat is called an *exothermic reaction*. Reactions that absorb heat or need heat to drive them are called *endothermic reactions*. Burning fuels is an exothermic chemical reaction.

Heating value is the relative heat energy that is liberated when a particular fuel is burned. There is an upper heating value and a lower heating value. The lower heating value does not take into account the water vapor that is generated in the combustion reaction. As an example, natural gas has a heating value of about 1000 Btu per cubic foot. Propane has a heating value of about 2.5 times this, or 2500 Btu per cubic foot. Knowing a fuel's heating value can allow you to estimate fuel flow to a burner. For example, if a burner is rated at 2 million Btu per hour and is burning natural gas, there is a design peak fuel flow rate of 2000 cubic feet per hour of natural gas (2,000,000/1000 = 2000).

The *lower flammable limit* (LFL) is the lowest concentration of a combustible substance in a gaseous oxidizer that will propagate a flame under defined test conditions (NFPA 68: 3.3.12.1). The LFL is sometimes called the *lower explosive limit* (LEL). The *upper flammable limit* (UFL) is the highest concentration of a combustible substance in a gaseous oxidizer that will propagate a flame (NFPA 68: 3.3.12.1). The UFL is sometimes called the *upper explosive limit* (UEL).

The *flammability range* of a fuel is defined on the lower end as the lower flammable limit (LFL) and on the upper end as the upper flammable limit (UFL). LFLs and UFLs are usually corrected to 32°F and 1 atm of pressure (14.7 psia). Hence, even at normal room temperature of 72°F, they differ from what is usually published. Increases in temperature and pressure usually result in reduced lower flammable limits and increased upper flammable limits. Upper limits can approach 100% in the case of high

TABLE 2.1 Properties of Natural Gas and Propane

Property	Natural Gas	Commercial Propane[a]
Minimum ignition temperature (°F)	~1100–1200	~920–1020
Lower flammable limit (%) at 25°C	5.0	2.15
Upper flammable limit (%) at 25°C	15.0	9.6

Source: NFPA. Properties of Natural Gas and Propane. www.nfpa.org.
[a]From NFPA 58, Table b12(b), 2011 edition, Copyright (©) 2011, NFPA.

temperatures. A decrease in temperature will usually have the opposite effect. You should practice caution when using published LFL/UEL data such as those in Table 2.1.

Flame speed is the rate at which a flame travels through an explosive mixture.

Ignition temperature is the minimum temperature at which a fuel will ignite spontaneously in a normal atmosphere without an external source of ignition. Propane has a lower ignition temperature than that of natural gas: about 920°F versus approximately 1100 to 1200°F. In Table 2.1 we compare some of the properties of natural gas and propane. Although combustion equipment is usually designed to burn one fuel at a time, burner designers must be able to anticipate the need to combust a range of fuels and properties in a safe and effective manner. Each fuel's unique properties are important to the design of the burner, ignition system, combustion air system, and even heat transfer equipment inside a boiler, furnace, or oven.

Real-Life Story 6: Propane and Natural Gas Systems Don't Mix Well

During a plant boiler safety inspection visit I noticed that there was a backup or standby propane system on the site. It happened to be a pure propane vapor system rather than a propane–air mixture system. Propane–air systems are designed to blend propane and air to get a fuel gas that has properties similar to those of natural gas, allowing it to be used interchangeably. I was told not to pay any attention to this system since it was used only once a year and was not the primary focus of my visit. I explained that my overall mission was safety everywhere on the site. I eventually got approval to take a quick look at the system. Upon review I saw that three-way valves existed throughout the natural gas piping system tied into the propane vapor lines. Later I found out that these valves allowed specific areas of the plant to run on propane vapor when there was a curtailment of natural gas. That's right—straight propane vapor with no blending, no changing of burners or nozzles, no separate fuel trains. There was nothing to account for the vastly different properties of these two distinct fuels. Simply moving a three-way valve released a fuel into the system that's heavier than air instead of lighter and that's 2500 Btu per cubic foot instead of 1000 Btu per cubic foot.

When asked how this could possibly be done safely, the plant's maintenance staff leader said: "Yeah, we don't like to run that way very often as the flames are real orange and smoky." As you read on, you'll discover just how tragically this scenario could have ended. The client was advised to stop this practice immediately. The plant agreed and had the entire direct propane system removed within months.

Lesson Learned Propane vapor systems cannot be used interchangeably with natural gas systems. A propane vapor system must be part of a specific fuel train and burner system designed for propane vapor if it is to be used for backup or standby purposes. Engineered propane–air systems are available that create fuel gas mixtures that can be used interchangeably with natural gas. Whenever you hear, "Don't worry, it might not be quite right, but it works, or it is an infrequent event," it's time to worry. This attitude does not work when it comes to safety.

2.3 HEAT/IGNITION

Next, let's discuss the heat or ignition part of the fire triangle. To cause combustion, heat has to be added to raise the temperature of some part of the fuel–air mixture above the ignition temperature of the specific fuel that is involved. This is usually done with a spark igniter. Such systems take a lower voltage, perhaps 120 volts, through an ignition transformer to produce thousands of volts that can arc across an air gap and generate a high-temperature spark. Another type of popular igniter in widespread use today is the hot surface igniter, made of resistive ceramic that is heated electrically to above-ignition temperatures.

Once a flame is lit, combustion requires ongoing ignition energy to make the flame stable and self-sustaining. This can be accomplished in a number of different ways. In some cases, it comes from reflected heat from refractory (burner block) that surrounds the burner nozzles. Flames can heat the refractory, and this heat, well above the required ignition temperature, can serve as a means to keep the new flowing fuel–air mixture ignited.

Obscure ignition sources sometimes make for unexpected explosion risks. We all know that "no smoking" signs should be posted near possible sources of fuel. Here are facts for you to consider on the consequences of smoking near natural gas. The ignition temperature of natural gas is usually given as between 1100 and 1200°F. A lit cigarette's temperature[7] (free burning) provides a heat source of 1292 to 1472°F. When a smoker is puffing the cigarette, the temperature can be 1472 to 1562°F. It's not hard to see how just carrying a lit cigarette into an area with the right mixture can cause a disaster.

2.4 OXYGEN/AIR

The third leg of the fire triangle is oxygen. Most combustion systems get their oxygen from the surrounding air. The air that we breathe is about 21% oxygen and 78% nitrogen. Some special high-temperature and metal melting or cutting applications use pure oxygen. Much higher flame temperatures (5000°F or more) can be reached with pure oxygen–enriched combustion systems.[8] Typical combustion processes using air can only achieve flame temperatures of 2800 to 3500°F because of the nonreactive 78% nitrogen that gets dragged along when we burn air.

Each type of fuel must be presented to the burner at the right fuel/air ratio for proper combustion. Complete combustion occurs when all of the fuel combines with all of the oxygen it needs and nothing but carbon dioxide, heat, and water vapor are released. When exactly the right amount of fuel is matched with exactly the right amount of air for combustion, "on-ratio" or *stoichiometric combustion* has occurred. Stoichiometric combustion doesn't often happen in the real world because it's difficult to provide the perfect amounts of air and fuel to a burner on a continuous basis. A good rule of thumb to remember about burning natural gas is that it takes about 10 parts of air to burn 1 part of natural gas stoichiometrically.

Real-Life Story 7: No One Expected a Hydrogen Explosion

Even the most experienced boiler veterans can make mistakes if proper safety protocols are not followed. That's a lesson that was learned in the maritime world when an explosion ripped through a liquefied natural gas tanker in 2003.[9] The following account is taken from the United Kingdom's Marine Investigation Branch report of this incident.

A ship's starboard-side boiler was undergoing repairs at a Caribbean shipyard. Two experienced chemical cleaning specialists were aboard to oversee the removal of boiler scale and corrosion as part of an extensive boiler tube replacement program. On the morning of the incident, the boiler was started up as usual, and an acidic chemical compound was added to the boiler water to remove the unwanted sludge. Later that day, during a test of the water–acid mix, the cleaning specialists found that the acid concentration was too strong and had begun dissolving the boiler steel.

The two workers stopped the cleaning process and opened up the boiler's steam drum hatch to investigate. One of the men grabbed a nearby halogen lamp and placed it just inside the steam drum. A small flame sprang up and an explosion followed. The two men were thrown back, knocking them unconscious and causing severe burns. One man lived; the other did not.

What they hadn't realized was that the acid-based cleaning products had caused an accumulation of hydrogen to build up in the steam drum. When the hatch was opened, the hydrogen combined with oxygen to create a flammable mixture. The halogen light's temperature was just enough to ignite the mixture and cause the flash flame. Flames don't have to last long for damage to skin to occur. In the case of electrical panel arc flash explosions, extreme temperatures are created for only milliseconds, but the damage to flesh can be permanent and devastating.

Lessons Learned Fuels can be derived in unusual ways and be in places where you don't expect them to be. It's possible that the workers had no idea that hydrogen was being generated and, if so, they may not have known that it had a wide range of flammability (generally, 4 to 75% concentration in air[10]). Any processing or use of acids has a potential for hydrogen release. There have been numerous explosions in lead–acid battery-charging areas used for forklift trucks when ventilation was not adequate. Had the boiler steam drum been made properly inert or purged with nitrogen, the hydrogen buildup might have been avoided. Be aware of gases that may be leaking or exist as part of a process. Understand their properties. When in doubt, test atmospheres first and create a safety plan that may include ventilation or making inert.

> **IMPORTANT**
>
> It is worth remembering that the wider the flammable range, the greater the likelihood of a mixture coming into contact with an ignition source, and the greater the hazard that the fuel represents. The flammability range of hydrogen is about 4 to 75% concentration in air. This means that almost any concentration can be ignited.

2.5 COMBUSTION CHEMISTRY

There's a little more that you should know about what goes on at a molecular level when combustion takes place. When you understand some of the simple chemistry behind combustion, you will better understand what happens when things go wrong. This will help you to operate more safely.

When a fuel burns, the reaction occurs according to a scientific principle called *conservation of mass*. This says that matter can be neither created nor destroyed. You can, however, add energy and rearrange the mass into different components and change their form. For example, when a cigarette is burned, much of the tangible form held in someone's fingers apparently disappears. But if we look at it from a molecular level, we see that an equivalent amount of other materials were formed when the paper and tobacco were burned. They became ash, carbon dioxide, water, carbon monoxide, and other particles.

When methane is combined with oxygen and burned, the result is carbon dioxide, water, and heat. What follows is the chemical equation for this reaction. This equation is known as a *balanced equation* because it has the same number of atoms on each side of the equal sign. This is stoichiometric combustion since it is the exact amount of oxygen the fuel needed for complete combustion.

$$CH_4 + 2O_2 = CO_2 + 2H_2O + \text{heat}$$

On the left side of the equation, one C (carbon) atom splits off from the methane molecule and combines with two oxygen molecules to form carbon dioxide (CO_2). CO_2 is the gas that is injected into ordinary soda pop to make it fizz. The hydrogen atoms combine with the oxygen atoms and form water. The reaction is exothermic, so heat is also released.

2.5.1 Applying Combustion Chemistry to Burner Systems

We have noted that in industry, combustion rarely occurs stoichiometrically. Let's consider the case when excess fuel is provided, that is, when we have a fuel-rich condition. When this happens, there is not enough oxygen to combine with or oxidize all of the carbon molecules. Many unwanted compounds can be formed when other than stoichiometric combustion occurs, including carbon monoxide, alcohol, ammonia, and formaldehyde. The most common indication of a too-rich mixture is grayish or even black sooty flue products leaving the stack as carbon separates from other molecules and is released in pure form.

The maximum amount of heat is released from the fuel when combustion is stoichiometric; that is, when the only combustion products are carbon dioxide and water. When combustion is not stoichiometric—for example, when carbon monoxide is formed—there's still 27% more heating fuel value available in the carbon monoxide if it can be converted to carbon dioxide by adding one more oxygen atom.

When we have very lean combustion, we also do not optimize the combustion process. When there is too much air, flames can become unstable. The air tends to have a cooling effect and quenches certain parts of the combustion process that have already begun to take place. This stops the reaction in some partially completed form.

If you're ever at a site where someone is experiencing strange smells, headaches, or soft-tissue irritation (e.g., the eyes or sinuses), it could be a sign of very dangerous conditions. The strange smells and discomfort may be coming from the ammonia, alcohol, or formaldehyde formed from incomplete combustion and flue products that are somehow making it back into an occupied area. If these compounds are present, it's likely that considerable carbon monoxide is also present. Carbon monoxide is odorless and tasteless. It has a specific gravity very close to that of air. Hence, if it's released during combustion, it will remain around the release point unless there is an air current or density difference. Carbon monoxide has a cumulative effect on the human body. It attaches itself to hemoglobin in red blood cells. Hemoglobin carries oxygen to vital parts of the body, including the brain. Hemoglobin bonds to carbon monoxide much more strongly than to oxygen. As the hemoglobin in the red blood cells accumulates more and more carbon monoxide, there is less and less oxygen-carrying capacity in the red blood cells. A person who is affected can have severe headaches, loss of consciousness, and ultimately, death.

If someone experiences such smells or irritations, you need to get him or her to fresh air immediately and check for carbon monoxide. Provide medical treatment as soon as possible, and call 911. As a practical matter I have always requested that wherever flue systems are connected to positive-pressure fire boxes, carbon monoxide detectors be placed in the mechanical equipment rooms where these are located. Similarly, all occupied boiler control rooms should have carbon monoxide detectors. You won't find this a requirement in any code, but it should be.

Real-Life Story 8: Boiler Operations, Headaches, and Being Left Alone Don't Mix Well!

Consider the case of a boiler operator in a large manufacturing plant who said that he wasn't feeling so good. He had a headache and dizziness near the end of his shift. He convinced everyone that he was just fine and could make it home on his own. He left the plant in his car and within 30 minutes was pulled over by a state highway patrol officer for driving over 90 miles an hour. When the officer approached the car, he found that the boiler operator could not speak and was disoriented. The officer could not even conduct a field sobriety test. He only stared at the patrolman with the very dazed, glassy-eyed look. An ambulance was called, and the man was rushed to the hospital, where he recovered after several days of special treatment for carbon monoxide poisoning. This situation could easily have ended in a tragic loss of life: his or that of some other innocent party.

Lessons Learned Combustion processes create relatively harmless by-products when everything goes well. When things don't go well, deadly carbon monoxide can quickly and easily poison anyone within its reach. Don't let anyone be left alone who you suspect has exhibited symptoms of carbon monoxide poisoning. Make sure that the person gets safely into the hands of medical professionals at once.

There are many reasons that flue gases can end up where they are not expected or supposed to be. For example, flue gases can enter occupied spaces when exhaust fans are turned on in boiler rooms with no corresponding supply air. This can cause negative pressures and back-drafting down flues. Other sources could be leaks in boiler casings or flue duct systems, changes in wind direction that blow flue gases downward into building openings or air intakes, or even because flue dampers or barometric relief devices have not been set right.

2.5.2 Burner Fuel/Air Ratio Operating Conditions

Now that you understand the fire triangle and issues related to fuel–air mixtures, you might be questioning the best conditions under which burners should operate. Should burners be set to burn stoichiometrically too rich, or too lean? The answer is that most commercially available burners are intentionally set up to have a slight excess of air (i.e., to run lean). The excess air provides a little extra oxygen, to allow for the slight variation of conditions that might occur at the burner. The consequence is that excess air wastes a bit of fuel but maximizes safety. For many burners of the type on smaller packaged boilers, the goal is to have between 2 and 6% excess oxygen. In other words, if we put a flue gas analyzer instrument into a stack, we would measure between 2 and 6% oxygen after the entire combustion process has taken place. The right amount of excess oxygen or air for an application depends very much on the style and type of burner you have and how it operates. Guidance on this can be provided by the burner or equipment manufacturer.

2.6 ENVIRONMENTAL EMISSION ISSUES

Certain combustion by-products are undesirable for reasons other than safety. Do you remember the 78% nitrogen in air? High-temperature combustion causes a process called *dissociation* which occurs when some of the nitrogen reacts with oxygen molecules to make nitrogen oxides. These oxides of nitrogen react in the atmosphere to make weak nitric acids (acid rain), known in the field as NO_x (pronounced "knocks"). This is why they are tightly regulated.

Sulfur compounds may contribute to acid rain as well. Generally speaking, natural gas and propane contain very little sulfur. It's usually when heavy oils and coal are burned that sulfur is an issue. The sulfur combines with oxygen to form sulfur oxides (mostly SO_2 and SO_3), which also cause acid rain.

In some cases, special burners and/or control systems may be installed to minimize unwanted elements such as nitrogen oxides. You may hear a term like "low-NO_x burners" and maybe even "ultralow-NO_x burners." This type of technology comes in many forms, so you'll have to be especially careful if you own these or are having burner service work done so that you don't compromise the technology in some way. There are sometimes special flame stability considerations and environmental performance considerations with these technologies. You may also have a U.S. Environmental Protection Agency (EPA) operating permit that dictates emissions limits. In many cases if you exceed these, you are legally obligated to turn yourself in. In the case of low-NO_x burners, you need to have either factory service or very experienced burner technicians setting fuel/air ratios. These are not pieces of equipment to learn on or experiment with. There may also be performance guarantees that are voided if anyone other than the manufacturer or original installer services the burners.

To understand some of the ways in which lower No_x is achieved, let's look into some of the burner technologies that are used. At least two different popular types of burner technologies are used to achieve low-NO_x conditions: staged combustion and flue gas recirculation. Staged combustion uses specially designed burners that allow combustion to take place partially in certain sections or zones of a firebox in a slow and staged or controlled manner. This reduces the peak flame temperatures and therefore reduces NO_x formation. Remember that natural gas ignites at approximately 1100 to 1200°F and that flame temperatures in commercial burners are generally 2800 to 3500°F. A lot more NO_x is formed at 3500°F than at 2000°F. Also, more NO_x is formed when there is more excess air (because with more air, more nitrogen is present). If you minimize excess air and stage the combustion by lowering flame temperatures, you can also lower NO_x formation. Fuel and air staging can reduce NO_x levels by 30 to 60% over those of conventional burners.

Flue gas recirculation is the other popular method of NO_x control. Flue gas recirculation systems come in two forms: internal and external. Internal systems redirect a small amount of flue gas back toward the combustion zone through special nozzles and mixing strategies within the burner system. External systems take a small amount of flue products out of the flue stack and puts them back into the combustion air. The resulting dilution and reduction in available oxygen decreases the flame temperature and thermal NO_x formation. When flue gas recirculation into the burner is used, the amount of flue gas volume is subject to the operational constraints of flame stability and impingement as well as flow-induced vibration.

2.7 BASIC BURNER DESIGN ISSUES

Basic burner design incorporates the three T's of effective combustion: time, turbulence, and temperature.

- *Time.* The time during which the combustion process is taking place is determined in part by the shape and volume of the combustion chamber and flame.
- *Turbulence.* Turbulence is the device's ability to get good mixing of the fuel and air.
- *Temperature.* Temperature is related to the ignition energy provided for the fuel–air mixture for sustaining the combustion reaction. Ignition energy is often influenced by the shape and design of the refractory material that surrounds the burner fuel–air mixture release point.

There are many different types and styles of burners. Although we cannot discuss them all here in detail, the goal is to give you the ability to recognize some of the most basic types and to understand what's going on inside a device. In the following sections we introduce you to airflow, nozzle mix, and premix burners.

2.7.1 Airflow Burners

Airflow burners (Exhibit 2.2) are used in many applications to heat relatively large quantities of air flowing past them. They are popular for direct-fired ovens, dryers, and makeup-air heating applications.

These burners generally consist of some type of aluminum or cast iron fuel manifold with holes drilled along the entire length. These fuel release holes (number, pattern, and size) are specific to the gas pressure being delivered in the fuel manifold for the design burner heat release. Air diffuser wings are then added. These are usually made of stainless steel. They have holes that are smaller near the manifold opening and progressively larger away from it. These holes allow some of the passing air to

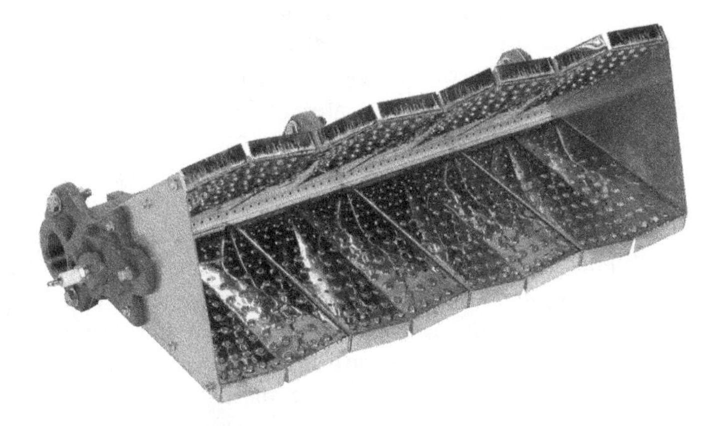

EXHIBIT 2.2 Eclipse AHMA-style burner.

enter and mix with fuel as primary air, depending on the firing rate. Secondary air (air provided after ignition) then comes from the mixed airstream past the burner where the combustion process is completed.

Igniters and flame detection are usually provided in the side of the burner near the point of connection with the fuel gas piping. These burners usually come in modular sections that can be added to for the overall heat release desired and the configuration of the air supply duct work. Airflow burners have a very robust flexible design and have been in use for many years. Failure modes for these burners can include corrosion and fouling, due to a lack of cleaning and maintenance. Overheating and destruction of diffusers can also occur if they are not cleaned properly, because debris gets lodged in them, preventing air from passing through the holes. The passing air cools the diffuser plates. These modes of failure, corrosion and overheating, are easy to see when a unit is running. If gas distribution holes are plugged, you can usually observe that the burner is not lit the entire way across. In some cases you can see the flame color change or be unstable at certain points, or there will be glowing red spots on the diffuser if the air holes are plugged and impingement is occurring. If you see any of this, shut the burner down and look closely at the gas manifold or air diffuser holes for plugging or cracking and impingement of diffusers.

2.7.2 Nozzle Mix Burners

Nozzle mix burners are used in many boiler and process heat applications. They generally consist of a series of fuel nozzles in a circular arrangement, with air provided behind them from a fan. In many cases there is a circular pipe or fuel manifold, providing constant pressure to serve these nozzles or orifices. The size of the nozzles or orifices is directly proportional to the desired heat release from the burner.

The fan can be integral to the burner casting and mounted directly behind the nozzles, or it can be remote through the use of ductwork. When the fan is integral, the burner is said to be *packaged*. Air diffusers are usually provided directly adjacent to the fuel nozzles to promote mixing. These are usually made of carbon or stainless steel sheet metal.

The term *nozzle mix* is used because fuel and air are introduced into a combustion zone separately immediately prior to ignition. Failure modes for nozzle mix burners can include air diffusers getting overheated and disintegrating if clogging occurs or if firebox pressure conditions change unexpectedly and move the flame to an undesirable position. Gas orifices can also become compromised causing uneven burning.

2.7.3 Premix Burners

The term *premix* does not describe a specific burner design. Instead, it refers to the concept of mixing raw fuel with air and creating a flammable mixture prior to introducing it to an ignition source. (If there is not enough air for a flammable mixture, the term *partially aerated* is typically used to describe the system.[11] In some cases this involves a remote mixing device, and in other cases the mixing can occur immediately prior to the burner nozzle release point. Premix burners are available in both inspirating and aspirating designs.

Inspirating premix burners induce or pull air into the mixing chamber at a low-pressure zone created by fuel gas that is discharged toward a nozzle. As the gas expands through the nozzle, this low-pressure area draws in the air. An air blower is not used. The resulting mixture moves toward the burner tip for ignition.

Aspirating premix systems work similarly, but they depend on forced airflow for mixing. In an aspirating burner, combustion air from a blower is released toward a nozzle. The resulting low-pressure area draws fuel gas to the mixing tube. Again, this mixture moves toward the burner tip for ignition.

Premix burner technology applications are very broad. They exist in everything from residential furnaces to refinery crude heaters and everything in between. Premix technology is important because it allows for flame release patterns that are different from those of raw gas or diffusion mix burners. Premixed burners often produce shorter and more intense flames than those of raw gas diffusion mixed burners.[12] This can produce high-temperature regions in the flame and opportunities for flame shaping that can be used for special furnace designs.

Dangerous conditions can occur for premix burners if the burner nozzles become clogged. If this happens, unignited flammable mixtures can back-flow out of air/gas mixers. There can then be an accumulation of flammable mixtures in areas such as rooms or spaces where the mixer is located, and explosions can and have occurred. I am aware of this happening with premix burners in heating, ventilation, and air-conditioning (HVAC) direct-fired makeup-air applications where moisture caused corrosion and plugging of burner holes. If you smell gas in areas where this equipment is located and no piping leaks are found make sure the mixer system is verified to be operating correctly.

2.8 DRAFT SYSTEMS

Draft is the difference in pressure that pushes or pulls the fuel–air mixture and flue gases through a combustion system. To understand combustion equipment air and combustion product flows, you first have to understand draft systems. The combustion equipment you will encounter uses one of four different draft systems: natural draft, forced draft, induced draft, or balanced draft. A description of each of these types, as well as some operational implications of each, follow.

2.8.1 Natural-Draft Systems

Natural-draft systems have no fans. The three other types of draft systems (forced draft, induced draft, and balanced draft) involve the use of at least one fan. Natural draft is the simplest type of system. It relies on hot flue gases being less dense and rising through a flue system to the ambient environment. You can encounter natural-draft systems anywhere from the hot water heater in your house to large refinery process heaters. The negative draft pressure that exists depends on many things, including the height of the stack, temperature of the gases, ambient environment temperature, and the relative pressure of the surrounding spaces.

Special Considerations for Naturally Drafted Equipment If naturally drafted equipment is located within a building, the draft pressure can be very sensitive to pressure conditions within the building and the facilities for providing combustion air. Conditions need to be correct for flue products to leave the device properly and thus make for a continuation of proper burner operations. Whether you're replacing an existing piece of equipment or installing new equipment, this is an area to which you'll want to pay close attention. There have been many instances of natural-draft equipment being installed into negatively pressured spaces where exhaust fans have drawn flue products down the stack and into the space. On very cold days and during startups, natural-draft equipment may be susceptible to transitional conditions where the negative draft pressure is not adequate. Be aware of this possibility and consider installing CO detectors and alarms, in mechanical equipment rooms where this has a potential to happen.

In many cases the amount of draft is controlled or limited by a device installed on the flue pipe called a damper. This is simply a piece of metal that can change position and restrict the flow out of the flue pipe. Relief or barometric dampers consist of what looks like a loosely hung damper blade with some type of weight or adjustment. These work differently than traditional dampers in that they admit air into the stack to reduce the draft when called upon to do so. There is an adjustment is to set the draft desired on the system. If the desired negative pressure or draft is exceeded, the relief damper is supposed to swing open and allow room ambient air into the flue, to dissipate or reduce the negative pressure. A relief damper should never allow flue gases back into a room. I have often found relief dampers to be broken, stuck, or not set correctly. Make sure that you check these to verify flow direction and proper operation. Manufacturers of these devices provide extensive information on how to set them and verify proper operation. These are often installed on venting systems in situations where stacks are very tall or with older masonry chimneys.

NFPA 54 provides detailed guidance on the installation of naturally drafted equipment, venting issues, and combustion air. In my opinion, one of the most important things in NFPA 54 regarding this equipment that is often overlooked is the need to do a performance test once the installation is complete. This is identified in NFPA 54 (2012 edition), Section 11.6, which requires you to verify that no draft hood spillage is occurring after 5 minutes of operation. You should also consult appendix section notes that define important parameters for conducting this test.

This performance test can help you to find concerns that were not anticipated, or perhaps building conditions you inherited that affect your installation in a negative way. Don't skip this step. Be aware that in practice, after the equipment is installed, people can do things like prop doors open to other areas of the building that are connected to the equipment room. They might also operate exhaust fans or block combustion air intake louvers with boxes of supplies, and barometric relief dampers can be compromised just from people bumping into them or hitting them with things like mop handles. There's no substitute for a regular review of this equipment once it's in service.

Combustion Air Considerations It is vital to provide enough air or pathways for air to get to burners without compromising the design intent of the burner or its flue system. This is important wherever fuel-fired equipment exists. Providing the right amount of combustion air is often an overlooked design issue. NFPA 54 calls for four ways to provide combustion air: room volume, forced air from outside being brought in, louvers or ducts from the outside connected to the room, or some combination of these methods. If a mechanical system brings in the required air, it must be interlocked to the burner safety system.

I have found many cases where although provisions had been made for code-required combustion air for the original equipment installation, the amount of air provided in practice was now compromised. Following is a list of safety considerations for providing the proper combustion air and making sure that draft controls function correctly. This partial list is not a substitute for reading and applying all of NFPA 54's requirements.

- If motorized dampers exist for bringing in combustion air, assure that they are interlocked to the burner firing controls as required by NFPA 54.
- Flues should not discharge near operable windows, building openings, or air intakes.
- All combustion air intakes should be clean and clear of debris and obstructions.
- Verify proper sizing of louvers and ducts according to NFPA 54 and equipment manufacturer requirements, and don't forget the provision that calls for discounting of the gross louver size to a net free area based on the louver design.
- Be sure that exhaust fans or other appliances that draw air out of a building are not communicated to the room where natural-draft appliances are located. If they are, verify that adequate ventilation air is brought in and interlocked (above and beyond the exhaust requirements) so that adequate combustion air can always be maintained.
- Air handlers should never be taking in return air directly from a room that has fired appliances, especially when they are naturally drafted appliances.
- It is a best practice to install carbon monoxide detectors in mechanical rooms that have fired equipment.
- Make sure that a performance test of naturally drafted appliances has been conducted to verify that all appliances are venting properly. This test should reoccur on a regular basis. Simulate conditions in the building that could represent actual seasonal conditions like exhaust fans operating.
- Verify that all barometric relief dampers are functioning properly.
- Verify that all flue piping sizes and arrangements are as required by NFPA 54.

IMPORTANT

Natural-draft-fired systems require careful consideration of their environment. Remember that backflow of flue products could easily occur into the space where they are located if just a few simple things go wrong. Verify proper drafting regularly, maintain CO detectors in these spaces, and always be on the lookout for signs of flame rollout (burn or scorch marks on equipment).

Real-Life Story 9: A School System in a Fog

Consider the case of a school system whose boiler room was said to have a substantial fog whenever it got very cold outside. Reports were that sometimes you couldn't even see through the boiler room because of the fog. In my investigation I found that the fog was really flue gases being drawn back down stacks and into the boiler room. This could have made for conditions that jeopardized the safety of the school children.

The facilities heating system consisted of six natural draft hot water boilers. Each of these was 8 million Btu per hour input with 36-inch-diameter flues. It was obvious just by looking around that the room had a substantial deficiency in combustion air. There were some louvers on an outside wall, but experience told me that they were not nearly enough. Later calculations would show that these provided far less air than was required by NFPA 54. There were also air handlers that were supposed to be interlocked to the boilers that should have been bringing in some fresh air for combustion whenever the boilers fired. The interlocks were not installed properly or functioning. The design of the boiler controls allowed for boilers to be fired in a lead–lag order that could change. This was never considered in the fresh air interlock system. It made for another way that firing could take place without the proper amount of outside combustion air being available.

The flue system on the roof consisted of short stub stacks that were not far apart. The boiler room was also in an area that was near several taller walls, which obstructed wind flow. This meant that in certain wind conditions, clouds of flue products hung near the roof level and near boiler stacks that were not firing.

I took a handheld anemometer, used for measuring airflow, and held it between a locked out boiler flue gas discharge and its draft collar outlet. There was obviously airflow back down the stack to make up for what was needed by the boilers that were firing. I watched carefully while three boilers were running and 2 others were made to fire. For some time, before a good draft was established, flue products from these additional boilers did not go up and out. These gases came straight into the room along with cold air down the stacks. The cold air caused the moisture in the unvented products of combustion to condense. This and some amount of flue gas recirculation from boilers firing nearby created the fog that everyone talked about.

It's amazing that no one was killed or injured. Besides the water vapor from the combustion process, there were substantial amounts of carbon monoxide from incomplete combustion in this room. It's horrifying to realize that this went on for months and possibly years in a school building.

Lessons Learned Natural-draft combustion equipment requires very careful consideration of combustion air design. This can include special considerations for boiler and supply air fan controls, flue stack installation, wall louvers, and the potential for flue products to back-draft into a space. It's why I have always recommended that carbon monoxide sensors be placed in these spaces. This is especially important if the room where this equipment exists is staffed (control room) or if there are handlers that can recirculate the rooms contents to other parts of the building.

I have found many combustion air deficiencies in my career. It's clear that many architects and building designers do not understand how to apply the codes and standards related to combustion equipment properly to satisfy this issue. It's an inconvenient truth that the combustion air requirements could create a need for a lot of wall louver area, which does not look pretty. You then also have the situation where in very cold weather, all this cold air comes into the room and can possibly freeze nearby water lines. Yes, it can be a design challenge, but it must be dealt with.

In many cases exhaust fans exist that are not accounted for. In some cases this equipment is in areas that are adjacent to and connected with manufacturing spaces that operate under negative pressures. I have seen many boilers and hot water heaters back-drafting under these circumstances. Evidence of back-drafting could be scorch marks (flame roll-out) on painted equipment panels that look burned. If back-drafting is suspected, equipment should be shut down immediately until conditions can be corrected.

2.8.2 Forced-Draft Systems

Forced-draft systems use a fan to push the fuel–air mixture through a burner and into and out of the firebox. In some cases the burners used for these systems are referred to as *power burners*. The firebox or combustion chamber pressure is positive in these systems. This means that any leaks in the combustion chamber or duct leading from the furnace will allow flue gases to come out into the building or space surrounding it. This scenario can be dangerous if these combustion products contain carbon monoxide, unburned fuels, or other combustible materials. Forced-draft systems are by far the most common type of combustion systems.

A comprehensive evaluation of combustion air requirements for power burners is now available from NFPA.[13] It reviews combustion air requirements for power burners from many different standards organizations, NFPA 54, and manufacturers. This can be a good reference for helping you to better understanding the needs of your equipment and facility.

2.8.3 Induced-Draft Systems

Induced-draft systems have a fan on the outlet of the firebox or combustion chamber. This fan draws the fuel–air mixture through the burner into the firebox for combustion and then pushes the combustion products out of the flue stack. The combustion chamber is usually controlled at some slight negative pressure relative to the environment outside the fired device. If a leak occurs in the combustion chamber or flue system, the incoming air can dilute combustion products and confuse instruments that may exist to measure flue gas properties and control fuel/air ratios. This has happened on many occasions and has led to serious problems.

2.8.4 Balanced-Draft Systems

Balanced-draft systems, a combination of forced- and induced-draft systems, have a fan on both ends (burner and flue). In these cases the draft control objective is usually to maintain a slightly negative firebox pressure. Many solid fuel boilers, such as coal stoker boilers, use balanced-draft systems.

Stokers are devices that bring in and feed such solid fuels as coal, wood chips, or even trash. Traveling-grate stokers have chain grates that move, carrying fuel into the firebox. The grates change speeds to accommodate different firing rates. In some traveling-chain grate designs, air is delivered under the grate for cooling. In other cases, air is delivered at the point of fuel entry, or over-fire is used to accommodate unburned gases leaving the grate area. Firebox volumes are often very large, and fans are required on the discharge end to help move the combustion products out.

2.8.5 Draft Controls

No discussion of draft systems would be complete without identifying how air, and consequently draft, are changed and controlled. Draft can be controlled directly with automatic or manual dampers or with registers at a specific burner. Draft can also be controlled at a flue outlet with dampers. In some cases supply or induced-draft fans are supplied with variable-frequency drives, inlet guide vanes, or outlet dampers.

Changing or adjusting a draft needs to be done carefully and in very small increments as it can have a dramatic impact on firebox pressure and, subsequently, on firing rates and burner operating conditions. In some cases, with sophisticated heat-treat equipment, draft control of large oven zones is very critical and small movements of a damper can change temperatures and process requirements.

2.9 UNDERSTANDING AND EVALUATING FLAMES

Once a burner is lit, it is monitored by a flame detector as part of the flame safety system. The flame safety system's job is to shut down the fuel supply immediately if the flame is extinguished. The flame safety system cannot, however, identify many flame quality issues, which can be creating dangerous conditions inside a combustion chamber. Nothing can replace the added element of safety that can be obtained by a periodic review of flame conditions by a skilled and trained operator. In this section guidance is provided on some of the factors that you should be aware of that can identify dangerous flame conditions.

2.9.1 Where To Look

Viewing flames is done through either a burner or an equipment sight port. Most burners have some sort of sight port built into the burner housing or casting. This is generally a tempered glass port $\frac{3}{4}$ to 2-inches in diameter that provides a view of the flames as they enter the firebox. This particular port offers little useful information because it's at the front of the burner looking at the back of the flame as it travels away from the burner.

For many styles of burners, depending on what you are trying to evaluate, the best place to observe a flame is with it coming in your direction. Hopefully, there's a sight port in the firebox that allows you to view the flame this way. This would normally be a port in the back door of a fire tube boiler.

IMPORTANT

Never look into a sight port during a light-off. Many people no longer have sight in one or both of their eyes and have experienced many cosmetic surgeries because they tried to watch a light-off and ended up with the sight port blowing off directly into their faces. Always wear safety glasses and flame-retardant clothing when you approach a sight port. You must first take a gloved hand and feel the glass to make sure that it is secure and that the port is not a source of uncontrolled flue gas releases that could cause you harm. Check out these factors before putting your face near a sight port.

You may sometimes find a sight port glass cracked or completely opaque. You should check this before you begin, and either clean the glass or replace it with glass of the proper strength and type so that it is clear, safe, and functional. You should always be able to examine a flame visually, especially before and after adjustments, if at all possible. Never put your face next to an open furnace or combustion chamber door or port associated with any boiler or combustion equipment. Negative pressure fireboxes can suddenly go positive, spewing out hot gases and flames within seconds for no reason that may be apparent to you. Some facilities have special glass, mirrors, or viewing equipment and protocols to enhance the safety of this activity.

With some fuels (e.g., oil), special filter glass (usually, cobalt blue) is sometimes used because the glow from the flame and internal refractory is so bright that it can hurt your eyes. If it's too bright, turn away immediately and protect yourself with filter glass the next time you look in.

Remember that you will be observing a snapshot in time. If you want to get a complete and full assessment, you'll need to watch the burner unit transition from low fire to high fire. Conditions can be very different as a burner transitions to different firing rates.

2.9.2 What To Look For

When observing burner flames, there are at least four important characteristics that can provide insight into flame quality and safety issues: flame color, shape, symmetry, and stability. Every site should have a daily flame observation program that has a person actually reviewing flames and noting their condition. There is no safety or risk management substitute for this that is available with sensors or other equipment that can be installed. In some cases, for large boilers or furnaces, cameras are installed in fireboxes so that operator visual monitoring can be continuous.

Flame Color This discussion applies only to standard burners (not to low-No_x burners) that burn natural gas. The color of other fuels and those associated with low-NO_x burners can be completely different. Hydrogen flames, for example, provide pure ultraviolet light and are almost invisible to the naked eye. Low-No_x flames are usually at lower flame temperatures and often appear yellow. Make sure that you learn from the manufacturer at commissioning, if possible, the proper flame color and shape for your burner and fuel type.

Generally speaking, the closer the fuel/air ratio is to stoichiometric, the more transparent a natural gas flame usually gets. The proper flame color for standard burners on natural gas with a slight amount of excess oxygen (2 to 6%) is blue, with yellow or orange flickers coming from the tail ends of the flame. The flame should also have movement from the burner airflow.

In the case of a traditional nozzle mix burner on a boiler with a circular burner ring, the flame should have a round swirl. You don't want to see a lazy orange or yellow smoky flame, as that would be fuel rich. You also don't want to see a very pale blue high-energy fast-moving sharp-edged flame, as that would be a flame that is too lean.

In the case of fuel oil, one usually also reviews stack opacity. If an oil flame is bad, you'll see smoke or opacity. Opacity can be measured with instruments and also by a Ringelmann spot test, which is more subjective. This test is performed by someone specially trained to look at the stack discharge and compare it to a color chart.

In the case of fuel oil, when you look into the firebox you might also see what are called *fireflies*. Fireflies are actual droplets of fuel oil that move past the initial burner zone and are transported to the rear of the firebox before they can become vaporized and consumed. They will generally look like sparks from fireworks flying at you. This is not a good situation. It means that some improper burner adjustment or problem with the oil nozzle system has occurred, since droplets are forming and not being burned completely in the right place. Issues to check include the atomizing medium and fuel pressures and oil gun nozzle tips for fouling or plugging.

Flame Shape Flames should generally be shaped to fire into the center volume of a firebox, have movement, and not be touching anything. This is especially the case for boilers. Flames should never be directed onto tubes. When a flame touches something continuously, it is called *impingement*. If the flame occasionally touches something and backs away, it is called a *lick*. You might say that the flames are licking the left-hand wall of the boiler tubes if they are touching them occasionally.

The properties of carbon steel begin to change as temperatures increase. If flames touch tubes or another part of a furnace firebox, the design temperature of whatever is touched is likely to be exceeded. In the case of some steels, a phenomenon called *creep* can begin to occur at temperatures above about 800°F.[14] Creep is the tendency of a solid material to move slowly or deform at elevated temperatures.

When a flame touches a steel object, it begins to change color as the temperature rises, and then it starts to oxidize. For example, you might notice parts of a burner diffuser glowing red if it gets overheated. Overheated steel (chemically changed) can also have a bluish tint when it cools. As it is heated further, it can lose strength and disintegrate. Steel also expands when heated, and this can lead to bending and warping of air diffusers and even furnace or oven structures. Boiler tubes that experience creep might also eventually blister and then rupture. Once a burner or air diffuser starts to overheat and change shape, it can lose its ability to mix well and to shape the flame as the designers intended. The burner can then get into a death spiral in which the rate of performance degradation escalates with time. It is for this reason that if hot spots are noted on burners or air diffusers, they must be investigated and repaired quickly.

Many nozzle mix burners are circular. You can describe and record issues with circular nozzle mix burners by mentally superimposing the face of a clock and identifying areas to be investigated or monitored by using the time on the clock face as reference. For example, you can write down in a log that there is a hot spot at 2 o'clock.

Flame impingement can be caused by refractory debris lodged in or around the burner, redirecting or deflecting the burner flame. Other causes of flame impingement could be carbon buildup on the burner ports, broken or missing air diffusers, burner misalignment, or deterioration of all or part of a burner tile block.

Real-Life Story 10: Flame Impingement Can Be Deadly

A terrible flame impingement–related refinery furnace tube failure and explosion happened in the summer of 1998 at a Canadian refinery.[15] A public report described the following account of this incident. It all started when a smoke plume and an oxygen-level problem were discovered in a refinery heater furnace unit that helped convert secondary crude into high-octane gasoline. This type of furnace typically has a firebox with many rows of tubes that contain some intermediate hydrocarbon liquid product that is being heated. The flow of the hydrocarbon product in the tubes removes heat and keeps the tubes themselves from overheating.

A furnace operator was sent to investigate the issue. He was joined by a team of several other workers. No one realized the severity of the conditions. The operator was attempting to correct the smoke plume and swirling yellow flame that had been detected inside the firebox along with oxygen levels in the flue gas that the control room saw to be dropping. Other team members were observing and helping to advise from a control room, and one was on the way to assist when a violent explosion occurred in the east cell of the furnace, sending a huge pressurized fireball through parts of the refinery. The operator was killed in the explosion, and another worker was injured. It took rescue teams more than two hours to recover the body, due to the intense heat and danger in the area.

The cause of the explosion was not immediately apparent. Staff responding to calls about the oxygen-deficiency problem saw it as a routine issue. They did not initially perceive themselves to be in danger, but later investigations found that tube hot spots caused by flame impingement had occurred. These had been detected and reported by operators previous to this incident. They made adjustments to the burners to reduce the firing rate of some burners, and the problem seemed to be resolved. However, the refinery inspectors and engineering staff did not realize the impact on tube life caused by even temporary overheating of the tubes. The excessive heat made for creep failure conditions and weakened tubes that eventually failed. The sudden rupture created a deadly mix of released hydrocarbons and air that caused the explosion.

Lessons Learned If tube or structure flame impingement is recognized, even for a short time, the equipment should be shut down, cooled off, and inspected for damage.

The damage might need to be assessed by a qualified metalurgist or through special testing. Remember that confined-space protocols, possibly those mandated by OSHA, might be required for firebox entries. Once even short-duration overheating has occurred, the damage is done, and tubes, burners, or even structures cannot heal themselves. Flame impingement issues are unlikely to be fixed without removal of the burner or making repairs inside the firebox. Some causes of flame impingement are fallen refractory, corrosion or plugging of burner ports, air diffusers that are compromised, or fuel/air ratio controls that are not functioning correctly.

Symmetry The burner flame should be present and fully lit in a consistent and solid fashion throughout the entire burner geometry. In the case of a round nozzle mix burner, the flame should be lit and stay lit the entire way around. It should not be flickering on and off in different sections around the circumference of the burner. In the case of airflow burners, the burner housing can be several feet long and in various shapes, such as an H. All of the burner surface in this case should be lit and show even consistent flames.

In a furnace or oven you should also look for symmetry from similar burners in similar zones. For example, burners at entrance and exit doors may be set up differently from interior-zone burners, but most entrance-zone burners for common equipment should look similar. Make sure to verify this with the furnace oven designer or manufacturer.

For individual burners, you should not see glowing red spots or dark spots. If you see a dark spot, it generally means that you are seeing a cold area of the firebox. This is an area where the refractory has possibly been compromised. In some cases, if you can get a view of the entire firebox, you might be able to see hunks of refractory spalling off or lying in the bottom of the firebox. Look for things like this and bricks moving out of place if they are visible.

You should examine the boiler or furnace wall on the outside of the combustion chamber if possible where this dark spot is occurring. Inspecting the outside shell or wall of a dark spot may reveal a hot spot on the outside. This is usually characterized, at the very least, by burned-off paint. In some cases overheating could make for a buckled outer skin or even a glowing red spot. Localized refractory damage and overheating could eventually make for a burn-through and fire.

Stability Flame stability is another important parameter that you must assess. An unstable flame is a flame that is close to being extinguished. Flame stability is influenced by many factors, including fuel/air ratios, burner conditions, and even fuel quality. Many burners are designed for flames to start $\frac{1}{4}$ to $\frac{1}{2}$ inch from the face of the burner nozzle. Sometimes, when conditions are not correct, the flame is pulled off the burner and is several inches to several feet into the combustion chamber. This condition, known as *blow-off*, is generally noisy, and the flame often appears to alternate being on and off. The sound, which may be similar to woofing, is caused by the fuel–air mixture searching for air and signifies a very dangerous condition. This sound often immediately precedes an explosion.

The phenomenon opposite to blow-off, *flashback*, occurs when a mixture does not leave the burner faster than the flame speed. This means that the flame burns back into the burner nozzle. This causes impingement, destroys the burner tip, and can make for incomplete mixing, flame stability issues, and explosions.

Combustion rumble, a sudden vibration and rumbling of the equipment, can be another sign of flame instability and can occur at almost any firing rate. It can get quite severe and scary. Sometimes, as the firing rate is changed, it goes away. At other times you may hit a resonant frequency where the vibration will start and then get worse. Combustion rumble is sometimes a flow-induced resonance that is caused by the design of the unit and a combination of fan, airflow, or flue gas flow issues. Combustion rumble flow issues may also be due to issues with linkages or regulators, where the flame is being extinguished and reigniting in rapid succession because of fuel/air ratio problems. I have experienced combustion rumble when fuel/air linkages have fallen off. Combustion rumble and flow-induced vibrations can be very dangerous. Don't operate equipment this way. If this occurs, shut the equipment down immediately and have it investigated by an experienced burner technician.

2.10 FUEL/AIR RATIO EVALUATIONS

You have learned about the importance of proper fuel/air ratios to safe and clean combustion. Here we provide information about how this is done in the real world for different burner systems. Fuel/air ratios are usually evaluated or verified in one of two basic ways (depending on the type of combustion system, open or closed). In the case of closed combustion systems serving vessels such as boilers or indirect heat exchangers, the burner technician measures the composition of flue gases with a flue gas analyzer that looks at the amount of certain unburned fuel by-products and excess air in the combustion products. For some industrial ovens, and other combustion equipment, many burners may be firing into a common chamber. In these open system situations, the technician uses ports into the burner that indicate the pressure of fuel and air delivered at different firing rates. These provide an indication of the ratio of fuel and air being delivered. This information can then be used to calculate the fuel/air ratio.

In the case of open systems such as makeup air heating equipment or grain dryers, there is so much dilution air going past the burner that you can't tell much from using a combustion product analyzer in the resulting hot airstream. Consider the case of an airflow burner on a grain dryer. A burner could be passing 20,000 cubic feet per minute (ft^3/min) of ambient air to be mixed with 600 ft^3/min of hot flue products to dry grain. Meaningful information cannot be gained from trying to assess with an analyzer what is happening downstream of this type of burner because of the dilution of the flue products from the burner. In airflow burner and other direct-fired open chamber applications (where there is no intermediate heat exchanger), you can make fuel/air ratio judgments only from direct flame observations and from an assessment of the gas and air pressures and subsequent flows being delivered.

2.10.1 Fuel/Air Ratio Evaluation of Closed Systems

Closed firing system fuel/air ratios are identified most commonly through the use of flue gas analyzers. These complex devices require skill and experience to use properly.

There are many different styles and brands of flue gas analyzers, and some technicians are fiercely loyal to a particular type or brand. Flue gas analyzers can cost from as little as \$1000 to more than \$10,000, depending on the features and accessories. Generally speaking, you get what you pay for when you buy a flue gas analyzer. Most facilities buy an analyzer only if there are a considerable number of pieces of combustion equipment and if there are people who can be trained to use them properly.

Flue gas analyzers, in this context, are handheld survey devices. They are an instrument with a pump and hoses connected to a probe. The technician inserts the collection probe to a hole in a section of flue pipe. The instrument typically draws a sample into the unit and reads the flue gas temperature, oxygen, carbon dioxide, and in some cases volatile organic compounds and potential pollutants like NO_x and SO_x. There are a number of important considerations in using combustion analyzers properly: waiting long enough for a piece of equipment to be at the proper operating temperature, making sure that probes are of the correct length to get to the center of a stack, and making sure that equipment has been calibrated recently are just a few.

Manufacturers of burners on smaller equipment usually do not provide specifications for ideal flue gas compositions, but only general guidelines. It's important that flue gas compositions be somewhat consistent throughout the entire firing range. There should not be step functions in the composition of flue gases as firing rates change.

For packaged boilers where NO_x is not an issue, it is generally desirable to tune burners to produce below 100 ppm of CO throughout the firing range. There are EPA testing guidelines for obtaining this information if it is to be used for compliance or permitting purposes. Before you start taking readings, make sure that you understand whether or not your equipment has environmental compliance requirements and what the consequences of nonconformance with these requirements will be. As stated earlier, in some cases you have a legal obligation to report yourself if you are not in compliance.

The most common closed systems are boilers. With boilers, the technician usually also verifies that the boiler system has the ability to fire up to full load. This may not seem as if it should be a problem, but consider trying to do this on a space heating system in the summer. In many cases boiler steam systems are designed with a manual vent valve that runs a full-sized vent pipe out to the roof and through a silencer, as venting steam at high flow rates can be very noisy. This vent gives the steam that is generated at high-fire a place to go if you choose to set the burner for a high fire level and evaluate it in the summer nonspace heating season or when the steam plant has insufficient load to absorb the steam generated during testing. Think about the possible need for a steam vent before starting testing and burner adjustments.

2.10.2 Fuel/Air Ratio Evaluation of Open Systems

In open systems, where the use of an analyzer is not practical, evaluation of fuel/air ratios usually includes measuring air and fuel pressures, converting them to flows, and comparing the readings to manufacturers' specifications where possible. Orifice plates are a common means of measuring fluid flow. They are widely used for identifying fuel and air flows for combustion equipment. Special orifice plate holders in common use for heat-processing equipment are often used in fuel and

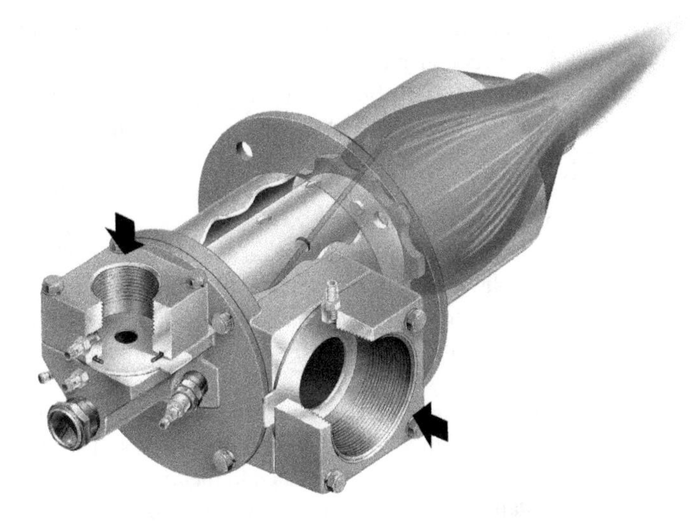

EXHIBIT 2.3 Eclipse Thermjet burner with pressure taps and integral orifice plate.[17]

air piping systems for this purpose. These orifice plates have holes that provide a specific pressure drop proportional to a given flow.

The pressure of fuel and air delivered to burners is usually measured in very small increments, usually in inches of water column, designated as inches water column "W.C." A manometer is often used to measure pressures from flow orifice taps. Popular models of digital and U-tube liquid-filled manometers are made by Dwyer Instrument[16] and others.

Some burners come with integral air and fuel flow orifices and pressure measurement taps (Exhibit 2.3). Sometimes these taps have small valves built into them to access the pressure-measuring ports through the taps. When these are provided, the manufacturer usually gives information on flows that correlate with the pressures that are read. These are usually also a function of firebox pressure, which you might have to read at another sample port (not on the burner) that extends to the firebox.

Be careful when using electronic digital manometers for measuring pressures because they can be so sensitive that the signal bounces all over and getting a stable reading may not be easy. Also be aware that some digital manometers have been approved for use with natural gas and some have not. When using liquid-based manometers, incline units that use special oil can show very small changes in pressure. Manometers also come with a direct-reading gage face. These can also be very sensitive to fluctuating pressures. Whatever type you choose, make sure that it's accurate, rated for the intended purpose, well maintained, and calibrated.

Usually, the orifice plate or holder will have a stamping or marking on it that indicates the manufacturer, who can provide you with the size or even with a table that indicates flow through the orifice based on pressure differences. Flows through orifices are usually calculated by applying the square-root law. The basis for the square-root law is derived from Bernoulli's equation for incompressible flow.[18]

Once you take readings and know the flows of gas and air at different firing rates, you can easily determine the fuel/air ratio at different loads and compare this to the manufacturer's recommendations. If the burner has a sight port, you can also examine the burner and flame and verify that all is correct and optimized with respect to color, shape, symmetry, and stability. If you're lucky enough to have a system with a gas flow meter, you can clock the meter and see if your flows are close, as an extra check on the fuel flow rates that you calculated.

Whether a system is open or closed, measuring fuel and air flows can be especially important when setting up burners for the first time. Technicians usually establish airflows, first setting what's called a *cold air curve* to provide airflow to the burner manufacturer's requirements. This is done before fuel is brought on and firing attempts are made.

Remember, the adjustment of fuel/air ratios is a high-risk process. It should only be done by very experienced and highly trained burner technicians. The goal of this chapter is to raise your awareness of the issues related to fuel/air ratio setting and give you a better basis to understand what is happening when this process is occurring, and what the critical success factors are. Finally, I want to leave you with a saying among burner technicians regarding fuel/air ratio adjustment. If the mixture is too rich, always look to pull fuel first slowly; don't add air. If there is an accumulation of unburned fuel (being too rich) and you add air too quickly, you can have an immediate explosion.

Real-Life Story 11: Black Smoke from a Stack Is Never a Good Thing

I was invited to assist a client in a developing country who was switching from manufactured gas (gasified coal) to natural gas. The heating value of this type of manufactured gas was about 500 to 600 Btu/ft^3, and the heating value of the newly supplied natural gas was about 1000 Btu/ft^3. The manufactured gas included a lot of carbon monoxide and hydrogen. The client wanted a review by someone who had experience with the burning of what the client believed to be that "dangerous carbon-rich gas" that we call natural gas in the United States. The change was plantwide. There had been two explosions and several other close calls over a six-month period during the fuel transition project. The plant facility covered a large area with many buildings and processes.

Upon driving into the site with my host, I saw a couple of stacks that were bellowing black smoke and said: "Hey, what's that building—the one where black smoke is pouring out of those tall stacks?" Yes, black smoke was literally pouring into the sky. I asked to go there first. I seemed to be the only one seriously concerned.

The building had long rows of radiant tube furnaces. Each radiant tube had a burner firing into it. These are high-temperature alloy tubes meant for use in a special atmosphere for heat treating. Flue products come out the other end to collectors, where they are eventually routed out of the building to tall stacks.

The facility had been operating this way for weeks. It was thought that this was one area that was operating correctly. Upon walking down the line of some of the furnaces inside the building, I was astonished to see piles of black soot (raw carbon) under some of the tube burners and collector boxes. You could walk up to any burner and see orange and bright yellow flames, which indicated very fuel-rich conditions. Later it was found that orifices in the burners were not changed out to accommodate the lower flow requirement for natural gas versus the manufactured gas.

In another area of the facility where ladles and their lids were relined, one of the workers commented: "This new natural gas fuel sure is smoky. It gives me a heck of a headache, too." I went over to review his refractory dry-out burners and found that no one had changed orifices in this area either. After this was done, I discussed with him what good and bad flames looked like so that he could avoid this condition in the future.

Lessons Learned It's important that fuel/air ratios be set properly when equipment is commissioned. You should understand what the flue gas composition should be for different firing rates and have documentation from a calibrated combustion analyzer to prove what you have. In addition, where possible you need to confirm things visually at a sight port for different firing rates. It's important that proper fuel/air ratios be verified immediately whenever a fuel switch occurs.

NOTES AND REFERENCES

1. U.S. Chemical Safety Board, www.csb.gov, completed investigations, Little General Store, West Virginia.
2. *North American Combustion Handbook*, 3rd ed, vol. 1, p. 283, definition of combustion.
3. Praxair, MSDS for methane, MSDS, http://www.praxair.com/praxair.nsf/AllContent/9837F06E014AAA0D85256A860081A239/$File/p4618f.pdf.
4. Ibid.
5. Praxair, MSDS for natural gas, http://www.praxair.com/~/media/Files/SDS/p4627e.ashx.
6. Praxair, MSDS for propane, http://www.praxair.com/praxair.nsf/AllContent/5CFAA6A6BF-7073C685256A860081E832/$File/p4627e.pdf.
7. R. Baker, Temperature Distribution Inside a Lit Cigarette, the British-American Tobacco Co., Southampton, England, presented November 1996, at the Tobacco Chemists Conference, Winston-Salem, NC.
8. U.S. Department of Energy, www1.eere.energy.gov/manufacturing/tech_deployment/pdfs/oxygen_enriched_combustion_process_htgts3.pdf.
9. Hilli incident, Marine Accident Investigation Branch, March 2007, Report 4/2007, http://www.maib.gov.uk/cms_resources.cfm?file=/hilli.pdf.
10. Praxair, MSDS for hydrogen, http://www.hydrogenandfuelcellsafety.info/resources/mdss/Praxair-H2.pdf.
11. William F. Raleigh, Premix Burners: Technology Advancement and Engineering Challenge, Tekniththerm Engineering, American Society of Gas Engineers 2008 Technical Conference.
12. Charles E. Baukalik, Jr. (Ed.), *The John Zink Combustion Handbook*, 2001, CRC press.
13. Yunyong P. Utiskul, Neil P Wu, and Hubert Biteau, *Combustion Air for Power Burner Appliances*, NFPA Fire Research Foundation, January 2012.

14. NBBI, Creep formation, http://www.nationalboard.org/Index.aspx?pageID=181.
15. Accident Investigation Report on the Explosion and Fire at the Irving Oil Refinery, Saint John, New Brunswick, Workplace Health, Safety and Compensation Commission of New Brunswick, June 11, 1999, http://ncsp.tamu.edu/reports/WHSCC/irvingreport.pdf.
16. Dwyer Instrument website, www.dwyer-inst.com.
17. Eclipse website, www.eclipsenet.com.
18. Francis S. Tse and Ivan E. Morse, *Measurement and Instrumentation in Engineering: Principles and Basic Laboratory Experiments*, Marcel Dekker, New York, 1989.

3

Natural Gas Piping Basics

The Veins and Arteries of the Fuel and Combustion Systems World

Combustion systems start with fuel systems and fuel systems, start with piping. By far the most common fuel burned throughout the world is natural gas. Natural gas use is growing even more in popularity as the United States develops shale gas deposits. For this reason the primary focus of this chapter is piping related to natural gas systems.

Before we discuss advanced gas piping concepts it's important to review the basics. In this chapter we attempt to discuss the most basic natural gas–related piping concepts starting with the piping itself, how it's made, and how it's installed.

Real-Life Story 12: Wrong Valve Pressure Rating Could Have Led to Many Deaths

I arrived at a site to find the office areas looking like the aftermath of a cheap action movie. The building walls had literally been blown out and cement blocks were everywhere. Ceilings had dropped along with light fixtures and their grids in many office areas. You have to believe that if the explosion had occurred during working hours, someone would at least have been injured, if not killed. Fortunately, the incident occurred over a holiday weekend.

The site was a 20,000-square-foot warehouse facility with an attached 5000-square-foot office area. The warehouse building had low-pressure gas piping throughout to serve natural gas unit heaters hanging from the roof trusses. A small mechanical room had furnaces for the office area and a small hot water heater. The building was fed by a natural gas utility service located on an outside wall. A relief valve had been

Fuel and Combustion Systems Safety: What You Don't Know Can Kill You!, First Edition. John R. Puskar.

installed after the utility's regulating station discharge. The relief valve discharge line had a tag that read "10 psig." This meant that the utility service was capable of delivering up to 10 psig of gas pressure into the building.

NFPA 86 and other codes require that all fuel train components must be rated for the maximum pressure that can be delivered to a fuel train. This is the pressure that can be "seen" by the fuel train if the closest upstream regulator were to fail. In this case the delivered pressure could be up to 10 psig, limited by the relief valve setting, if the main service regulator were to fail.

It was obvious from the fact that there was devastation in several distinct parts of the building that there were gas releases in multiple areas. All of the fuel trains to space heaters and the main furnace and hot water heater room were found to be piped with the same manual equipment shutoff valves. These were light-duty manual isolation valves that were found to be rated for only 2 psig of pressure. In fact, the rating was cast into the body of the valve. I believe that the pressure normally delivered was less than this. However, since the relief valve on this system was set at 10 psig, a service regulator failure could have easily fed at least 10 psig.

After the incident, a number of the 2 psig isolation valves were tested at higher pressures. It was found that if these valves were in the open position, a pressure of about 7 psig would cause the stem to pop up out of the valve body and release gas. The investigation found that the utility did have an overpressure situation that weekend. The time-stamped real-time pressure recordings from the utility matched the timing of the incident. Upon closer examination, several of the isolation valves on equipment were found to have their stems in the "popped up" position.

Lessons Learned Gas utilities provide some maximum allowable operating pressure (MAOP) to each site. In this case it was clearly 10 psig—not what the site was operating at. It's important that you ensure that your facility's gas distribution systems and fuel train component ratings exceed whatever the MAOP and relevant upstream regulators are capable of providing. You can see that it may be different from the pressure at which your systems operate today. Call your local utility to discuss what MAOP your facility was designed for so that you can compare this to operating conditions.

There are a number of ways that fuel train components can be protected from the possibility of regulator failures for code compliance: including having robust high-pressure designed components, adding gas pressure relief valves at fuel trains, or using special lockup or overpressure-compensating regulators with internal relief valves. An experienced combustion engineer should be consulted on this if you think you have a problem.

You'll find pressure ratings stamped on the side of many components or on manufacturers' tags together with model and serial numbers. When in doubt, call the manufacturer to discuss the rating. When you review ratings of fuel train components, you must understand what some of the terminology means. For regulators and pressure switches, there is burst pressure and maximum pressure. *Burst pressure* relates to the integrity of the device housing. *Maximum pressure* is usually the most pressure a component can "see" and still provide repeatable results. For example, a regulator may be overpressured momentarily to something below the burst pressure of the casing. It may, then, never be able to regulate pressures accurately because of damage done to internal components. In the case of valves, this can mean maximum closing pressure, which indicates the greatest pressure difference against which the valve can operate. Other pressure ratings related to valves might speak to the ability to hold and not leak through when the valve is closed.

3.1 NATURAL GAS PIPING CODES AND STANDARDS

Natural gas piping is a very broad topic, and the codes and standards that apply are diverse (see Table 3.1). The major applicable codes and standards that apply are NFPA 54, the National Fuel Gas Code; NFPA 56, the Standard for Fire and Explosion Prevention During Cleaning and Purging of Flammable Gas Piping Systems; ASME B31.1.1, the Power Piping Code; and ASME B31.3, the Process Piping Code.

NFPA 54, called the National Fuel Gas Code, is intended to cover fuel gas piping up to 125 psig but does not cover installations where fuel gas is not used as a fuel or piping in electric utility power plants; see the NFPA 54 Scope for a complete list of installations covered. Fuel gas has a specific definition and is considered to be commercially available gaseous fuels. That's a unique distinction because it means that intermediate noncommercially available hydrocarbons piped around a facility such as a refinery would not be within its scope.

ASME B31.1, the Power Piping Code, covers the design, materials, fabrication, erection, and testing of piping systems typically found in electric power generating stations, industrial plants, and other installations. See the ASME B31.1 Scope for the document coverage. The fuel gas piping requirements in NFPA 54 are consistent with the requirements in B31.1 for pressures under 125 psig.

ASME B31.3, the Process Piping Code, focuses on piping systems that may be associated with refineries and petrochemical plants. In many cases this piping could be related to fuel gas systems. See the ASME B31.3 Scope for the document coverage. The requirements provided for gas piping systems in B31.1 and B31.3 are not dramatically different. There may be elements in both codes that support the types of systems that you need to know about.

TABLE 3.1 Partial Summary of Applicable Codes and Standards for Natural Gas Piping Systems

Piping Section	Applicable Code or Standard[a]	
Up to gas meter discharge (typically, the utility's distribution piping)	U. S. 49 CFR Part 193	
Meter to appliance shutoff valve	NFPA 54/56 ($\leq$125 psig)	ASME B31.1/3 (>125 psig)
Appliance shutoff valve to appliance burner	Boiler	NFPA 85; ASME CSD-1
	Furnace/oven	NFPA 86

[a]Be sure to verify your project's requirements within the scope of these documents.

NFPA 56, the Standard for Fire and Explosion Prevention During Cleaning and Purging of Flammable Gas Piping Systems, covers all flammable gases, not just natural gas. It also covers natural gas used in power plants and other industrial applications over 125 psig. I have had the privilege of being one of the founding members of this standards committee. The standard is discussed in more detail in Chapter 5.

Before you can understand issues of repair and maintenance, you must understand the basic elements used in piping systems. Pipe, valves, and fittings are each discussed here to provide a basic understanding of how these can be selected and specified for a gas piping project. Remember, this is not meant to be an in-depth instructional design guide, simply to make you aware of what exists and why it's used.

3.2 GENERAL INDUSTRIAL UTILITIES PIPING FUNDAMENTALS

This discussion of piping basics common to many types of industrial utility systems includes specifications regarding piping thickness and lengths as well as pipe mill operations. Most gas piping systems are constructed with steel pipe, although copper is also an acceptable material in certain applications. If copper is used, the joints must be brazed with materials that have a melting point of over 1000°F (most commonly used brazes, but not solders, meet this requirement). This is different from the typical lead-based solder used for water pipes, which has a melting point below 400°F.

Let's start with how steel pipe is made. There are three primary manufacturing methods: the electric resistance welding process, the continuous welding process, and the seamless process. The *electric resistance welding* (ERW) process and the *continuous welding* (CW) process both start with steel in a precision strip form. In the ERW process the strip is fed cold (unheated) toward rolls that slowly form the pipe into an oval and then a precision cylinder.[1] A continuous electric resistance weld is applied at the joint that is formed. Excess metal "flash" that evolves from the welding process and the pressure of the two pieces being pushed together is removed with a special cutter.

In the *continuous welding* (CW) process the strip is first heated in furnaces and then to form the pipe is shaped into an oval and precision tube, just as in the ERW process. The difference is that the temperature and pressure make a forged weld.

To put it in its simplest terms, seamless pipe is produced by piercing a solid billet of deoxidized and conditioned steel which has been heated to a required temperature. It is then processed through a series of mills and rolls until it is finished to prescribed dimensions.

Pipe mills usually provide pipe in 21-foot lengths, although specialty lengths of up to 40 feet can be obtained in some sizes. Pipe can be provided with plain ends, beveled ends, grooved ends, threaded ends, or threaded and coupled ends. Pipe is also available in a number of different surface finishes. If a threaded and coupled arrangement is purchased and the intended use is natural gas piping within the scope of NFPA 54, you must make sure that the threads are tapered threads, not straight threads. If straight thread couplings are used, they may be mixed with tapered thread fittings, which will undoubtedly leak over time. NFPA 54 requires that only tapered thread fittings be used on gas piping systems.

Steel pipe can be purchased that conforms to many different specifications and metalurgical grades. NFPA 54 identifies pipe for gas lines up to a maximum allowable pressure rating of 125 psig. It states that ASTM[2] A53A or A53B is suitable, along with ANSI B160 seamless pipe. ASTM A53B is stronger than ASTM A53A.[3] The A and B designations denote the grade of the pipe. Seamless B160 is even stronger, but it usually costs much more than the others. ASTM A53A and A53B can be bought either longitudinally welded or seamless.

After determining the designation, you would then specify a schedule or thickness of pipe: schedule 10, 40, 80, or other. The higher the schedule number, the thicker the pipe wall and the smaller the inside diameter. For example, a 2-inch schedule 10 pipe has an approximate wall thickness of 0.109 inch, a 2-inch schedule 40 pipe a wall thickness of 0.154 inch, and a 2-inch schedule 80 pipe a wall thickness of 0.218 inch. Guidance on pipe thickness calculations and designations can be found in ASME B31.1 and B31.3 for natural gas applications and in ANSI specifications.[4] I have often specified schedule 80 for fuel gas piping systems that may be installed in areas where corrosion is possible (even though it was not needed from a pressure rating perspective) to provide an additional element of safety. Adding material for corrosion considerations is a traditional practice in process plant design.

3.3 MANUAL ISOLATION VALVES

Manual shutoff isolation valves play an important role in combustion systems and equipment safety. Ball valves and plug valves are the most popular styles valves installed in fuel systems for shutoff purposes. Regardless of the style of valve used, most NFPA codes call for components to be listed for use by a nationally recognized testing agency. Two common approvals for valves are Factory Mutual (FM) and Underwriters' Laboratories (UL). Both FM and UL have guides available where you can look up and verify components and their ratings. There are often markings on the body of the valve or on its handle with specific approval codes you can research. You may also find pressure limits marked on the valve body.

Manual shutoff valves for equipment shutoff are required to be installed in an *accessible location*, defined in some codes as being less than 6 feet above the finished floor, and the placement must not be encumbered by any sort of access door or panel, which must first be removed to get to the valve.

Ball valves are constructed with a chrome-plated or specialty-material ball that has a hole through the center. The ball is designed to seal against some type of Teflon or other durable seating material. Ball valves usually have an indication of their suitability for use in natural gas systems and their pressure rating stamped or cast into the side of the body or denoted on the handle. Many of these valves show the term "WOG" (water oil gas) somewhere, which will be followed by a number, such as 300 psig. However, the gas referred to is not natural gas but an inert gas. These valves should be used only with inert gases and are not permissible in flammable gas systems. If you are in doubt as to a valve's capabilities or want to understand these risks better, check with the valve manufacturer.

Unlike plug valves, ball valves do not require annual maintenance. Ball valves can also be purchased with locking handles, which make them preferable to plug valves in many applications. Ball valves come in the form of either a floating ball or a trunion mount. In the floating ball style the ball has the capability to move slightly in the valve body so that the pressure in the pipe can help to force it against a seat to seal. In trunion mounts, which are generally in larger sizes, (i.e., over 6 inches), the ball is held in place by precision bearings and a variety of sealing mechanisms are used. In fact, trunion mount ball valves are available with double seals and even in double block and bleed arrangements. Many trunion mount ball valves have what appear to be sealant injection fittings. In most cases manufacturers' instructions indicate that these are to be used with special sealing materials and injection devices only if there is a need for the valve to be sealed on an emergency basis.

Plug valves are quite simple in design. Usually, there are only four main parts: the plug, the body, the bonnet, and the button head fitting. The two major types of plug valves are lubricated and nonlubricated. It's easy to tell the difference. You will find a button head fitting (for injecting sealant) only on lubricated plug valves.

Two primary body/plug types are offered in today's marketplace. One is a straight plug and the other is tapered. On a plug valve, the plug and body are not in intimate contact. Lubricated plug valves require a special a sealant to be injected that fills the space between the plug and the body. It's the injection of this sealant that actually provides the seal. There is no seal without it. This is an often overlooked but vital part of lubricated plug valve maintenance. This sealant is specific to the type of service the valve is fulfilling. The sealant is not common grease. It is a specially engineered sealant and lubricant that must be installed with a special high-pressure sealant injection gun. Valve manufacturers and standards like NFPA 86, the Standard for Ovens and Furnaces, require reinjection of this sealant on an annual basis.

The plug has a hole in it so that when it is aligned with the piping, flow can take place. If little or no sealant is present, the main hazard is gas leaking through the valve, around the plug, when you think it is in the closed position. Also, inadequate sealant promotes corrosion to set up between the plug and the body. This creates a condition where the plug can seize in the body, making the valve no longer operable.

Although they require at least annual sealant injection, lubricated plug valves are durable and widely used. In smaller ball valves, if you scratch the ball or the seating surface with pipe debris, you must discard the valve. It will leak through in the closed position if this happens. Lubricated plug valves, on the other hand, can take small amounts of debris and continue to function well as long as the sealant is maintained. Non-lubricated plug valves may not be rated for fuel gas service. Also, the lubricants/sealants required can change with the fuel gas service. Make sure that you understand the types of valves installed at your facility, their ratings and whether (and how) they need to be serviced.

3.4 BLANKS OR BLINDS

There is the possibility of a leak in any valve, and for this reason, valves should never be trusted completely, especially during maintenance shutdowns or when someone has to be inside a firebox (always be in compliance with OSHA 1910.147 for achieving zero-energy states when locking out utilities). Blinds are the only sure protection for isolating any piping system. The terms *blanks* and *blinds* are used interchangeably. The word *blinds* is used in the balance of the book to describe isolation equipment consisting of a flat piece of steel of the appropriate thickness installed between two flanges to prevent the possibility of any flow.

Blinds come in many forms, including thin sheet metal slip blinds meant for temporary use to thicker higher-pressure-rated permanently installed blinds. The thickness, gage, and strength of blinds depends largely on the pressures that are involved. Blinds are also available in a number of surface finishes. These include being painted or having a root mean square (RMS) designation such as RMS 150. RMS is a machining tolerance that relates to how rough a surface is: The lower the RMS value, the smoother the blind surface. I have found that, typically, RMS 150 blinds work well with traditional gaskets.

Blinds may come in the form of a simple circle, a paddle, a figure 8, or in a host of other shapes. The blind may have bolt holes that align with a standard flange, usually class 150. If the blind is to be reused several times at a location, consider a spectacle blind for permanent installation (Exhibit 3.1). One part of the blind has a solid disk to totally block the line; the other half of the blind has a ring with the inner diameter equal to the inner diameter of the pipe. When the blind is positioned for flow, the solid disk part shows. When it is in the isolation position, the open circle shows. There are also mechanized line blinds that use hand wheels or motors to move the blind plate between the open and closed positions. In the steel industry, large-diameter blast furnace gas blinds called "goggles" are common. Automated blinds are usually quite expensive and are used only when there is a need for frequent blind removal and reinstallation.

Many designers don't remember to design their systems to accept blinds. Remember that as a practical matter, pipe does not stretch. You don't want to force joints and misalign existing flanges in places where room to install blinds was not provided. In some of these cases, piping designers add spacers (a ring in the piping system the same thickness as the intended future blind) so that there will be adequate room to install a blind easily in the future.

EXHIBIT 3.1 Spectacle blind installation; opaque end is in, no-flow isolation condition.

When installing a simple slip blind between two flanges, you can use a tool called a flange spreader. These usually consist of some type of wedge that gets clamped onto the flanges and is then turned with a wrench to spread the flanges apart slowly. Use of a flange spreader may or may not be successful, depending on the condition of gaskets between the flanges. In many cases, blinds need to be installed on both sides of a valve. Spreading pipe for these purposes may require loosening hangers or friction clamps and/or installing come-alongs (winches) to pull joints apart.

When installing a blind, remember to ensure that the side of the valve being opened up (for a valve/blind installation) is completely purged with nitrogen and that all line-breaking safety protocols are observed. Make sure that the valve has been sealed off and is verified not to leak. Bleed the other side of the system/valve to reduce any possibility of the existence of gas on the side of the valve to be blinded. Refer to Chapter 5 and to NFPA 56 for more information on other processes for conducting this type of work, to make sure that the work is done safely.

IMPORTANT

It's always better to plan ahead and make space for blinds. Even on retrofits it's important to cut and weld on new flanges if you have to, rather than using portable winches to pull pipe apart and try to force-fit the blinds into place. Forcing blinds can create misalignment and leaking flanges.

NFPA 54 states that you cannot reuse gaskets with any flanged gas piping repair. On a recent project I saw blinds installed that would not seal. It was suspected that heat from welding either warped the blinds or surface defects allowed leaks. The most common problems identified during the final pressure test of that facility was leaking gaskets and warped blinds. Make sure that you have spare blinds and gaskets of all anticipated sizes available when planning a project.

3.5 STEEL PIPE JOINING METHODS

Pipe joining methods for steel gas piping systems include welding, flanges, and threading. Each of these is discussed below.

3.5.1 Welded Fittings

Fittings to connect piping lengths use either welded or threaded joints (flanges are connected to pipe by welding or threading). Available welded and threaded fittings include T's, elbows, reducers, couplings, unions, and 45's. Welded fittings are made available as standard and high-strength forged steel. Welded piping connections occur as butt welding (end-to-end prepared pieces fitted together), slip-on fittings, and socket-welded fittings.

In butt-welded "weld neck" flanges, the two pieces (flange and pipe) are prepped and then clamped together. The prepping is very important, as the objective is to make a somewhat precise V-groove to accept the welding material once the two pieces are butted together and tacked. Most welding fittings come prepped with half a groove. Tacking is the process of making a temporary small weld to hold things together while alignments are checked. Pipe fitters also have to be aware of proper holing alignment. *Holing* is the process of lining up the holes in the flange before it is installed permanently so that when it mates up with an existing flange, valves or connections on both ends of the pipe can be made and fit with other mating flange holes. Once aligned, holed and tacked, a welder or pipe fitter, if credentialed for that type of welding, would then lay in the first welding pass, called the *root pass*; subsequent passes are *filler passes*. Different welding rods, materials, and welding machine current settings can be used for these, depending on the piping system requirements. These variables and materials and methods all constitute welding specifications which are critical for a projects success. Welding specifications can also include a requirement for periodic weld inspection and testing.

Other examples of butt-weld fittings are Bonney Forge's Weldolets and Threadolets.[5] These are an economical branch connection designed to minimize stress concentrations and provide integral reinforcement for branch connections and take-offs. Manufactured in a number of different pressure ratings, they provide a safe alternative to what is called a fish-mouthed or field-fabricated connection, where one pipe is just welded over a hold in another pipe to make a connection.

Slip-on flanges and fittings are just that: They are slipped onto the end of the pipe instead of butted up to it. In the case of a flange, the pipe is welded up around the contact points of the pipe and the flange. Slip-on flanges are not considered to be as strong a joint as weld neck or butt-welded flanges and fittings.

Socket welding is like a slip-on for small fittings. Elbows, for example, are available with a socket into which the pipe fits. The welder then seals the contact point with weld materials. The primary precaution to consider is that the pipe cannot be bottomed into the socket weld fitting. Socket welding is popular for fuel oil piping since threaded fuel oil piping tends eventually to leak.

3.5.2 Hot Taps

Hot tapping is a process by which a branch from a main line can be installed on a piping system without a shutdown and removal of the pipe contents. Making hot taps is a specialty that should be undertaken only by very skilled contractors experienced in carrying out such procedures with very special tools.

After careful planning and a safety review, the hot tap process usually starts with a Weldolet or Theadolet being attached with a careful weld. The procedure might include having flow in the pipe to remove heat from the welding process. Then a fitting and attachment are installed that allow a special drill bit to be sent down through a special valve. The drill cuts the pipe and captures the slug simultaneously. The drill is removed from the hole through the valve and the valve is closed without a release of what's in the pipe. The new takeoff is then ready to have pipe attached to it.

This has been done safely thousands of times and carries little risk when done correctly by qualified people.

3.5.3 Welding Certifications

Pipe welding can be complex and requires both knowledge and skills. Welding on gas lines must follow ASME B31.1 and B31.3 guidelines. The welders must also have the appropriate certifications as defined by ASME because the word *appropriate* can mean many things to many people. ASME welding certifications are well known and well respected throughout the world. Other proprietary welding schools and organizations certify welders for a variety of tasks. Job scope, including line size and pressures being used, will determine what welding certifications are required. Each person who is going to perform the work will need to have a document, such as a diploma or certificate. A great resource for understanding the world of welding certifications is the National Certified Pipe Welding Bureau (NCPWB; www.mcaa.org/ncpwb/). People with the proper certifications know what rod and welding machine settings to use for a given joint. They also understand how best to prepare and position an area to be welded and how to follow detailed welding procedures.

3.5.4 The Integrity of Welds

Another consideration is weld integrity testing, which can be destructive or nondestructive. Each of these has certain capabilities, advantages, and limitations. There are many nondestructive techniques, including visual examination, radiography or x-ray, magnetic particle, dye penetrant, and ultrasonic.[6] In most cases a nondestructive testing organization is looking for cracks, voids, or porosity: that is, air holes or gaps that cannot be seen by the naked eye. In addition to the nondestructive tests identified above, there are code requirements for destructive tests, such as impact, bending, cross tensile, corrosion, and V-notch.

It is usual and customary that pressure piping welds be inspected according to code requirements and welding procedures/specifications for the intended service of the piping system. Testing is usually done by a third-party contractor who acts as the neutral observer. In most cases a certain percentage of welds are chosen to be checked according to a specification. If failed welds are found, they must be repaired. This often means grinding them out and redoing them.

3.5.5 Threaded Fittings

Threaded fittings are available in both malleable and forged steel. Malleable fittings are used for low-pressure systems and small-diameter piping (usually up to about 4 inches). Forged steel threaded fittings are heavier and thicker and are used for higher-pressure ratings, (see ASME codes and standards to verify requirements).

All fittings must be checked carefully before their use. Common details to check include thread alignment with the body, thread condition, and even the number of threads. In addition to these details, malleable or cast fittings must be checked for sand holes that might have caused porosity in the casting process.

Threading of pipe is done using pipe threading machines. The most popular ones are made by Ridge Tool,[7] headquartered in Elyria, Ohio. These machines are designed to cut pipe to length and then turn the pipe while a set of fixed dies cuts threads into the end. The dies are changeable according to the pipe size by the pipe fitter operating the machine. There are also portable motorized threaders and hand-rotated threaders.

It is usual and customary to weld over a 2.5-inch pipe size instead of using threaded connections. It's much harder to thread consistently and cleanly above 2.5 inches and that the chances of leaks goes up dramatically as the diameter goes up. In my experience, most natural gas–related jobs do not use 3- or 5-inch pipe.

3.5.6 Flanged Connections

Flanged fittings are generally used with welded systems and larger pipe sizes (generally starting at 4 inches and over). They come in raised-face and flat-faced formats. Make sure that you match raised face to raised face and flat face to flat face.

Just like valves and other components, flanges have pressure ratings of which you need to be aware. The ratings are usually stamped on the circumference (e.g., 150 psig). Once a piping system rating (or class) is determined from, for example, ASME B31.1 or B31.3 (i.e., 150 psig, 300 psig, etc.), all piping system components, including flanges, should be specified for that rating. You can also often tell the pressure rating by the number of holes in a flange, but this is not always an accurate method. Bolt-hole patterns usually follow ratings (e.g., 4 bolt, 6 bolt).

Flange materials can also be tricky. You'll always want to make sure that you're using flange materials that follow code requirements. In most cases this means carbon steel not cast iron flanges for natural gas applications.

It's also important that one consider what's called *holing and alignment*. Remember, a flange can be installed in any number of rotated positions. If one is not careful with the alignment, a valve can be installed at an angle instead of straight, or one end of a valve's bolt holes may line up but not another. Experienced pipe fitters are trained how to handle these types of issues properly.

3.6 FASTENER ISSUES: WHEN A BOLT IS NOT SIMPLY A BOLT

To ensure leak-free flange joints, it is important not only that new gaskets be used but also that the bolts used in the flanges be properly rated and torqued. An excellent resource is ASME PCC-1, Guidelines for Pressure Boundary Bolted Joint Assembly. A lessons-learned technical bulletin prepared by the National Certified Pipe Welding Bureau is also a good reference.

Fasteners have grades and specifications that are indicated in a language all their own with markings and stamped information. ASME fastener guidelines[8] (i.e., ASME B18.2.1 for fasteners and B18.2.2 for nuts) can help to put you on the path for using the proper studs, bolts, and nuts. These must also be installed with the correct torque to ensure proper holding force. ASTM International (formerly the

American Society of Testing and Materials) also has lots of information and standards related to fastener materials and is available at www.astm.org.

Also, when installing studs, be aware of the need to obtain the proper material for the service. Studs that are simply made from an off-the-shelf threaded rod cut to size are not adequate for pressure piping. Real rated studs are usually stamped to indicate that they are a special grade and type of material that is of sufficient tensile strength. Make sure as well that proper nuts, washers, torque, and installation techniques are used.

Finally, be careful that you purchase fasteners from reputable sources. Many counterfeit fasteners have entered the market in recent years. These might have the right markings but are not the actual materials required. Fasteners are not a place to try to save money on a project.

When reviewing existing installations, be aware of fasteners that appear corroded. It may be time for a replacement. Metal loss from fasteners indicates that their strength has been compromised. You may need to remove pressure from a system or complete a formal nitrogen purge and inerting procedure to safely replace compromised fasteners.

Another issue to be aware of regarding bolts and studs is *short studding*. ASME B31.3 allows threads to be only one thread short of being fully engaged. Good practice dictates that fasteners be flush with a nut or have at least one thread protruding.[9] Where there is no protrusion or, worse yet, there are obviously fewer threads than the design holding force requires, the fastener is compromised, as is the integrity of the flanged connection.

A great example of how fastener issues can lead to tremendous devastation and death is provided in the opening real life story at the start of Chapter 7 involving the US Navy ship Iwo Jima tragedy.

NOTES AND REFERENCES

1. Pipe manufacturing, http://midstate-steel.com/how-pipe-is-manufactured.html.
2. American Society for Testing and Materials, www.astm.org.
3. US Steel, Product specifications, http://www.ussteel.com/uss/portal/home/products/tubular.
4. American National Standards Institute, ANSI B36.10, Standard for Steel Pipe.
5. Bonney Forge, Weldolet/Treadolet Fittings.
6. *NDT Net*, The ABC's of Non-destructive Weld Examination, Charles Hayes, Vol. 3, No. 6, June 1998, first published in the *Welding Journal*, May 1997.
7. Ridge Tool, www.ridgid.com.
8. ASME B31.1, Power Piping Companion Guide, www.asme.org.
9. Walter J. Sperko, ASME PCC-1 Guidelines for Pressure Boundary Bolted Joint Assembly: Lessons Learned, NCPWB Technical Bulletin, National Certified Pipe Welding Bureau, 2010.

4

Gas Supply System Issues

Understanding Main Incoming Services

Once natural gas piping is inside a facility, it is pretty easy to look up, see it marked, and understand what it is. Many people don't quite understand how the gas might have gotten there. It's important to know where the gas came from, who owned it and at what point, how the pressure got controlled, and how to shut it all off if necessary. In this chapter we also discuss alternative fuel considerations, such as propane, landfill, or digester gas service issues.

Real-Life Story 13: Never Disrespect a Gas Leak

The evening of June 9, 1994, was a harrowing night for the residents of the John T. Gross Towers retirement home in Allentown, Pennsylvania.[1] Just outside the eight-story building, contractors were working to remove soil contaminated with heating oil that had leaked from an 8000-gallon storage tank removed a day earlier. A backhoe struck a 2-inch-diameter steel gas service line that had been exposed during the excavation, causing the line to separate at a compression coupling. Over the next 15 minutes, contractors and residents began to smell a strong odor of natural gas. The escaping gas flowed underground along the pipe and into the retirement home. The construction crew found the gas line's shutoff valve but did not have the proper tools to close it. The gas company was called, but not as an emergency 911 call.

Just before 7 P.M., the natural gas in the building ignited, causing a small explosion. A second explosion occurred about 5 minutes later, setting portions of the building on fire. More than 300 residents were evacuated, many of them having to be carried out.

Fuel and Combustion Systems Safety: What You Don't Know Can Kill You!, First Edition. John R. Puskar.

In total, 82 people were injured and one person was killed. The fatality was a 73-year-old occupant of the third floor who died when he was struck by a door propelled by the force of the explosion.

The building suffered more than $5 million in property damage, as the cinder block wall enclosing the tower was blown outward, fragmenting the adjacent exterior walls. The first-floor concrete slab over the crawl space was pushed out. Several boilers were damaged and several windows were blown out. It was nearly a year before some residents were able to move back. An investigation from the National Transportation Safety Board concluded that the damage might have been avoided had the proper authorities been notified in time. If the foreman had reported the broken gas line immediately as a 911 emergency, crews would have been alerted almost 15 minutes before the explosion, giving them enough time to respond, initiate evacuations, provide building ventilation, and shut off the flow of gas into the building. This would have prevented the explosion or reduced its impact significantly.

Clearly, the crews lacked the proper tools, training, and respect for the explosive properties of natural gas. Fifteen minutes of inaction made the difference between life and death for one person. When the smell of gas is present, it shouldn't take an explosion to spur you into action.

Lessons Learned You must have a response plan in your facilities for people who smell gas at any time. Near the top of your priorities must be evacuation plans as well as calling in the gas leak as a 911 emergency call to the utility. In the case of any excavations, make sure that the local utility's locating service is called to locate the underground utilities before you dig. Every local area has such a service; ask your local utilities for contact information; the service is usually available at no charge.

It is also important to know where gas shutoffs are and to verify that valve handles exist and that the key isolation valves actually function. I have found that in at least one out of 10 incoming natural gas service entrances, the main shutoff valves have seized in position and cannot be closed because they have not been exercised or cared for properly through the years. As you can see from the story, a couple of simple things can be the difference between life and death.

4.1 INCOMING NATURAL GAS SYSTEMS

Before one can begin to understand gas piping systems safety and risk management effectively, it is important to understand the basic design elements of gas transmission delivery piping systems, gas yards, and gas conditioning systems. The scale of this discussion is somewhat macro, as these types of service entrances are usually associated with electrical utility plants, petrochemical facilities, or large manufacturing complexes. However, there are many elements to discuss that are also commonplace for smaller facilities.

The world of gas piping systems for larger facilities usually begins with a transmission line some distance off the property or battery limits. In some cases this line is owned by the entity being serviced or a subsidiary. However, in most cases this line is owned and operated by a third party that provides service to a gas yard somewhere on the plant property. There is also some important delineation or demarcation point where a change of custody occurs for the gas. This point could be identified by contract documents as a flange or pipe section or a valve or metering station. This section of piping, from the transmission line to the gas yard, is usually designed according to the U.S. *Code of Federal Regulations* (CFR) Title 49, Part 192, Transportation of Natural and Other Gas by Pipeline: Minimum Federal Safety Standards. The sections of piping covered by these standards usually include considerations for corrosion protection and for future line cleaning (e.g., pig receivers and launchers).

4.1.1 Gas Yards and Fuel Conditioning[2]

In large facilities there is likely to be an area, called a *gas yard*, that has a number of important pieces of fuel conditioning equipment. The gas yard is the heart or nerve center of all the incoming gas for the facility. What is shown in Exhibit 4.1 is typical for many but not all gas yards. The fuel gas from the service line passes through the main shutoff valves and, in some cases, through a dew point water bath heater and a pigging station. Depending on the equipment installed and pressure needs (i.e., some combustion turbines need high pressure gas to operate) there may also be a fuel gas compressor. The gas could then flow through pressure regulators to a coalescing filter and in some cases an odorant insertion station before being distributed throughout the facility. This simple schematic is provided to show how major components might be interrelated.

4.1.2 Main Shutoff or Isolation Valves

There are usually several isolation or shutoff valves at a gas yard. Some of these may be part of the gas utility or gas transmission company's piping, and others may be part of the customer's plant piping. The demarcation point is typically the discharge of the gas meter (make sure that you verify where this is with the supplier for your facility). It's important for the owner of a facility to know and designate which manual valves are designated as the customer's main isolation or shutoff valve or valves and which of these would be used in an emergency to isolate the plant.

In addition to a manual isolation or shutoff valve, many sites also have in place an automatic valve that has actuators and can be actuated remotely. Many automatic valves are actuated by system natural gas pressure, compressed air, or a nitrogen bottle and have manual hydraulic pump backup systems. There are many styles of manual shutoff valves and automatic valves. It is important that all valves be maintained and function-tested periodically.

4.1.3 Pig Receiver and Launcher

Many gas yards contain a pig receiving and launching station. Pigs are devices that are inserted and moved through the pipe to some other location, where they are removed. They are the most common means of cleaning major pipelines while they are in

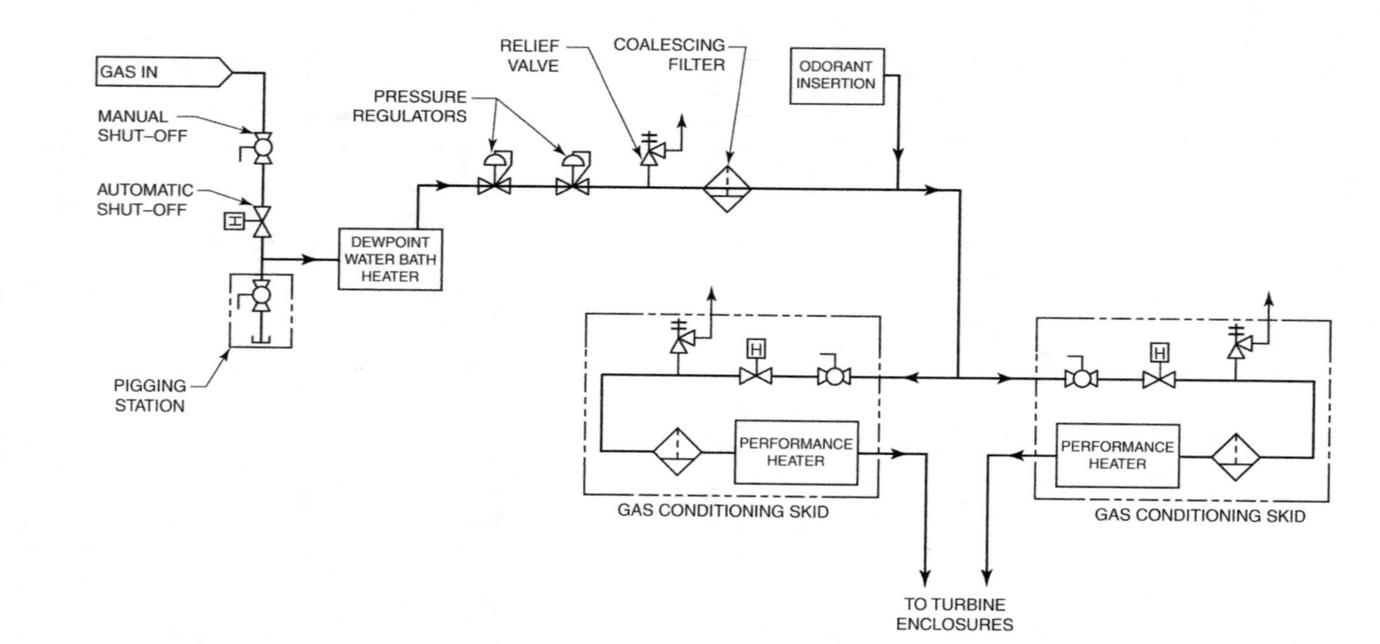

EXHIBIT 4.1 Typical schematic for gas yard conditioning equipment, combustion turbine systems.

service. (Pig types are discussed in Section 5.10.6.) There are many styles and types of pigs and receiving and launching stations. The design of the pigging station is usually commensurate with the design of the piping system and pigging that may need to occur. U.S. DOT regulations require that all new and replaced transmission piping systems be "designated and constructed to accommodate the passage of instrumented internal inspection devices," except for station piping including "compressor stations, meter stations, or regulator stations" (49 CFR Part 192.150). Most facilities require inspection of the transmission service at some time in their operating life, and smart (instrumented) pigging is the most common method of inspecting transmission piping.

4.1.4 Pressure Regulation and Relief

Interstate transmission pipelines usually operate at much higher pressures than those used within a major industrial or power generation facility, although some power plants operate turbines at transmission-line pressures. The pressure delivered to a gas yard from an interstate pipeline can be 1000 psig or more. This gas is usually dropped in pressure through a series of regulators to between 500 and 700 psig for combustion turbine installations, and sometimes much less, possibly 100 to 200 psig, for distribution around an industrial site. In most cases when gas is dropped in pressure, some type of overpressure protection is provided downstream in the case of a regulator failure. This is usually a relief valve or a series of relief valves. The discharges from these high-pressure relief valves must be placed carefully because the high pressure and considerable volume release can create a large flammable cloud if there's a problem. Considerations for discharges from these valves must include the surrounding areas' electrical hazard classifications.

The act of reducing the gas pressure drops the temperature. There is about a 1°F drop for every 15-psig drop in pressure (1°C drop for every 186 kPa). Gas that is very cold can make for cold wet dripping and even icing of piping and components. In cases where the pressure drop is large, hydrocarbons and water vapor can also be condensed inside the pipe. Hydrates and liquids inside gas lines can clog instrument sensing lines. Wet dripping pipe accelerates external corrosion. For all of these reasons, dew point heaters are sometimes provided.

4.1.5 Dew Point or Water Bath Heaters

Dew point or water bath heaters are fired heaters that heat natural gas to prevent it from being too cold (below design conditions) as its pressure is reduced and it is distributed throughout a facility. These are usually a consideration only where the starting gas pressure is at least several hundred psig. These heaters can be either before or after the main regulator stations. These heaters are usually shell-and-tube heat exchangers with a series of small-diameter tubes through which the natural gas passes. In a water bath design these tubes are submerged in a glycol–water mixture. The water bath is usually kept at about 160 to 180°F. Typical gas delivery temperature from these heaters is 80 to 120°F, depending on the system design and heater placement. Non–water bath direct-fired designs are also available but are used primarily for relatively small systems.

4.1.6 Particulate and Coalescing Filters

Particulate and coalescing filters may also be an important part of the natural gas–conditioning systems for a site. These can be installed permanently and located in the main gas yard or specific to important pieces of equipment such as combustion turbines or at both locations. Particulate filters are usually a mechanical cyclone, mesh, or filter element, or a combination of these that trap debris. Most equipment manufacturers have specifications for the particle size they do not want to see and sometimes for the maximum recommended dirt loadings per volume of fuel gas consumed that could cause equipment damage.

Coalescing filters are designed to bring together small liquid droplets to coalesce or form larger droplets, which would then remain within the vessel due to a velocity drop or impingement, so that they can be drained and removed. It is not uncommon for natural gas to contain hydrocarbon liquids that get transported with the gas and sometimes come out of solution with the gas. These can damage fuel train components and equipment such as combustion turbines.

4.1.7 Metering

Some facilities meter at the gas yard for custody transfer and then again at individual units or buildings to permit an evaluation of performance and costs. Metering is usually with temperature- and pressure-corrected turbine meters, Coriolis effect meters, orifice plates, or annubar metering technologies.

4.1.8 Odorization (an important issue)

Natural gas and propane have no odor when they come out of the ground. When odor is required or desired, a chemical odorant called *mercaptan*, a sulfur derivative that has an intense skunk odor, is added by pipelines or utilities. This chemical can be added upstream of the gas yard, in the transmission system or at the gas yard. Odorization must be applied to very carefully controlled specifications. U.S. DOT regulations, 49 CFR Part 192.625, have requirements for gas odorization. For natural gas, the standard calls for about a 1%, mercaptan concentration, and for propane, about 0.4%. These levels have been deemed to be thresholds that allow people with a normal sense of smell to detect the mercaptan. Some plants with very large services (e.g., power plants, chemical plants) use gas that is not odorized. You should find out whether or not your facility uses odorized gas. If not, you should understand what the consequences might be for identifying leaks and vent releases and for managing gas piping leaks. Facilities without odorant should consider the use of fixed methane monitors near vent release points and other strategic areas so that rapid detection of a release is possible.

Many factors can affect one's sense of smell, especially when it comes to detecting mercaptan. Mercaptan can be absorbed by new steel pipe and other materials, a phenomenon known as *odor fade*. Not all persons are able to detect the odor, and there are conditions that make it less detectable by certain persons, including the aged and those with medical issues such as sinus problems. It's also possible for chronic low levels of mercaptan to desensitize a person's sense of smell, called *odor fatigue*. If someone is exposed to a low-level mercaptan environment over time, they may not be able to detect it effectively in the future or be able to discern different mercaptan levels. This is one of the reasons that one can ever rely only on the smell of odorized natural gas to determine its presence or concentration. Instead, properly calibrated meters or detectors should always be used as part of a safety process for determining the presence or concentration of mercaptan. Make sure that the gas detectors used are intrinsically safe or explosion-proof so that their use does not create an ignition source. Not all meters use detection technologies that are intrinsically safe.

4.2 PIPING CORROSION PROTECTION

The fuel distribution piping at your facility may be in danger—not because of terrorism, vandalism, or other human cause but from the naturally occurring process of *corrosion*. Corrosion can occur from either inside or outside a pipe. It can also occur on both above- and belowground (buried) piping systems. Corrosion is the cause of about 35% of all fuel pipeline accidents according to the U.S. Department of Transportation's Pipeline and Hazardous Materials Administration.[3] Luckily, a method of halting the process of underground corrosion, *cathodic protection*, was discovered in 1824. An excellent resource for corrosion engineers, training, and general help on this topic is the National Association of Corrosion Engineers, website, www.nace.org.

Underground piping can be painted and coated to prevent corrosion; steel pipe with a thick plastic coating is also available. Unfortunately, paints, coatings, and other barrier methods will eventually fail. Failure can occur within months of installation. When the coating fails, it is usually a small opening in the coating called a *pinhole* or *holiday*. When a pinhole develops in underground piping or other underground metal object, a galvanic cell is formed. This is essentially a wet cell battery where the metal is the cathode. As current flows through the pinhole, metal is lost or "sacrificed" there. As metal leaves the pipe and a hole develops in the container, liquid or gas will flow out.

Cathodic protection is the process of negatively charging the underground fuel piping (the cathode). There are two ways to apply the voltage needed to make a cathodic protection system work:

1. *Active cathodic protection systems*, sometimes called *impressed current systems*, supply direct current to the underground piping system or metal object being

protected using a rectifier converting alternating-current (ac) to direct-current (dc) voltage. Active systems are most commonly used in transmission pipelines and storage tank installations (below- and aboveground tanks) to protect their bases.

2. *Passive cathodic protection systems* use sacrificial anodes buried near and connected to the pipe to complete the circuit. Oxidation or corrosion will tend to occur in the anode rather than the pipe (the cathode) and can maintain its integrity. Passive systems are very common for underground piping in use at utilities and industrial plants for buried distribution systems. Passive cathodic protection is also used in residential water heaters, where a sacrificial anode protects the water tank from corroding.

Eventually, the sacrificial metal in a passive system will be depleted and leave the piping vulnerable to major damage. That is why cathodic protection systems must be monitored regularly to determine the rate of anode depletion. The test or monitoring should be done when the system is installed and periodically thereafter. U.S. DOT regulations call for annual testing for piping under this jurisdiction. Most industrial systems should be monitored at one- to three-year intervals (depending on soil conditions and the rate of decay). Readings must be archived and compared from year to year and this information maintained on site. A corrosion professional can then make a judgment on the readings taken based on soil conditions, the type of pipe, and other factors and let you know if there is a problem. Most plants are not equipped to make this measurement or to interpret the results. This is where the services of a qualified corrosion engineer can be very important.

IMPORTANT

Cathodic protection systems contain test wires that may be located near gas services in small enclosures at ground level or wrapped around pipe risers. Make sure that a competent corrosion engineering firm is engaged to read signals from these and interpret the results on a regular basis.

Another important passive cathodic protection system issue to consider is proper electrical isolation. Underground services need to be isolated electrically from building mains or other systems that are not protected. Corrosion protection systems are designed for specific lengths and segments of pipe and are usually installed in discrete cells. There are usually isolation gaskets between flanges and even special bolt isolation kits for flanges. All of this is to separate the protected sections electrically. If those removing and replacing piping or conducting service work do not reassemble things properly or compromise the integrity of electrical isolations, the cathodic protection system can be rendered useless.

4.2.1 Corrosion Concerns Where You Would Not Expect Them

Often, little attention is paid to exterior corrosion. Sometimes industrial piping is not well protected when it's installed outside or in a corrosive environment. Organizations such as NACE, the Society for Protective Coatings, (SSPC; www.sspc.org), and ASTM have information and specifications for coatings, including types of coatings and recommended thicknesses based on the application.[4] It is also important to check coating thicknesses periodically and reapply them to retain full protection. Paint gages are a simple and inexpensive tool that can allow one to determine the coating thickness left on a piping system. Piping systems installed outside lose a certain amount of paint or coating thickness every year, depending on where they are located. It's very important to check coating thicknesses on new systems to make sure that they meet specifications and then to maintain minimum thicknesses once they are installed.

It is common in the boiler world to have piping run in floor trenches. In many cases such trenches allow for water drainage from the boiler room floor. These are usually wet or humid environments. I have removed floor plates and grating from many trenches and exposed a lot of piping in really bad shape. If your boiler room has piping systems that run in trenches, make sure that you open them (remove the trench plates) carefully to examine the condition of the piping. None of this piping should have obvious wall loss or excessive pitting. In one case I found a trench half full of mud with gas piping running through it. Once the gas piping was secured and locked out, I removed some of the mud and found that I could dig my fingernails into the gas line and pull away pipe wall. It just kept falling off. I requested that the system be shut down and the pipe replaced. The pipe wall had in some places lost half of it's original thickness.

In another boiler room, we examined a steel propane line with threaded fittings in a trench. Once the trench access plates were lifted, there was an obvious odor of propane. When handheld leak detection gas sniffers were used, the threaded pipe in the trench was found to be leaking at several connections. Since propane is heavier than air, it was accumulating in the trench. We were lucky that when we lifted the plates we did not bring in enough air to make this a flammable mixture and cause an explosion. Try to avoid running fuel piping in trenches. It's just not a healthy environment.

In yet another boiler room trench at a facility in Europe, electrical conduit came down a wall to the piping trench. Once in the trench the conductors were run unprotected and just laid on the piping in a dangerous hot wet steamy corrosive soup. You never know what you will find in such places, but it is often not good.

In some cases it's not trenches that are the problem, it's the environment. In a food-waste rendering plant, trace amounts of hydrogen sulfide gases were generated that hung in the air throughout the boiler house. Over time these gases caused deep pitting in everything they touched. This was especially true of the gas lines that ran outside, where the weather and gas temperatures made for a dew point on the gas piping and condensation. It was scary to look at some of the deep pits on these gas lines. Special coatings and paints had to be used on this piping to keep it intact for any length of time.

4.2.2 Conducting a Gas Piping Survey

Conducting a comprehensive gas piping risk survey is an excellent way to identify steps that can be taken to mitigate risks. Following is a list of things to consider when conducting such a survey. Before starting, make sure that updated drawings of the natural gas piping systems exist. Make sure that principal shutoff valves are identified and marked according to whether or not they are accessible and functional. Note key regulators and vent lines on these drawings as well.

1. *Verify the integrity of service delivery piping from providers*. Federal pipeline regulations in the United States require utilities to do periodic surveys for leakage and for cathodic protection system integrity, but is your site really being surveyed? Ask the provider for records related to your site. If you ever happen to notice signs of dead vegetation over the pipeline, bubbling of puddles after it rains, or occasional gas odors, these could be signs that there is an underground service leak that should be reported immediately. Repairing an incoming service could interrupt a facility's operations for days.
2. *Conduct a risk-based gas leak survey for piping inside the facility*. Use a handheld combustible gas indicator to attempt to identify leaks in high-risk areas such as those associated with fuel trains and rooms that may be small in volume or that might have little ventilation. Remember, there is no such thing as an acceptable small gas leak. No matter how small or seemingly minor, all gas leaks need to get on a list and be prioritized for repair.
3. *Conduct a review of piping systems for external corrosion*. The piping most vulnerable to external corrosion is that installed outside, in trenches, or in special processing areas. Be especially aware of and review piping coming up through the ground at soil-to-air interfaces (where it goes in and comes up). Soil-to-air interfaces are notorious for corrosion because of the availability of free oxygen to the pipe at that point and the change of protection systems needed.[5] If obvious pitting or wall loss is discovered, contact a nondestructive testing (NDT) provider to evaluate the damage. These service providers are experienced at evaluating remaining pipe thicknesses. Also make sure to review and document the remaining thickness of coatings on piping and critical structures installed outdoors. Handheld paint thickness gages are inexpensive and easy to use. You should create a coating maintenance plan that includes a recoating schedule for all critical systems.
4. *Find and note the condition of cathodic protection systems*. Verify that underground piping systems have cathodic protection systems. If they do not, consider adding protection. If cathodic protection systems exist, verify that their effectiveness is being documented and that readings taken are within the ranges recommended.
5. *Review the condition of bolts and nuts holding flanges together and verify their condition*. Verify that fasteners are in good condition, are of the proper type by their markings, and are not corroded excessively. Verify that they are fully engaged and not short-studded.

6. *Verify the MAOP from the local utility or gas service provider and make sure that downstream components are protected.* Once you know the MAOP, you will need to determine if components like valves and regulators have the proper pressure ratings. Components must be compatible with the pressure that could be seen if the nearest upstream regulator were to fail.
7. *Identify abandoned gas piping sections or equipment and reduce their risk for leaks.* Abandoned piping systems that are still energized or those no longer used can pose an unnecessary risk. Consideration should be given to removing flammable gases from abandoned or unused piping to render it safe. In some cases, the installation of nitrogen to unused sections of piping can be appropriate. Manual isolation valves still open to equipment or systems that are abandoned create a risk for gas leakage through closed manual or automatic valves. This can allow an accumulation of gas in abandoned equipment and an explosion. Closing off manual valves to unused or abandoned equipment is a help. However, the most positive means for removing the risk of leakage is the installation of a blind downstream from a closed valve.
8. *Review the condition and spacing of pipe hangars.* ASME B31.1 has a table for suggested pipe support spacing. You should verify that pipe supports meet these criteria. In some cases you may find that they may have fallen or been removed. You should also verify their condition; that is, they should actually be touching and supporting the pipe, taking the load, and be installed securely.
9. *Review the physical condition of piping for damage.* Is any of the piping obviously damaged, bent, or no longer plumb? Is some of it in high-traffic areas by docks, where, for example, hitting a unit heater means a mass evacuation? Note piping that has been damaged and may be vulnerable to further damage.
10. *Verify that key isolation valves exist and are functional and accessible.* Are key distribution valves in place and accessible, functional, and serviced in case things need shutoff for different areas of the site? Are these marked? Do they need to have chain operators, hand wheels, handles, or geared operators? Do some underground valves require church key–type handles that are not immediately available? Are some valves in locked fenced-in areas? Should they be? How is access controlled?

Problems Related to Shutting Down for Repairs In the course of conducting your survey, you may find issues such as leaks, obvious external corrosion, or valves that look like they may not work. It has already been stated that the responsibility for fuel distribution piping at your site usually starts with the discharge flange of the utility company's meter (call them to verify this). If problems are identified on piping systems that you own, the fuel supply may need to be shut off at the street or main incoming supply to make repairs. Once a utility shuts off service, they have the right to dictate under what conditions it will be brought back online. This usually necessitates that your piping systems pass some type of pressure hold leak test. In some cases this can be extensive and problematic since a very small leak that has existed for years outside in a low-priority location might now cause days of extra work to find and fix. The following real-life story takes you through just such an incident to give you an idea of the type of planning that may be required.

Real-Life Story 14: A Small Gas Smell Uncovers Big Problems

One warm summer day about 15 years ago I had the pleasure of visiting a manufacturing plant that had a gas service entry point to the property in a small block building in the parking lot. This is the case with thousands of other manufacturing plants. Inside I found a gas line coming up from a gravel floor that went to a pressure-reducing and metering station. It then went back into the ground and on to various parts of the complex. In many facilities the discharge flange of the meter, located in this building, is the point of custody transfer for the fuel from the utility to the customer. I always like to start my review of a site with this point since it's where the customer actually first owns the gas.

When standing in this little building, I noticed a strong gas odor. I thought it might have been caused by leaking flanges or fittings. The customer had also noticed this smell prior to my arrival and had spent a lot of time and effort making sure that all the flanges and fittings were not leaking. Duct tape had been wrapped around the flanges, as it is customary to use to find leaks on flanges. By poking a small hole in the tape wrapped around the mating flanges, one can then check the hole with leak-detecting fluid or a combustible gas indicator to see if there is a leak somewhere on the flange. The leaked gas, trapped by the duct tape, would make its way to the hole and make it easier to detect. Otherwise, it's very difficult to find gas leaks on flanged connections.

After speaking with a representative of the utility company about the status of their aboveground piping and watching them run some simple checks, it was clear that the leak was not in the building from the aboveground piping but in the utility's underground line serving it. The utility representative agreed to send someone over during the customer's traditional two-week summer shutdown and hand-dig around the entering pipe. He said that more than likely a pinhole leak right near the soil-to-air interface needed to be repaired.

When the digging started, it was discovered that the underground pipe was like Swiss cheese, with numerous holes caused by underground corrosion. The entire service out to the street, 800 feet long, had to be replaced under emergency conditions during a two-week window to avoid an interruption in the plant's production schedule after the annual summer outage.

Every plant with underground gas piping has this risk. The integrity of service piping owned by the utility company is supposed to be monitored and reported on regularly according to U.S. DOT regulations. You can ask your utility to show you evidence of the integrity of the line serving you. They should have historical information on the condition of the cathodic protection on this piping and leak surveys that have been done.

When the line goes underground after the meter, it's usually the site's responsibility for integrity verification. In many cases I have found that these lines have no cathodic protection and that the sites are at risk of a major service interruption. Plants that use propane have a similar concern. In many cases there is an underground line connection from a truck unloading station to the propane storage tank, or the service from the tank goes underground to the main facility.

Lessons Learned Don't wait until you have a verified underground piping problem. Take steps today to understand whether or not your underground systems are cathodically protected. A competent corrosion engineering firm can stop by for not much money, take some soil-to-pipe readings, and tell you if your corrosion protection system is actually in service and providing protection. Make sure that annual corrosion protection records are kept for reference. The preventive fix in this case is a lot safer and better than the surprise fix.

Create a schedule for gas piping surveys for your site. The scope defined above is very ambitious, but once done and documented, a slimmed-down walk-around version for assessing some of the more obvious issues can be carried out annually.

4.3 CONSIDERATIONS FOR LIMITING ACCESS TO SERVICE ENTRANCES

In some cases, incoming natural gas services should be protected from general access and the public. In today's world of potential terrorism, or sabotage from a disgruntled employee, why allow casual access to something that can cost you lives and millions of dollars if it is tampered with?

If your service entrance or critical gas-related equipment is vulnerable to tampering, you may want to consider a chain-link fence, with the approval of the utility, which includes locks—one for you and the other for the utility company. These must be installed such that a common key will open either lock. Natural gas service entrances should also include signage for no smoking, emergency call numbers for both the utility company and your site, and an emergency shutoff valve. *Make sure that emergency shutoff valves have handles*. Offer your local fire department a key so they can shut off the gas if they respond to a fire at your plant.

4.4 GAS SUPPLIES FROM DIGESTERS AND LANDFILLS

In some cases fuel gas is available from landfills or digesters. Digesters are often found at wastewater treatment plants, and these have been the site of many explosions. Digesters are large vessels with specialized bacteria that consume organic materials and give off methane as a by-product. The methane is collected and either used in a process or flared (controlled burnoff) for disposal. The green revolution that our country is undergoing is making digester installations ever more commonplace in such facilities as large farms and other nontraditional settings.

There are many issues related to fuel trains, piping and special flame arrestors, and flares associated with systems that burn gases from these processes. These are not addressed here. Please understand that these systems are not "plug and play" compatible with natural gas systems. There are issues with moisture, other mixed-in gases, heating values of the fuels, contaminants, pressure fluctuations, and in some cases, supply interruptions.

If you operate or expect to operate a digester or a landfill gas system you will also probably have a flare. There has to be a place where the gas is sent if the user cannot take it. Flares are an entity unto themselves. Flares burn off safely gas that cannot otherwise be used. Simply venting large quantities of gas without burning it can create a large explosion hazard, and storing the gas is not usually an option. Steel plants have tried to store things like blast furnace gas in gas holders (tanks) for years. After several horrific explosions and fires of historic proportions, the use of gas holders is no longer commonplace. To better understand the issues involved with using these gases, it is important to understand where they come from and how the gas would get to you.

4.4.1 Digester Gas Supplies

Digesters are used in wastewater treatment plants, both public and private, for getting rid of organic wastes. Digesters use a pool of liquid waste that is saturated with specialized bacteria that consume specific material. In most cases, the materials fed to a digester are consistent in makeup, and the gas coming off is consistent in composition. However, the gas can vary in Btu value and moisture content because of the weather, the health of the bacteria, the waste temperatures, the waste makeup, and other factors. Most facilities use some sort of gas conditioning equipment (could be filters or moisture seperators), prior to routing it to a compressor or blower to make it available to be transported in a pipeline. In most cases special burners are required to use the gas because the fuel/air ratio requirements are very different than for natural gas, due to the low heating value and composition of digester gas. It's also important in these applications that valve train components such as automatic shutoff valves have stems and seats that are compatible with the gas to avoid corrosion, moisture, and other contaminants. In many cases, flame arrestors are prevalent in different parts of the system to prevent flashback into piping systems.

NFPA 850, the Recommended Practice for Fire Protection for Electric Generating Plants and High Voltage Direct Current Converter Stations and NFPA 820, the Standard for Fire Prevention in Wastewater Treatment and Collection Facilities, may be helpful if you have this equipment. The 820 standard addresses some of the safety issues related to methane hazards at these facilities (Both provide some information on biomass and digester gas plant gas use). Also remember that digester gas would not be odorized with mercaptan.

4.4.2 Landfill Gas Supplies

Landfill gas wells are drilled down into the landfill. This is usually done in some sort of geometric grid pattern. Pipes are placed in the wells and connected to a manifold system that has a blower attached. The blower draws gas from the well pipes and sends it to the consumer. In many cases, some gas conditioning is also done (separation, filtration, or treatment). There are usually many wells, only some of which are in service at any time. Just as in the case of digesters, bacteria in contact with the organic materials in the landfill are decomposing and consuming the organic materials and giving off methane as a by-product. In many cases, as a bank of wells deteriorates in Btu capacity, another bank is switched into service. The partially depleted wells are periodically given a chance to regenerate, to be put back into service at a later time.

By their very nature, landfill gas wells have a greater variability of Btu content and contaminants than that of natural gas wells. At times, landfill gas can contain foreign materials such as refrigerants, mercury, and even silica, which can cause damage to combustion equipment. Some of these problems can be overcome by cleanup equipment installed on the delivery side; however, the user may have to negotiate with the landfill gas producer to include specialized cleanup equipment installation and maintenance in the service. Just as with digester gas, burner and fuel train components have to be selected carefully to be compatible with the low heating value and composition of the landfill gas, and flame arrestors are not uncommon.

While a significant number of landfills produce fuel gas, there are no known standards for these installations. This gas would not be odorized with mercaptan unless an odorization system were to be installed.

4.5 INCOMING PROPANE SERVICE CONSIDERATIONS

Propane is a commonly used fuel that is available at many sites. It can be used as straight propane vapor through special burners and nozzles or as a mixture of propane and air through a commercial mixing and blending system with a heating value similar to that of natural gas. When providing an industrial facility with propane, special consideration needs to be given to the storage tanks and the possible need for vaporizers to get adequate volume flow. There are important criteria in designing and locating storage tanks identified in NFPA 58, the Liquefied Petroleum Gas Code. If your facility has propane storage tanks and distribution systems, you need to be aware that NFPA 58 has had several recent retroactive changes. Owners should verify that tank and piping systems have been reviewed for the latest compliance requirements. Nothing presented in this section or anywhere in this document is a replacement for obtaining this code, reading it, and understanding it. It provides safety and risk management knowledge that you must be familiar with if you own or operate propane systems.

Propane vaporizes at about −40°F. In other than very large applications, propane is stored in a pressure vessel containing propane liquid and vapor. If propane vapor is withdrawn from the pressure vessel, some of the liquid will vaporize, cooling the liquid. Heat enters the tank through the tank shell to warm the liquid to ambient temperature. If vapor is withdrawn faster than heat can transfer through the pressure vessel walls to the liquid, the tank pressure will drop. This condition is worst at low ambient temperature, and in many industrial applications a vaporizer is required to maintain an adequate flow of vapor to meet process demands. Vaporizers are usually fired devices that either heat a small tank of propane directly or heat a larger water bath that has propane tubes running through it. Electric and direct-fired vaporizers are available for smaller vaporization rates.

Like methane, propane has to be odorized if this is a desired property. Propane is normally odorized before it arrives at propane company storage or at the industrial facility. Verification of the presence of odorant is required when you receive a shipment directly from a wholesale supplier. The truck driver must verify the presence of odor when full truckloads are delivered to industrial plants, and the plant should also verify that it is receiving an odorized load. A sniff test is required as a minimum, but a quantitative method using "stain" tubes is available if documentation is required. Make sure that the person doing a sniff test has a normal sense of smell and does not have a cold or other condition that prevents detection of odor.

There are several special things to be aware of with propane piping systems that are not typical of natural gas systems. Parts of large propane handling systems probably handle propane liquid but not vapor. Piping systems where propane liquid can be trapped between valves are required to have small relief valves installed. Storage tanks must also have special relief valves to guard against overpressurization. Remember that the presence of these piping and tank reliefs means that unexpected heavier-than-airpropane can exist from releases at or near ground level at many different locations within a bulk propane installation.

Finally, be aware that propane specifications vary widely. The quality and consistency of the product have been an issue at times. Grade HD-5 propane[6] is supposed to be at least 90% propane by volume and up to 5% propylene. There can then also be up to 5% other gases. Normally, this is not a problem; however, if propane is used as an engine fuel, excessive propylene can cause knocking.

NOTES AND REFERENCES

1. National Transportation Safety Board, PB96-916501, NTSB/PAR-96/01, Pipeline accident report, UGI Utilities, Inc., Allentown, PA, June 9, 1994.
2. Electric Power Research Institute, Document 1023628, Guidelines for Fuel Gas Line Cleaning Using Compressed Air or Nitrogen, September 2011, www.epri.org.
3. U.S. Department of Transportation, Hazardous Pipeline Safety Administration www.phmsa.dot.gov.
4. Joint Standard SSPC-CS 23.00/AWS C2.23M/NACE No. 12, Specification for the Application of Thermal Spray Coatings of Aluminum, Zinc and Their Alloys and Composites for the Corrosion Protection of Steel, Society for Protective Coatings, American Welding Society, and NACE International.
5. PHMSA, Corrosion Control Issues and Regulations for Natural Gas Systems, http://www.state.tn.us/tra/gaspipefiles/TSIInfo/PHMSAPart192ExternalCorrosion.pdf.
6. Propane 101, Propane Grades and Quality, www.propane101.com.

5

Gas Piping Repairs and Cleaning

Applying Key Piping System Safety Concepts to the Repair and Cleaning of Gas Lines

In this chapter we provide advanced concepts for facilitating the safe repair and cleaning of gas piping systems. Some of the most significant and horrific tragedies have come about from mistakes made in preparing gas piping for maintenance, bringing gas piping back into service, and trying to clean gas lines. The concepts presented in this chapter need to be made the subject of policies and practices with both designers and maintenance staffs. A section at the end of the chapter highlights a relatively new standard, NFPA 56, Standard for Fire and Explosion Prevention During Cleaning and Purging of Flammable Gas Piping Systems, which is central to this topic. It took many months of meetings with contributions from over a dozen experts to write NFPA 56. This is a very important and ground-breaking piece of work that applies directly to many of the concepts presented in this book. Anyone who does or oversees activities related to gas line repairs and cleaning must become familiar with this standard. This chapter is not a design guide or a "how to" for gas line purging and cleaning. Each site and its circumstances and conditions are different, and nothing here should be seen as a replacement for sound engineering judgment and the requirements prescribed by applicable codes.

Real-Life Story 15: My Friend's Changed Life

In 1975, a catastrophic workplace explosion left 19-year-old steamfitter Sean Patrick George with burns over 65% of his body and one of his co-workers dead.[1] I have come

Fuel and Combustion Systems Safety: What You Don't Know Can Kill You!, First Edition. John R. Puskar.

to know Sean through committee work as part of NFPA 56. The things he has gone through as a result of his injuries are beyond comprehension. I will provide some of the highlights here of what happened to him in an effort to reinforce the topics presented in this section. However, you really need to go to his website, www.seanpgeorge.com, and purchase his DVD, where he provides many more first hand details and presents them much more effectively than I can here.

Sean came from a family of pipe fitters. His dad was a steamfitter at the local union in Pittsburgh and so was a cousin, John Rogers. Sean was proud be carrying on the family tradition. It was his first week on his first job. He would be in the company of his cousin John for this project. They would be working together at the West Allegheny Middle School near Pittsburgh.

They were installing a new gas line coming into the basement of the school. There was a takeoff soon after the line entered the basement that led to a unit heater. The line was installed and they were moving to the process of energizing the line with gas for the first time. A foreman had told them that there was a sense of urgency to get this done. The process laid out was to bleed the line of air and get a pilot light lit in the basement unit heater. The two worked for hours with small plug valves, releasing from the pipe what they thought was air into the basement space. The unit heater pilot would not light. They went to lunch and came back and opened valves more fully, when suddenly the discharge from the valves burst into flames. Sean indicated that before he knew it, a thunderous explosion occurred, knocking him off a ladder and to the ground. He tried to stand and was suddenly engulfed by a huge fireball. He ran out of the room through the only exit he knew and threw himself down into a puddle of water outside.

The building was in total shambles. Neighbors thought an airplane had crashed since the school was not far from the Greater Pittsburgh Airport. Sean's cousin John died within hours of the incident. Sean underwent many weeks of treatments, defied the odds, and survived. In his DVD he describes the treatments and the pain that he went through and how the incident affected his entire family.

Lessons Learned Remember, this was not a project for a high-pressure gas feed to a sophisticated process plant. It was a relatively small low-pressure system for a school. The hazards related to bringing gas piping systems into and out of service are real and they can be deadly, regardless of the size of the system or the pressure. Gas from energized piping systems should never be released indoors. Hoses and/or temporary piping should be used by experienced trained personnel using carefully drafted plans to discharge the piping contents to safe areas. NFPA 54 and 56 provide guidance in these matters. These need to be read and understood by anyone who deenergizes and reenergizes flammable gas piping systems.

5.1 KEY STEPS TO SAFE GAS PIPING REPAIRS

There are six key steps to conducting gas piping repairs safely, each of which we discuss in detail: (1) creation of the job plan, (2) isolation of the piping system, (3) prerepair venting and purging of flammable gases, (4) pressure testing/leak checking, (5) postrepair purge of air from the piping system, and (6) gas reintroduction (reenergizing the piping system).

5.2 PLANNING THE PROJECT

The first part of the process for safe gas piping repairs, and probably the most important, is proper planning. This involves a thorough survey of the work to be done and the creation of a purge plan for de-energizing the piping and making it safe to work on. This could be a simple document for small and routine projects or a multiple-page detailed description, including photographs and drawings for a more complex, somewhat unique installation of piping or process vessels.

5.2.1 Routine and Small-Volume Low-Pressure Venting and Purging Projects

There are fuel system–related aspects within industrial and combustion turbine operations that are considered somewhat regular and routine. For example, in the combustion turbine industry, fuel gas conditioning skids containing filters need regular service and attention. Best-practice facilities have written protocols for routine purging-related work that are usually in the form of checklists. These provide a consistent basis for careful screening of the work to be carried out related to hazards. Checklists can form the basis for a group discussion with all of those who are to perform the work.

The 2012 edition of NFPA 54, Section 8.3, has provisions for purging small-volume and low-pressure (design operating pressure of 2 psig or less) gas systems. This information provides a number of options and safety steps for removing residual gas and for purging these systems.

5.2.2 Nonroutine Venting and Purging Projects

Project specific purging plans should be developed for unique and nonroutine gas piping projects. These specific plans should provide a step-by-step process using piping drawings and pictures of the work areas and include major valves and vent points if possible. The plans should also include consideration of general conditions, valve and blind lineups for isolation, training for nitrogen safety, prerepair purges, pressure testing, postrepair purges, and gas reintroduction steps. An example of a site-specific purge plan is included in the NFPA 56 appendix. Many industries do not currently approach the purging and release of natural gas in a cautious or planned manner. Recent disasters and changes in codes and some state laws (e.g., Connecticut's) are evidence that it's time for long-held paradigms about gas piping to change.

Real-Life Story 16: Nonidentical Twins

A set of nearly identical gas turbine combined-cycle power plants owned by the same company were separated by about 300 miles. But even though the facilities were technically the same, the operating and maintenance policies couldn't be more different. I had done consulting work related to purge planning at one of the sites during a maintenance turnaround on one of the turbine fuel conditioning skids, reference site A. I was at the sister site, reference site B, along with several staff members from the plant I had worked with previously. It was startling to see personnel from the two plants interact and discuss what was to be done, and how and why. The plant I had worked with previously (site A) had extensive procedures for purging gas lines, pressure testing, and for reintroducing gas safely. Plant B had no procedures and no interest in changing. There was a misplaced sense of machismo at site B that led them to release high-pressure gas the same way they had done for many years.

The beginning of an outage that required gas lines to be opened started with removing high pressure gas. This venting of gas at site A used a vent that was distant from all personnel. This vent point went straight up into the air. Site B did its high-pressure gas release venting at a point where 1-inch piping pointed down from about 20 feet up. The gas pressure was about 550 psig before venting. Releasing this produced a very high velocity discharge toward the ground, allowing mixing with air and generating a substantial flammable gas cloud hazard. After venting, site A used nitrogen to purge and inert residual gas in the line; site B did nothing.

The lack of standardized policies, training, and procedures at site B was certainly troubling. Best practices shouldn't be kept confidential; they should be shared with the entire company for the safety of all.

Lessons Learned Formal processes and tools to share best practices and training can be invaluable. The most successful organizations have fuel and combustion system intranet websites specifically to share this type of subject matter. Organizations need to be structured to be able to recognize, screen, and implement new codes and standards that are related to their industry. Assignments should be made for personnel to monitor and periodically review relevant NFPA, ASME, and related industry standards and code organizations. Systems need to be implemented for accountability in these areas. In this case neither of the sites had adequate information about NFPA 56. The organization also had no comprehensive plans for implementing the standard at its sites. Claiming ignorance of relevant standards and codes is never an adequate defense.

5.3 ISOLATION

Actual work on implementing a purging process must start with isolating systems to be worked on from other systems that may remain in service or from the gas utility or pipeline service entrance. In many cases, when you evaluate the piping system for isolation, you will find that it was not designed to be sectionalized or purged. This is usually discovered during an emergency or when a new project requires considerable modification of the gas piping systems.

In the United States, federal worker safety laws under OSHA (29 CFR Part 1910.147) cover the control of hazardous energy, lockout, and tagout. These regulations require the system you are working on to achieve a zero-energy state. This necessistates isolating by creating a manual valve double block-and-bleed system or through the installation of line blinds.

Make sure that your facility has safe practices for leak-checking valves before flanges are separated for installing line blinds. If double block-and-bleed systems are to be used, make sure that there is a practice in place for continuously monitoring the bleed vent for leaks.

5.4 PREREPAIR VENTING AND PURGING OF FLAMMABLE GASES

Whenever gas piping in service is to be opened, it must be purged so that it is in a safe condition to be worked on. Purging is the process of using an inert medium such as nitrogen to remove the residual gas in the pipe and leave it in an inert state. Purging should not be confused with venting. The term *venting* is generally used to describe the release of pressure from gas piping systems so that only residual gas exists for purging. Venting usually needs to happen before purging to some degree so that flammable materials to be purged are minimized. The best way to reduce venting needs is first to safely consume as much of the gas as possible in the process, equipment, or appliances.

5.4.1 Venting-off High-Pressure Gas

Sometimes, gas to be vented is under considerable pressure. In such a case there is concern about how and where to relieve this gas so that no harm will occur. When discharging natural gas from any vent system, whether from a venting or purging operation, or even just a relief valve discharge, the velocity at the discharge point, the orientation, the height, the wind direction, the gas composition, and even the temperature of the gas are important in determining where it is going to go and what concentrations will occur. There are a number of commercially available software models that can provide information on the dispersion of gas in air, including ALOHA, available from the EPA at no charge (www.epa.gov). These models differ in their sophistication, but even the simple ones can provide useful information about threat zones. A *threat zone* is a defined gas concentration area that is in the discharge plume. For example, one might consider a plume discharge area that is at the LEL of natural gas to be a threat zone. Models such as ALOHA can predict where these occur within a range of probability depending on the input conditions.

When venting takes place, noise sometimes needs to be considered. Noise is directly proportional to the release pressure, the discharge height, the opening size, and even the wind direction. High-pressure releases are noisy and can sometimes be heard for miles. Neighbors, emergency personnel, and public affairs staff need to be notified about such events so that releases do not cause public concern.

Depending on wind conditions, people outside the perimeter of a site may smell a slight odor of gas from a release. This, too, is something that people answering phone calls from the public need to be prepared to address. NFPA 56, described in detail in Section 5.12, offers guidance and considerations for accomplishing venting safely.

5.4.2 Purge Points

A *purge point* is a fitting or series of fittings and pipe nipples in the piping system where a temporary hose or piping connection of some type can be made for inserting nitrogen or removing natural gas during a purging process. In some cases, these are actual connections in the piping with valves installed specifically for this purpose. In other cases, these are ports on a valve that can be used for this purpose. A lack of purge points will make purging considerably more difficult, as it will expand the scope of piping that needs to be taken out of service to do the job safely. In some cases, additional purge points may need to be installed (possibly with hot taps) so that safe purging can be accomplished and the smallest amount of piping can be taken out of service.

When purge point pipe nipples are added, they should be made from schedule 80 pipe. Experience has shown that purge point nipples can be a source of vulnerability when large pipe fitters with big wrenches get involved to make sure that things are tight, or even from careless operation of in-plant industrial trucks and scissor lifts.

Real-Life Story 17: Fixing a Leak Becomes a Big Problem

A new natural gas piping system had just been installed. It was a 6-inch line about 150 feet long. During the pressure test the pressure kept dropping very slowly over the course of an hour. An investigation found that a purge point threaded nipple was leaking. I was working with someone I called BPF (big pipe fitter) because he was 6 feet 4 inches tall and weighted well over 300 pounds. BPF grabbed a 36-inch pipe wrench to fix the problem. He grabbed onto the leaking pipe nipple, turned it, and then when it seemed that all was well, he kind of hung on the wrench just a little. That's all it took, that last little bit sheared the nipple, and all hell broke loose. The test pressure was about 50 psig. The sudden release of compressed nitrogen caused an extremely shrill noise. It required several hours of additional work to dig the end of the broken nipple out of the fitting and start the pressure testing process over again.

In another case, I watched a forklift truck in an industrial plant lift a load that snagged a purge point nipple on some overhead piping. It sheared the nipple right off on a 10-psig gas line. The entire plant's hundreds of employees had to be evacuated. All these people had to stand outside for an hour and a half waiting for the gas to be cleared and the situation to be called safe. This cost tens of thousands of dollars in lost production.

Lessons Learned In both cases the purge points were schedule 40 pipe. BPF might also have broken off a schedule 80 nipple, but it would have been more of a challenge. Obviously, great care must be taken whenever tightening things under pneumatic test conditions. If something shears or breaks off, the stored energy could send objects flying like bullets. There's also the issue of the unexpected release of nitrogen if that's the test medium, which can make for an asphyxiation risk.

Be careful when locating and installing purge points to be sure that they don't stick out into an aisle, where they could be damaged by a passing forklift truck or scissors lift. Another important factor to consider is the purge point's proximity to a safe place to discharge the pipe's contents. The best places are a considerable distance away from ignition points, mechanical equipment (air handlers/exhaust fans), traffic, windows, and/or personnel. However, you also have to consider the difficulty of trying to get purge point gas outside from the middle of a plant where no roof penetrations are nearby.

5.4.3 Vent Orientations and Clearances

Vents for the discharge of natural gas or for propane purging need to be designed and installed with careful consideration. Vents that point down toward the ground or toward a roof can be especially hazardous. For example, plume dispersion models have indicated that in the case of a 125-psig discharge, the natural gas LEL threat zone plume can extend over 200 feet. Even a 25-psig vent discharged through a 2-inch vent opening can have a natural gas LEL threat zone plume extending 120 feet. This factor, the length and breadth of the LEL threat plume, is often not considered when vents are installed and required clearances are considered.

5.4.4 Ignition Sources

Before a planned release of gas, the area needs to be reviewed for possible ignition sources. Natural gas in air has an ignition temperature between about 1100 and 1200°F, depending on the source of the gas and the pressure and temperature conditions. At any given work location, the number of possible ignition sources can be very high. There are some obvious ones, but you should never rely on thinking that you have removed all of the ignition sources. You should always assume that no matter how hard you try, you are always near an ignition source. This is the most conservative approach to take. Less obvious ignition sources include static electricity, electrical devices, lighting, combustion equipment, tools, communication instruments, and some pipe-cutting processes.

1. *Static electricity.* Static electricity is the electrical charging of materials through physical contact and separation. Static electricity is generated by the transfer of electrons (negatively charged particles) between bodies, one giving up electrons and becoming positively charged, and the other gaining electrons and becoming negatively charged. Static electricity is dissipated in humid air, so it is a greater issue in winter and during dry weather. For information on the mitigation of static electricity, see NFPA 77, the Recommended Practice on Static Electricity.
2. *Electrical devices.* Electrical devices include everything from motor starters and switchgear to light switches; any of these may produce high-temperature sparks. The act of shutting off a piece of equipment can cause a problem because when a device deenergizes, two contacts under power are quickly pushed apart by a spring. As these contacts separate, there is a spark. These turn-off arcs often contain more energy than do turn-on arcs because the equipment is under load.
3. *Lighting.* Lighting can be a source of ignition for a number of reasons: Some lamps can operate at high temperatures. Lamps can also break, exposing the high-temperature elements. Remember, too, that switching lamps on and off can make for contact arcs at the switch contact poles.

4. *Combustion equipment (heaters and boilers).* Turning off a piece of combustion equipment might leave refractory materials hot. Refractory is a ceramic high-temperature layer of protection found in most combustion equipment. Flame temperatures are usually 2500 to 3500°F. Once the flame is turned off, the refractory is still probably hotter than the ignition temperature of natural gas for some time.
5. *Hand and power tools.* Nonsparking hand tools are common for work around gas piping but are not required by codes. The use of nonsparking tools does not mean that it is safe to work around areas known to be at or near LEL. These tools might be considered for use *as an extra measure of safety*. Power tools, even battery units, can be obvious sources of ignition as they spin and engage and disengage. Plugging in and unplugging battery packs can produce arcs.
6. *Communication equipment and instruments.* Cell phones radios, and instruments, including combustible gas indicators and detectors, must be intrinsically safe if they are used near gas purging operations. Being rated intrinsically safe is a very special requirement and something that you will need to examine carefully. NFPA 70 the National Electrical Code, has more information on device ratings for hazardous locations.
7. *Pipe cutting.* Make sure that the site has line-breaking permit processes and that these provide for consideration of what's in the pipe and its condition before anyone even thinks about cutting it. There are many ways to cut pipe. Of course, from an ignition source perspective, the most hazardous ways to cut pipe are with a carbide wheel or with a torch. In many cases, gas piping to be cut can still have residual hydrocarbon fluids on the inside that can escape or at the very least form a coating. The coating can catch fire with the use of any cutting method, but some methods make this condition more possible than others.

   Cold cutting processes using manually turned pipe cutters can be used for small diameters. These types of pipe cutters can be advantageous since the rate of cutting and the generation of heat can be controlled. Other cold processes include pneumatically driven pipe bevel cutters that are available for larger pipe diameters.

5.4.5 Purging Processes

The three most popular purge processes are the displacement or slug purge, the dilution purge, and the trickle purge. The following is not meant to be an instructive guide to conducting these properly but is meant only as a way to increase your awareness of what these are and how they might be applied.

Displacement or Slug Purge A displacement or slug purge is an attempt to pass a large instantaneous volume of inert gas rapidly into a pipe so that very little mixing occurs. In the slug purge process there's the inert substance, a contact zone where some mixing occurs, and then a zone of "to be purged" material leaving the pipe. The objective is to have as small a mixed zone as possible. Once this type of purge starts, it should not be stopped. It must be continuous until the material is cleared.

If perfect slug flow purging conditions were possible, you could clear the entire line with one volume change, but this does not happen in the real world. It's very hard to send enough nitrogen through a long pipe and a lot of equipment without getting substantial mixing, no matter what calculations tell you. It has been my experience that at least three complete volume changes with an inert material should occur before sampling to determine the LFL.

Dilution Purge In this style of purge the objective is to force an inert gas under pressure into the piping system to mix with the hazardous gas being purged. In this purging process we want mixing to occur and in fact want to use time to help encourage it. The idea is to release the mixture down to a pressure that is still above atmospheric pressure (so that no air gets sucked in) and then to repeat the process. In each cycle of packing inert dilution material into the vessel or pipe and then releasing the resulting mixture, the original target gas is diluted more and more. The objective is to get the mixture down below LEL. The number of pack (pressurize) and release cycles for obtaining a desired gas endpoint are dependent on the system volume, residual gas pressure, nitrogen pressure, and the blowdown pressure. The calculations used to estimate the number of cycles are relatively straightforward ideal gas law calculations.

Trickle Purge The trickle purge process is sometimes used to keep air from getting into a part of a system where welding or hot work might be taking place. It describes the process of introducing nitrogen to a piping system in very small quantities in an effort to keep the piping system under positive pressure. In some cases—for example, if combustible residual liquids might be present once gas has been removed—a trickle purge (a small flow of nitrogen), during cutting operations could minimize the chances of a fire because oxygen would be displaced. Care must be taken to protect personnel who might be working on or near piping being trickle-purged because of the danger of nitrogen asphyxiation.

5.5 LEAK CHECKING AND PRESSURE TESTING

There are two concepts to grasp here: leak checking and pressure testing. *Leak checking* usually refers to verifying that the system has no leaks after some type of minor repair or disassembly and reassembly. A given pressure that you determine is held for a time that you also determine. You also decide what will be the criteria for success. *Pressure testing* usually refers to a more formal process that is done when the system is new and being put into service for the first time or possibly has just had a substantial repair. ASME B31.3 and NFPA 54 identify pressure test criteria for how much pressure a line should hold and for how long. Local or state requirements may also exist, and these may exceed NFPA or ASME requirements. Make sure you comply with the most stringent requirements that are relevant to your project.

It is a good practice prior to introducing natural gas to have documentation of successful pressure testing for new or repaired piping systems. Documentation provides a reference point in the event that leaks are found later. The results of these tests should be retained for the life of the piping system. I have had good success with the use of pressure chart recorders for documentation. These paper-and-pen battery-powered recorders can come with very small pressure increment gradations (1 psig or less) and provide excellent records of pressures and hold times of tests.

Pressure tests should always be witnessed by someone other than the contractor doing the installation work or piping repairs. Sometimes, the local permitting authority, utility or gas supplier, or insurance inspector will also want to witness the pressure test or see documentation.

NFPA 54, the National Fuel Gas Code, is the nationally recognized code for gas piping repairs in industrial facilities where operating pressures do not exceed 125 psig. This document identifies special conditions where pressure testing cannot be carried out against valves. Some parts of the piping system may have to have blinds installed for this requirement to be met.

You must also be careful that fuel train components and other piping devices are not exposed to excessive pressures that can damage them during pressure testing. You can minimize some of the risks of this through the use of blinds for more positive isolation than manual valves can provide. Valves can be left open or leak through in the closed position. Blinds provide positive isolation and eliminate the possibility of damaging devices such as regulators and pressure switches that are not rated for the elevated test pressures.

5.5.1 Pneumatic Pressure Testing Versus Hydrotesting

It has been my experience that pneumatic pressure testing is the norm for piping projects involving relatively small diameters, small volumes, or relatively low pressures (usually less than 125 psig). NFPA 54, and ASME B31.1 and B31.3 provide guidance on test pressures and hold times to use for testing, along with additional protocol considerations. In the case of large-volume or high-pressure projects, hydrotesting is commonly used, for safety reasons.

The pneumatic pressure testing process usually starts with filling the piping system to some relatively low pressure first to see if there are any gross leaks, such as valves left open, caps left off, or flanges not tightened. Once lower initial pressures are deemed to be holding, it is customary to start ramping up the pressure in slow increments, periodically allowing the pipe to stabilize until the desired test pressure is reached.

The main hazard for pneumatic testing is the tremendous amount of energy stored in the compressed gas in the pipe. If a fitting fails or there is a breach, the results could be catastrophic. Many injuries and deaths have been attributed to pneumatic pressure testing. On February 6, 2009, a catastrophic pneumatic pressure testing accident occurred at a Shanghai LNG terminal that was under construction.[2] The accident killed one worker and injured 15 others. The facility was pressure testing a 36-inch-diameter pipe section that was about 600 meters long. The explosion occurred when a flange at the end of the test section suddenly ruptured. The test pressure desired was 2262 psig. The rupture occurred when the pressure was about 1785 psig. A worker died when he was hit by a large piece of shrapnel.

He was some 350 meters away from the site when struck. The injured workers were about 100 meters away when they were struck by shrapnel. Because of the risks involved, ASME B31.3 requires an owner's permission to conduct pneumatic pressure tests.

Hydrostatic testing involves filling the piping system with water, which is incompressible, so that if there is a leak or a breach, the consequences are not catastrophic. The problem is that once you introduce water into a system, you must be concerned with removing the water and drying out the pipe before it can be used. Corrosion inside the pipe is also a consideration.

5.6 POSTREPAIR PURGE

The postrepair purge step is necessary to remove air from a piping system before flammable gases are reintroduced. This is again done as a slug or dilution purge using nitrogen. If you did not do a postrepair purge, you could send natural gas into a pipe containing air. This could create a flammable mixture in the pipe that can be ignited in a number of ways, including from debris traveling down the pipe, causing a spark. This can make for one very large pipe bomb!

5.7 REINTRODUCTION OF NATURAL GAS: STARTUP

The last step in the process is the reintroduction of gas. I believe this to be the most dangerous part of any purging and repair process. It's a time when you will be connected to the main service and have an unlimited supply of gas that can fill a building or be released and cause great devastation. Very simply put, you are trying to get the residual nitrogen out of the pipe and the natural gas back in up to the fuel train or burner (or as close as safely possible) so that a safe light-off can take place.

In higher-pressure natural gas systems the safest method for reintroduction, assuming that the system contains only nitrogen somewhere near atmospheric pressure, is to open the gas valve slowly and pack the line. In most cases, if you do the ideal gas law calculations, the pounds of nitrogen left in the pipe are insignificant compared to the pounds of natural gas that you will be installing. Once the gases mix and are at equilibrium, you may have over 90% natural gas and can usually carry out the combustion process, i.e. the burner will light.

With low-pressure natural gas systems you will need to be moving gas in at one end of the piping system (slug purge) and monitoring what is coming out the other end. This can create hazardous conditions at the purge discharge point, and releases here have to be managed very carefully. This needs to go on until endpoint conditions are such that the endpoint can be closed and the system packed, as for a high-pressure system.

During the reintroduction process it is imperative that the natural gas source valve be attended continuously. Communication with the person tending this valve should be maintained at all times so that if a problem is detected, the supply of natural gas can be terminated immediately.

5.7.1 The Utility or Pipeline's Role in Gas Reintroduction

You must consider the involvement of the natural gas utility or pipeline company for energizing new gas piping installations. They are likely to have requirements that must be followed before they allow service to be established. This is also likely to be the case if repairs required the utility to shut off service temporarily. The utility company may have to introduce gas from a street or pipeline service connection into the meter assembly and then set their regulation equipment. This could take days and may need to be coordinated with your facility.

Once the utility company shuts off service to a facility, they will return that service at their discretion, which usually involves a number of preconditions, including the possible need for an approval from a local building authority or inspector to certify the integrity of your natural gas piping systems. Then the utility company usually conducts some form of its own leak check to validate that the downstream piping does not leak. Communicating with them up front can allow you to avoid schedule disruptions.

5.8 GAS SAMPLING AND DETECTION

Gas sampling and detection instrumentation will be required to measure purge end-points to determine when some purges are complete (i.e., prerepair, postrepair, and reintroduction). There are many things to consider when it comes to gas sampling and detection. This includes the type of instruments required along with safe practices to prevent injuries to anyone using the instruments.

5.8.1 Sampling Instruments for Evaluating Gas Conditions

Many different metering technologies are used to evaluate gas conditions. NFPA 54 addresses two types of gas evaluation instruments: combustible gas indicators (CGIs) and combustible gas detectors (CGDs). This code requires that CGIs have a volume display scale from 0 to 100% in 1% or smaller increments. They must be listed for the intended use and calibrated as recommended by the manufacturer. Calibration of CGIs and CGDs must follow the manufacturer's recommendations and frequency. It is also good practice to make sure that multiple instruments are available during purging processes so that an instrument failure does not interrupt the work.

Some CGIs use a catalytic combustion sensor for determining LFL. Others use a catalytic combustion sensor for determining LFL (0 to 5%) and a thermal conductivity sensor for determining higher volumes of gas (5% to 100%). A catalytic combustion sensor will not detect fuel gas accurately at low levels with no oxygen present. This is important when purging gas piping out of service using nitrogen since it is unlikely that oxygen will be present. If you are not sure of the capabilities of your CGI, contact the manufacturer for information.

CGDs must be capable of sensing and indicating the presence of fuel gas; however, they are not capable of determining the volume or concentration. Indication may not be on a numerical scale, but instead may be by an audible "ticking," or visual blinking lights with more frequent clicks and blinking lights indicating that a higher concentration of fuel gas is present. Listing is not required, but they must be calibrated as recommended by the manufacturer.

A simple field check of a CGI or CGD can be done by placing the instrument's sensing probe in a small plastic bag and using a small cylinder of calibration gas to fill the bag. The instrument should indicate the concentration of fuel gas in the bag within the tolerance allowed by the manufacturer.

5.8.2 Gas Sampling Protocols

The taking of gas samples can be a very dangerous part of the purge process. It should only be done with great caution by trained and experienced people using the right equipment. Make sure that NFPA 56 protocols have been followed before attempting to locate vent discharges and take samples.

Before discussing endpoint targets for purging operations, it makes sense to discuss issues related to just how one would actually take samples in a safe manner. In my experience, samples are taken at a sample port in a piping system that is part of a discharging vent or at the end of a hose or pipe vent discharge. It's always better and safer to have a sample port in a piping system. A sample port is usually a drilled hole with a small nipple and valve connected to the pipe. Into this sample port would go tubing from a meter that would draw a sample.

> **IMPORTANT**
>
> You should never drill a hole in a vent pipe that is not known to be clear of residual gas, because the drilling process can create an ignition source.

I have used the following protocol to take samples safely from vent discharges involving natural gas purging operations from industrial facilities to remove residual natural gas from piping systems. First, a safety perimeter is established in the area of the vent release point using caution tape. Much consideration must, of course, be given to the vent release point to keep it away from traffic and ignition sources, building openings, and other hazards. The safety perimeter area required for the vent point depends, of course, on many factors, such as the volume of the release, pressure, and so on. In the case of an industrial site my experience is that this perimeter might be a 50-foot radius from the vent location with the vent pointing up. It is recommended that a person with monitoring equipment (a CGI, CGD, or four-gas meter) be stationed at all times on the outer bounds of this perimeter, oriented where this detection would make the most sense relative to prevailing winds and protection strategies. This person would be there to communicate a need to stop the purging process if a reading of 10% LEL or greater were indicated at the metered perimeter point or if the oxygen levels began to drop appreciably in the case of a nitrogen release. Although 10% LEL is not ignitable, it is an indicator that something is keeping a substantial part of the purged material near the ground level and it is accumulating. This would be a reason to stop the release and investigate the problem.

In no case should the person taking samples stay in the threat zone near the sample point waiting to take a sample. The actual sample taker should have personal protective equipment when in the safety perimeter taking samples, including safety glasses, flame-retardant long-sleeved clothing (similar to arc flash clothing), and possibly breathing air in addition to other equipment meeting that project's specific risks requirements. There should also be a rescue plan in case the person taking the sample is overcome. This may require personnel on standby who are equipped with breathing air and other equipment.

In my experience, flow to the discharge would be stopped or reduced substantially immediately prior to a sample being taken. Only then would the sample taker enter the safety perimeter, walking slowly with a four-gas meter toward the discharge point but with the prevailing wind at his or her back. If the sample taker identifies a hazardous condition on the gas meter, the person would retreat immediately. Once at the discharge point the sample taker would insert a sampling hose from the meter, wait the appropriate time for the meter to get the sample through the hose, record the reading, and then retreat from inside the safety perimeter. In all cases, one has to be sure that the length of tubing and its configuration do not exceed the meter's pump capabilities. Sample points should also be grounded so as to minimize the potential for a static discharge if tubing is inserted at the release point. In some cases, sample hoses can be grounded using wet burlap wrapped around the hose and in contact with the ground.

5.8.3 Purge Endpoint Targets

Three distinct endpoint targets will need to be assessed: prerepair purge, postrepair purge, and purging into service (reenergizing the piping).

Prerepair Purge Endpoint Targets The prerepair purge comes after venting pressurized gas. Nitrogen is made to flow into the piping to remove residual gas so that the piping is made safe to work on. It has been my experience that prerepair purges from natural gas piping systems can be stopped when the concentration of fuel gas at the endpoint is identified to be about 10% of the LFL or down to about 0.5% natural gas. This condition may depend on other factors at the site, and this is not a code-prescribed endpoint. It is conservative and, in my experience, warranted. If all nitrogen is flowing properly and the purge process is well designed, it is not difficult to get at least to this level.

Postrepair Purge Endpoint Targets The postrepair purge endpoint is the process of removing air or oxygen from a piping system. Nitrogen is made to flow into the piping system to push the air (oxygen) out. It has been my experience that the target endpoint has been to get to zero percent oxygen. Again, if the purge process is well designed and there is enough nitrogen flow, this is not difficult to achieve in practice.

Purging into Service (Reintroduction) Endpoint Targets Reintroduction of fuel gas, energizing the system, is the last step in the process before trying to light burners or start the process. NFPA 54 and NFPA 56 are specific in requiring that reintroduction provide at least 90% gas by volume into the system.

Remember that the meter used to measure this much methane is not a traditional four-gas meter. Most such traditional meters are designed to indicate only up to 100% LEL (LFL). This means that when the meter indicates 100%, you are at only 4.4% methane. It is also important that the meter used for this process and for the prerepair purge process be able to read accurately in an oxygen-deficient environment since you will probably be sampling a mixture of natural gas and nitrogen. In some cases, meter technologies cannot read accurately unless oxygen is available.

5.9 NITROGEN-HANDLING ISSUES TO CONSIDER

There are several decisions to be made about the use of nitrogen in a purging process. These decisions include how much you will need and in what form you will purchase it. You will need to perform calculations regarding the volume required. Nitrogen is relatively inexpensive compared to the cost of the people involved in the purge process and the lost production opportunity costs for your facility due to the length of time the purging operations require. It is much better to be conservative as to the nitrogen required than not to have enough. In some cases suppliers charge only for what is used.

Remember, you will need nitrogen at least three times [prerepair purge, pressure test (possibly), and postrepair purge]. It has also been my experience that at least three volume changes should occur when slug purging is taking place. Remember, too, that the leak checks or pressure tests may not go well. This may require several system discharges and refills to get all the leaks identified and fixed.

You will also have to put special emphasis toward training anyone who is involved with the purge on the hazards of nitrogen. Although the air we breathe is 78% nitrogen, breathing even a small amount of pure nitrogen can kill you. Even though nitrogen is inert, it is nothing to fool with.

IMPORTANT

Remember, one full breath of pure nitrogen can cause unconsciousness or even death. It's not flammable, but don't let your guard down when handling it. Also, be careful of the other hazards, including stored pneumatic energy risks and the low temperatures that may be involved.

5.9.1 Nitrogen Flow Issues

If there's not enough nitrogen volume flow capability, the purging process will take an extended time and may not occur in an effective manner. A number of things can compromise flow volumes. For example, regulators on nitrogen sources need to be "purge" regulators capable of high flow. You might have to specify this type of regulator when ordering from the nitrogen supplier.

If hose is used, its size, length, and configuration can affect flow capacities dramatically. For example, when extra hose is in a coil, it produces a considerable pressure drop since the hose behaves as a series of continuous elbows. Yet another important issue is the size of unanticipated orifices in the system. Orifices can be fittings on the ends of hose or additional regulators or valves through which the nitrogen has to pass. The size of the pipe or orifice at its smallest point is what ultimately limits the flow of nitrogen. This smallest orifice somewhere in the system can be the choke point and dictate the peak flow capacity.

Consider the example of 900 feet of 14-inch pipe serving a combustion turbine. You may want to purge through a fuel-conditioning skid at the end of this line. The skid may have 150 cubic feet of volume, counting vessels and heat exchangers. The total volume of this system could be 1000 cubic feet. If you are purging at only 20 ft^3/min, you might need almost 3 hours to get three volume changes if things go well. If you can flow 100 ft^3/min, you could be done in 30 minutes.

Other safety considerations for hoses include verifying that their pressure ratings are compatible with discharge pressures that might be capable from the nitrogen source. If high-pressure hoses are in use and the lengths are long, always consider securing the ends and all connections with whip checks (especially if a hose is used for the discharge of a purge).

Also be very careful with high-pressure vent discharges. In most cases, venting of high pressures is done with little or no discharge piping. If piping is used for the venting, it must be of sufficient diameter and must be secured very well. I have seen high-pressure gas discharge pipe several feet long whipping around like a noodle. This pipe could easily have been broken off and sent flying.

Real-Life Story 18: Nitrogen-Filled Vessel Work Ends in Death

In November 2005, contractors at a U.S. east coast refinery were working to reassemble a pipe on a reactor while it was being purged with nitrogen.[3] The U.S. Chemical Safety Board report indicated that a pipe fitter noticed a roll of duct tape lying inside the reactor about 5 feet below the opening and knew it had to be removed before work could continue (Exhibit 5.1). The employee and his foreman discussed the problem, and rather than delay their work, they attempted to fish out the tape with a wire hook. Nobody is certain, but it appears that the pipe fitter either fell or intentionally climbed into the reactor and was overcome by nitrogen. The foreman climbed into the reactor to help his fallen colleague and he too was overcome. Even though the foreman had emergency training, his desire to help his co-worker trumped his experience, safety training, and proper emergency response protocols.

Despite the quick arrival and actions of emergency responders, the two men were deprived of oxygen for nearly 10 minutes and could not be revived. In a confined space, a nitrogen-rich/oxygen-deficient environment can induce a coma in less than 40 seconds, followed by convulsions and death through asphyxia.

The U.S. Chemical Safety Board (CSB) later determined that contrary to the plant's safety procedures, a nitrogen purge warning and barricades were not in place at the worksite. The CSB also concluded that workers at the refinery were not properly

EXHIBIT 5.1 Vessel involved in nitrogen purging incident.

trained as to the dangers of low-oxygen atmospheres around the unsealed openings of vessels and equipment undergoing purges with odorless gases such as nitrogen.

"The CSB Case Study underscores the importance of strict safeguards when working around low-oxygen environments," CSB lead investigator, John Vorder-brueggen, said. "Workers are in danger not only inside confined spaces, but also around the opening where inert gases like nitrogen are flowing out." It's a harsh lesson to learn, but the danger of nitrogen cannot be underestimated. The CSB issued a Safety Alert and video on this incident. They are available at the CSB website www.csb.gov.

Lessons Learned It's not just being inside a confined space containing nitrogen that can hurt you. It's possible to be overcome by sticking your head into a manhole opening and taking a breath, or even being near a vent that is discharging nitrogen. Even being outside in open air, discharged nitrogen plumes can be hazardous. It's important to have hazardous areas barricaded off, personnel informed, watches established, and personnel properly trained whenever nitrogen purging work is being carried out on at a site.

5.9.2 Nitrogen Delivery Methods

There are a number of major producers of nitrogen in the United States, including Airgas,[4] Air Liquide,[5] Air Products,[6] Linde,[7] Matheson-Trigas,[8] and Praxair.[9] Nitrogen is also available from specialty gas and welding supply distributors. Nitrogen suppliers offer considerable technical support and can answer questions regarding nitrogen purging.

Nitrogen for purging use can be provided to a site in many forms including gas cylinders, liquid cylinders, and tube trailers. Prepiped nitrogen gas cylinder packs can be used where a small number of individual gas cylinders are not adequate. Liquid cylinders provide greater volume, but the pressure available is normally limited to about 100 psig. Higher pressure can be obtained by using a special regulator. Tube trailers usually provide the largest volume and the most pressure capability of any form of supply available.

5.9.3 Nitrogen Gas Cylinders

Nitrogen gas cylinders commonly come in sizes up to 300 standard cubic feet. Multiple cylinder packs contain more, depending on the number of cylinders. These volumes are available nitrogen at standard conditions, 70°F and 14.7 psig. The gas in these cylinders is usually provided at about 400 psig.

Using these cylinders is very simple: You connect the properly rated hose, open the regulator valve to some suitable discharge pressure, and start nitrogen flowing into your piping system. The discharge pressure is adjusted to the pressure needed using a pressure regulator. Remember to open valves slowly and build pressure in steps so you can better control what is happening.

Remember that cylinders can be vulnerable to improper handling. Keep protective head assemblies around and use them when cylinders are being stored. Also, make sure that cylinders are secured by chaining them to a wall or support, or by nesting (chaining a number of cylinders together, with each cylinder touching at least two other cylinders). A good reference for cylinder handling safety is P-1: Safe Handling of Compressed Gases in Containers, available from the Compressed Gas Association at www.cganet.com. Gas suppliers also have safety information and should be consulted on the proper handling, storage, and use of their containers.

5.9.4 Liquid Nitrogen Cylinders

A liquid nitrogen cylinder is a specially designed container that stores low-temperature (cryogenic) liquid nitrogen that changes phase to gas at room temperature. Liquid cylinders are available in several sizes. Note that different suppliers may have different sizes available. A 160-liter liquid nitrogen cylinder holds about 4000 standard cubic feet of nitrogen.

Liquid nitrogen starts to boil or vaporize at −320°F at normal atmospheric pressure at sea level. At any higher temperature the liquid will boil to make gaseous nitrogen available. The higher the ambient temperature, the greater the flow will be. Liquid cylinders are usually constructed with a delivery vaporization coil built in around the outside inner chamber of the tank. This allows the liquid to become heated by the ambient environment the tank is in, helping it to vaporize and become available as a gas.

Liquid cylinders are pressure vessels with design pressures typically between 250 and 350 psig. They usually have a pressure relief valve on top, which protects the unit from overpressure. Overpressure can be caused by over heating the contents of the liquid cylinder. As some heating is ongoing due to the limitation of the insulation, liquid cylinders that are not used for extended periods may have a discharge from the pressure relief valve. Be careful when approaching liquid cylinders since you could come into the presence of a nitrogen release. Keep these cylinders stored in well ventilated areas where a release of this type would not create a problem. Also, be sure to wear gloves when adjusting liquid cylinder valves, as they can be dangerously cold. Make sure to receive operational and safety training from the supplier, regardless of what form of nitrogen you procure.

> **IMPORTANT**
>
> If you overdraw gas from a liquid cylinder, you risk moving liquid up through the gas delivery line. This can be dangerous. If you notice frost on the delivery line or hose, on the regulator valve assemblies, or on the outside of the tank itself, you may be overdrawing the gas delivery capacity.

5.9.5 Nitrogen Tube Trailers and Pumper Trucks

Tube trailers (Exhibit 5.2) are an assembly of tubes or small cylinders under very high pressure, usually 2500 to 3000 psig. These tubes are all linked together through small-diameter tubing and valves. You open each tube's valve separately to connect them to a manifold and pressure regulator system. Because the pressures can be so high, you might have to rent hoses from the supplier that are rated high enough to be safe. It's not a matter of what pressure you are using so much as what pressure the hose can possibly "see". Make sure to consult the supplier about pressures you plan to use and hose capabilities.

EXHIBIT 5.2 Nitrogen tube trailer.

Standard tube trailers have a capacity of about 40,000 cubic feet. Jumbos have a capacity of 60,000 to 100,000 cubic feet. Remember that you won't be able to use all of this volume since some residual pressure and volume will always be left in the cylinder, depending on the process you use. Consult the nitrogen supplier to get an estimate of volume delivered versus the volume that is likely to be actually available for use.

Pumper trucks contain liquid nitrogen tanks and on-board high-capacity vaporizers. They can provide tremendous volumes of nitrogen and are often used to support oil drilling operations.

Real-Life Story 19: Only One Whiff of Pure Nitrogen Can Kill You

In this incident, one worker was killed and another severely injured when they were asphyxiated by nitrogen that was venting through a large open pipe where the men were working at a chemical plant on March 27, 1998.[10] According to the U.S. Chemical Safety Board (CSB; www.csb.gov), the two victims, both skilled and experienced workers, were inspecting the inside of a 48-inch open pipe end to gauge the effectiveness of an earlier cleaning effort during a major maintenance project. While the men worked, nitrogen gas was being used to purge air and moisture from the unit. The two men climbed into the pipe using a black light that causes grease, oil, and other contaminants to glow in the dark. But the midday sun made it difficult to see, so the workers asked two contractors to hold a black plastic sheet over the open pipe end while they crouched just inside. None of the workers were aware that the plastic sheet created a dangerous enclosure where nitrogen gas could accumulate, displace oxygen, and cause asphyxiation. After 15 minutes, the contractors pulled the sheet away and found one worker unconscious and the other in a daze. The first worker was pronounced dead on arrival at a local hospital. The second man was hospitalized in critical condition but survived.

Based on incidents such as this, in 2003 the CSB issued a Safety Bulletin[11] to draw attention to the ongoing problem of nitrogen asphyxiation. According to the bulletin, between 1992 and 2002 nitrogen asphyxiation caused 80 deaths and 50 injuries in industrial settings. Deaths and injuries occurred at chemical plants, food-processing facilities, laboratories, medical facilities, and other sites. The majority of incidents occurred during work in or near confined spaces, and many incidents were caused by the failure to detect an oxygen-deficient environment. The Safety Bulletin identified a number of good practices for preventing nitrogen-related injuries, including comprehensive worker training programs, warning systems, continuous ventilation, atmospheric monitoring, and planning for emergency rescue operations. Copies of the bulletin are available at the CSB website.

Lessons Learned Whenever nitrogen is in use for any line blowing or purging activity, no matter how minor or innocent the intended use is, there needs to be a safety plan that includes monitoring for oxygen in the environment. If confined spaces are involved, there needs to be a protocol that is in alignment with OSHA and other applicable industry criteria. If you're not really sure what is or is not a confined space,

OSHA 1910.146 can be a good starting point. Remember also that nitrogen asphyxiation can occur in open areas where a release is taking place. It's not just confined-space areas that present a hazard when nitrogen is being released.

5.10 THE WORLD OF GAS LINE CLEANING

Gas lines have many sources of contaminants when they are fabricated and installed. These contaminants can cause damage to combustion turbines and fuel trains. There are cleaning devices that can be installed like strainers and sediment traps in industrial and commercial systems and more sophisticated coalescing filters and separators in combustion turbine systems to remove debris. However, depending on the conditions related to the fabrication and installation of your piping systems, a preoperational cleaning may be in order. This is especially the case in combustion turbine systems, where the manufacturer will insist on it as a condition of any warranty. It is also recognized more and more that some type of cleaning of industrial and commercial gas piping systems can enhance the reliability of fuel trains and equipment for years to come. In this chapter we provide an overview of piping system contaminants and cleaning processes.

5.10.1 Piping System Sources of Contamination

The types of contamination observed in steel natural gas piping systems include iron oxide (rust), pipe mill scale, welding slag, other miscellaneous debris, and water vapor and free water from hydrotesting or other water-based processes.

Iron Oxide (Rust) Iron oxide (rust) is one of the biggest sources of gas piping contamination that must be removed so that it cannot be carried downstream, continually clogging filter elements and plant equipment. If enough iron oxide is present during a blow, it can appear at the end of the pipe as a large orange or brown plume during initial blows (Exhibit 5.3). Rust can occur within new steel piping if it is not carefully managed through the storage and installation process. Moisture can be introduced through such weather conditions as precipitation and rapid temperature changes where dew point occurs. These sources of moisture can make for immediate flash rusting inside and out.

Moisture can also be introduced through hydrostatic pressure testing processes and water jet or water aeration flushing where moisture is not removed completely. Flash rusting can occur quickly and spread if steps are not taken to include chemical inhibitors with water-based processes.

Pipe Mill Scale Mill scale is formed on the outer surfaces of steel sheets as they are produced by hot rolling steel. These sheets are then converted to piping with the inside of the pipe containing mill scale. Mill scale can be formed at a minimum temperature of about 900°F (482°C) and is composed mostly of iron oxides that are bluish black in color. This material is usually less than a millimeter thick and initially adheres to the

EXHIBIT 5.3 Plume from a pipe cleaning blow showing iron oxide and other debris being removed from the piping systems.

steel surface. Any break in the mill scale coating will cause accelerated corrosion at the break. Thus, mill scale helps to prevent corrosion until it breaks off the surface due to a mechanical cause such as expansion and contraction of the pipe from pressure or temperature changes and even scouring from fluids at high velocities. Newly installed pipe is vulnerable to this mill scale coming off over time. This is why cleaning processes seek to remove as much loose scale as possible before startup.

Pickling, acid removal of mill scale, is available for carbon steel piping. However, this is usually reserved for smaller sections within specialized applications such as lube oil systems for combustion turbine systems. Stainless steel piping is another approach to avoiding mill scale issues for lube oil and final fuel delivery piping sections.

Weld Slag and Spatter Weld slag is the residue left on a weld bead from flux. Flux is provided in the welding process to shield newly deposited weld metal from atmospheric contaminants that will weaken the weld joint. Spatter comprises globules of molten metal that are expelled from the joint and then resolidify on the metal surface. These materials, and particles left from pipe grinding and end preparation, are also sources of contamination.

Other Debris Many other things inside a pipe could contaminate a fuel gas system, including welding rods, pieces of gaskets, and even things completely unrelated to piping systems installations (e.g., soda cans). This is why foreign material exclusion covers are a necessity during pipe transportation, intermediate construction phases, and storage.

5.10.2 Removing Water and Vapor After Hydrotesting or Other Water Processes

Residual water and water vapors can be present from hydrotesting of piping or from cleaning processes such as water jet flushing. Combustion turbine manufacturers have specifications for maximum water vapor content in fuel gases. If water and water vapors are not removed properly these too can become sources of contamination which can fail a turbine manufacturer's gas quality criteria and cause damage to equipment.

Removing residual water is usually done with a series of pigging and blows using low-relative-humidity (dry) air through the piping system. Low-relative-humidity air is usually generated by heating air or using dried compressed air. Compressed air dryer systems are usually rented for this purpose, along with compressors for blows. Dryers are available as refrigerated and desiccant types. Desiccant dryers make for air of much lower relative humidity than do refrigerated dryers.

5.10.3 Build It Clean

The best way to avoid contamination is to "build it clean," a phrase used often when describing processes that are aimed at minimizing contamination. The following are key considerations in building it clean.

1. Design piping with a particular method of cleaning in mind so as to facilitate removal of debris upon installation. Try to avoid unnecessary or inaccessible dead legs, low points, and abrupt changes in flow path or diameter. Also be careful with sizing and the placement of vents and drains. Consider using removable spool pieces to accommodate venting points.
2. Specify and monitor welding techniques carefully and change the way that root passes are installed to minimize protrusion of weld beads and the presence of pipe bore slag.
3. Clean individual piping spools upon fabrication or in place as they are installed.
4. Consider chemical rust inhibitors and the laying up of piping with nitrogen instead of air prior to placing it into service.

5. Carefully control access to piping from foreign objects on job sites through the use of foreign material exclusion covers on piping sections as the piping is transported, stored, fabricated, and installed.

5.10.4 Overview of Pneumatic Line Blowing for Cleaning

Pneumatic line blowing involves the discharge of a high-pressure gas (compressed air or nitrogen) through the piping systems in a series of rapid bursts. The process starts with a volume of gas accumulated in some reservoir at some starting pressure behind a valve. Then the valve is opened quickly and there's a sudden release. The gas flows, scrubbing the walls of the pipe and transporting debris until the reservoir reaches some stopping pressure that represents an ineffective scrubbing velocity for the gas or the reservoir is discharged to 0 psig. Then the reservoir system is recharged and in some cases the process starts over. When considering pneumatic cleaning options there are conditions under which both compressed air and nitrogen have merit (but never natural gas). The issues to consider include whether air compressors of suitable capacity are readily available at the site, if storage in the form of a new out-of-service boiler or some section of pipe is available, and the arrangement of piping system segments to be cleaned.

Although the use of compressed air is most common, in some cases nitrogen can offer advantages. For example, portable nitrogen tube trailers or pumper trucks, can make available very high volumes and flow rates. This can be important if there is limited or no reservoir capacity or if the compressors required are not available. However, the trade-off is always the higher cost of nitrogen versus compressed air. Nitrogen also requires additional safety considerations because it is an asphyxiant.

Line-blowing effectiveness has been identified to be a function of the cleaning force momentum ratio that is achieved. Cleaning force momentum ratio is a product of mass flow and velocity. The fluid momentum of the air or nitrogen near a pipe wall entrains particles at the wall into the fluid flow through aerodynamic forces. It is the goal of the cleaning process to pass a fluid (compressed air or nitrogen) through the piping that makes for higher cleaning forces than can ever be achieved from the flow of natural gas through the pipe during operations. Cleaning force momentum calculations must take into account the density of the medium used and changes in geometry of the piping system as might affect the velocity of the medium during the pneumatic blow process. Care must be taken in establishing the desired flow path, the sequence of blows, and the treatment of dead legs, branches, and in-line elements to prevent damage. More detailed information regarding pipe cleaning processes is available from an EPRI document that I authored: Electric Power Research Institute Document 1023628, Guidelines for Fuel Gas Line Cleaning Using Compressed Air or Nitrogen, September 2011.[12] I was allowed by them to excerpt some sections of this document in this book. The EPRI document covers pipe blow cleaning force ratio calculations and more information for implementing pneumatic blows.

Although pneumatic blowing sounds simple, there are many safety considerations any time that pneumatic systems are in use. Hazards can include pipe explosions, damage to piping and instrumentation systems, and injuries from flying debris discharged out of the end of the pipe.

Real-Life Story 20: Seemingly Simple Gas Line Cleaning Costs Lives

It was supposed to be a beautiful new state-of-the-art combined cycle power plant nestled in the hills of New England. The plant was designed to have 620 megawatts of power capacity.[13] Two Siemens combustion turbines would discharge their excess heat energy into two heat recovery steam generators. Gas was being delivered to the site from an interstate pipeline through a 12-inch high-pressure distribution line. It had appeared that the plant could be started ahead of the original planned date. Startup and commissioning efforts were in full swing until February 7, 2010 the day that six people lost their lives and many others were injured.

I arrived at the site weeks after the incident as part of a team lead by attorney Robert Reardon. Reardon represented several plaintiffs (people killed and injured). I was shocked by the devastation (Exhibit 5.4). I had never seen anything like it, and nothing could have prepared me for what I saw. There was debris everywhere, including many pieces of metal hanging from upper stories that occasionally fell and rang out. It was clear that even walking around the site after the incident could be hazardous. The following is a description of the events that occurred, taken largely from the U.S. Chemical Safety Board's more detailed description, available at www.csb.gov.

Construction was nearing completion and the high-pressure gas piping that fed the turbines was being cleaned. Combustion turbines operate at high temperature, high speed, and close tolerances. The natural gas that is supplied must meet the turbine manufacturer's cleanliness specifications for particulates and other contaminants so that warranties will remain in good standing. The natural gas pipe cleaning process chosen for this project was pneumatic pipe cleaning or gas blows using the actual pipeline-supplied natural gas as the material to flush out debris. A total of eight blows were planned on the day of the incident: one at the main plant entrance, two on each of the two fuel conditioning skids, one on each of the two duct burner sections, and one for an auxiliary boiler feed line. Blows had already been completed the weekend before for the sections of piping that started off below the mountain and led up to the facility. These were carried out in areas that were somewhat wide open.

EXHIBIT 5.4 Power plant site postexplosion.

This second set of blowing procedures was going on, directed by a specialized commissioning contractor who had done this type of blowing many times. A local mechanical contractor who fabricated and installed temporary piping sections, such as the discharge vents, was also on hand. The blow gas source (again, natural gas) was controlled from the gas utility's metering and regulating station, some distance away, and from valves at the top of the mountain. There were teams communicating to open and close the large valves by hand.

If you understand the layout of the site, the conditions that led to the incident become clearer. The area where the deadly blow release occurred was between the two heat recovery steam generators (HRSGs). These were about 50 feet apart and about 120 feet away from the generator building. The gas skid being purged was outside but beneath a very congested section of pipe racks. Temporary scaffolding was installed in this area as well. Much of the gas released hit the pipe racks and scaffolding (Exhibit 5.5). This resulted in rapid mixing with air, making for larger volumes of a flammable gas–air mixture.

It was cold that day, and the gas coming out was even colder and denser because of its pressure drop. The cold dense gas was less prone to rising. The HRSG courtyard created a somewhat confined area. It's easy to visualize how a huge flammable cloud was created that did not dissipate well. Records indicated that about 400,000 cubic feet of natural gas discharged in just the last 10 minutes of the process. This would have made for over 10,000,000 cubic feet of flammable mixture. When natural gas is being vented or discharged, you have to remember that it is mixing with air and that 1 cubic foot of natural gas produces 23 cubic feet of flammable gas/air mixture.

EXHIBIT 5.5 Gas release vent point in a congested area.

The entire plant is nestled into the top of a mountain. When standing with your back toward the generator building and looking out through the HRSG courtyard, you are facing a couple of metal-sided buildings, a cooling tower, and a rock wall that was cut away to allow room to build the plant. The rock wall towers to almost the height of the cooling towers. It's easy to understand how the shock wave created between the HRSGs could have been channeled toward the cooling towers and even reflected and amplified off the rock wall. Explosions can become more devastating if amplification of the blast occurs after the initial explosion.

The source of the ignition is not exactly clear. Usually there is no lack of ignition sources, although some can be very obscure. A ground heater, a portable boiler that provides a source of hot water or hot water–glycol mixture, was about 75 feet from the skid where the gas discharge occurred, but it was said to be turned off. In a ground heater liquid is pumped through hoses placed in contact with frozen ground to thaw it to permit work such as pouring concrete bases. In this case the unit was oil fired and had its own oil tank and electrical generator, all packaged in a trailer. In some cases, even though equipment like this is off, refractory in the firebox can stay hot for some time (possibly hotter than the ignition temperature of natural gas).

The Middletown, Connecticut Police department confiscated what was left of the ground heater. Pieces of it were found throughout the courtyard. I was told that this was part of a criminal investigation at the time. Connecticut has criminal negligence statutes that provide a maximum of six months in jail. Engineers, contractors, and tradesmen rarely realize that their actions can result in jail time. Many believe that the worst that can happen is that their employer may be fined, they may be found civilly liable, or they may lose their jobs. I got a real sense of this during viewings of the confiscated pieces of the ground heater the police department made available in the police garage. It was intimidating to sit there and be interfacing with a bunch of guys with guns and badges when you're only used to being in an environment with other technical people.

Many people were involved with the gas line cleaning project. No one wanted to kill or injure anyone. There were good people involved who made tragic mistakes and errors in judgment. We may never know why these errors occurred, but they were very costly. The human trauma is still being tallied. There were six deaths, many injuries, and an economic loss in the hundreds of millions of dollars.

Lessons Learned Natural gas should never be used to clean gas lines. This practice was sometimes used in the past for convenience by power plant builders. Since the Middletown incident these practices have been outlawed in many jurisdictions. Codes have been changed to prohibit the use of natural gas and combustible gases in this way. NFPA 56 was created to provide guidance for these matters, prohibiting the use of natural gas for pipeline cleaning. Turbine manufacturers have also recommended against this practice.

There are many safe alternatives to the use of natural gas to clean natural gas piping systems. Pneumatic processes that use compressed air are by far the most popular choice. There are many hazards in performing pneumatic blowing processes that are not discussed here. Make sure that pneumatic cleaning processes are performed only by trained and experienced personnel.

5.10.5 Fuel Gas Piping Cleanliness Requirements

Cleanliness requirements are somewhat subjective for industrial gas piping applications other than combustion turbines. Combustion turbine manufacturers, on the other hand, typically require power plant owners to meet very specific fuel cleanliness requirements to maintain warranties. Turbine manufacturers publish overall contaminant loading guidelines where both air and fuel contaminants are considered. Combustion turbine internal piping is such that fuel gases are released through combustor cans with small nozzles positioned in a radial pattern around the hot end of the turbine. These nozzles have relatively small holes that can erode or become clogged with debris, compromising performance. Getting into the cans to clean them out can be time consuming. Debris can also erode and damage turbine blades. All of this affects plant performance, reliability, and maintenance costs.

Cleanliness requirements are usually agreed upon with and witnessed by both an owner's representative and the turbine manufacturer's site technical representative. When the cleaning process is a pneumatic blow, once several preliminary blows have occurred, targets are installed on the blow pipe outlet discharge. Targets can be many things, including polished copper plates, stainless steel, or even plywood painted white. Additional blows then occur and these targets are evaluated for markings that have occurred from debris coming out. There might, for example, be a requirement that there be no more than three marks on a target for the pipe section to be called clean.

5.10.6 Cleaning Processes Other Than Pneumatic Blowing

In this section we provide an overview of flushing and pigging technologies other than pneumatic line blowing used for pipe cleaning that can be applied to fuel gas lines. These topics are each introduced and described from an overview perspective.

Water Jet Flushing, Milling, and Aerated Water Blows of Piping Systems A number of water-based flushing processes are also commonly used for gas line cleaning. These are usually part of a family of processes that are deployed sequentially to get the overall cleanliness desired. These processes include water jetting or milling and aerated water jet flushing.

Water jet flushing or milling is a process in which high-pressure water jets connected to special hoses are moved through the piping systems. The jets can operate at pressures of 10,000 psig or more and are effective in removing pipe mill scale, weld slag, and rust from the inside gas lines. The debris is removed and flushed from the system as the jet nozzle head is retracted. Contractors operating these systems usually provide facilities to collect water and debris. The water can be disposed of or recycled through the special collection systems provided as part of the cleaning plan. Contractor equipment for these processes typically includes a trailer to house the control facilities somewhat near the operation, and space to locate the pump system skid. Pump system skids can be diesel operated or run from a rented generator. A source of water and space for water and debris collection equipment would also need to be provided.

Aerated water jet flushing is a similar process, in which highly aerated water is forced as a slug down a pipe at high velocities to dislodge from the pipe debris, weld slag, corrosion deposits, and other foreign objects.

Pigging Pigging is the most popular and widely used process for cleaning in the pipeline industry today. It is the process of moving a specially designed device through a section of pipe for cleaning and other purposes, such as pipe inspection. There are many varieties and levels of complexity of pigs. In this section we focus on pigs used for cleaning.

There are open and closed pigging processes. In an open pigging process a pig is inserted and moved through the system to some endpoint with compressed air and then blown out through an open end with debris. In a closed pigging process the pig and debris settle into a pig receiver at the end of the pig's travel. In closed systems the pig is moved with compressed air or online using the motive force of the product pressure in the pipe. There is little release of the compressed air or product gas that is used as the energy source to move the pig because of the design and operation of the receiver. In this discussion we do not address the design elements of receivers and launchers or any particulars regarding their operation but, instead, seek to inform the reader regarding the various types of cleaning pigs that are available.

Pig Types and Cleaning Applicability The following are pig types that are commonly used directly in support of cleaning processes.

Poly pig: flexible polyethylene pigs ideal for applications where the pipeline condition is not well known and the primary requirement is not to block the pipeline (www.pipepigs.com).

Pinwheel pig: a pig designed for aggressive removal from the walls of a pipeline of debris such as scale and hard wax. Hard steel pins burst and scrape debris from the pipe wall.

Pressure bypass pig: has a built-in pressure relief that allows for a burst of liquid or gas to exit the pig in the direction of flow to move debris and accumulation from in front of it when the pig becomes stuck.

Inhibitor spray pig: applies a corrosion inhibitor inside the piping system as it moves.

Magnetic cleaning pig: a preinspection pig that removes from a pipeline ferrous metallic debris such as welding rods. It uses strong permanent magnets to collect and hold these materials.

Cup pig: designed for use in long runs in which wear may be a special concern. They are specified for commissioning of pipelines as an alternative to bidirectional pigs and for separation of various media.

Brush pig: a bidirectional cleaning tool that cleans without scraping the interior wall of the pipeline. It uses both metallic and nonmetallic brushes.

Smart pig: has electronic data collection instruments that can find defects and determine pipe wall thicknesses. Some can also be tracked along their path.

Gaging pig: used to determine roundness and sag issues in piping systems.

Dual-diameter pig: seals tightly to the internal pipe wall and can be custom-produced for any diameter differential.

For pigging processes to be effective, pipe runs must be designed initially with pigging in mind. Temporary launchers and receivers can also be installed to accommodate pigging.

5.11 NFPA 54: CHANGES RELATED TO PURGING ISSUES

In light of the tragic accidents involving purging of natural gas, the U.S. Chemical Safety Board (CSB) voted in 2010 to make a recommendation to the National Fire Protection Association (NFPA) and the American Gas Association (AGA) to implement immediate measures to prohibit venting of natural gas indoors; NFPA and AGA jointly publish NFPA 54, the National Fuel Gas Code. The CSB is an independent federal agency charged with investigating industrial chemical accidents. CSB does not write federal regulations but makes recommendations based on investigation of the causes of accidents in the chemical and fuel industries. The NFPA 54 committee developed and submitted for a vote related tentative interim amendments (TIAs) in August 2010. NFPA also developed a new standard, NFPA 56, for the prevention of fires and explosions during the cleaning and purging of flammable gas piping systems. Each of these efforts is discussed in this section.

The NFPA 54 cycle for revisions is every three years. TIAs are amendments to the current edition that are usually incorporated into subsequent editions. The 2012 edition incorporates the 2010 TIAs. The 2012 edition of NPFA 54 was available in September 2011. The next edition, which will be published with a 2015 date, may contain further revisions that support the subject of purging based on more widespread use. Since 1988, NFPA has also published the *National Fuel Gas Code Handbook*,[14] which contains the full text of NFPA 54 and commentary (printed in a different color) to provide background, illustrate, and explain the code. Although the commentary is not mandatory, readers of the book will find it helpful in areas with which they are less familiar.

The revised requirements in the latest edition of NFPA 54 recognize that fuel gas is used in smaller residential and commercial piping systems, and larger systems, usually in industry. The requirements for larger systems mandate outdoor purging in terms of five significant recommendations on how to conduct outdoor purging. These are commonsense practices that can enhance safety during purging operations.

5.12 HIGHLIGHTS OF AND COMMENTARY REGARDING THE NEW NFPA 56 STANDARD

The NFPA acted on CSB's request by convening a group to create first a provisional standard for NFPA 56. In the NFPA provisional process, standards have the means to get fast-tracked. I was fortunate enough to be chosen as one of a handful of people to serve on this provisional committee. I am still on this committee as of the time of this writing. Once a provisional standard is completed, the document begins the normal NFPA cycle to become a regular standard. The provisional NFPA 56 standard has been in effect since August 11, 2011.

This standard is significant because it's about all "flammable gas" piping, not just "natural gas" or "propane." This means that it covers blast furnace gas, coke oven gas, and other process intermediates in many industries.

In my opinion, NFPA 56 is the most comprehensive source of information available today for conducting flammable gas purging work. The following is an adaptation of issues related to the standard. Remember, the standard is a living document. If you do this work, are about to do this work, or are affected by it in any way, you need to own the latest version and read it, including the appendixes, which have excellent reference materials that can be very helpful in understanding everything else in the document.

1. Applicability

NFPA 56 now brings all flammable gases into a standard or code, including natural gas over 125 psig operating pressure and other common industrial flammable gases (not just fuel gases). Adoption of NFPA 56 by users for liability protection is likely to be significant in the face of two significant gas piping incidents that occurred in 2009 and 2010) and the U.S. Chemical Safety Board's request for more comprehensive standards for purging industrial gas piping systems. This standard applies both to new facilities for cleaning and placing equipment into service and to existing facilities for purging operations.

1.1.1 Applicability. This standard applies to fire and explosion prevention during cleaning and purging activities for new and existing flammable gas piping used in electric generating plants, industrial, institutional, and commercial applications.

2. Nonapplication of Standard

The following are exempt from the standard. Remember, codes and standards suggest minimum safety requirements. Your organization can choose to adopt NFPA 56 for any flammable gas operations that you see fit.

1.1.2 Nonapplication of Standard. This standard does not apply to the following items:

(1) Piping systems covered by NFPA 2, the Hydrogen Technologies Code
(2) Piping systems covered by NFPA 54, the National Fuel Gas Code

(3) Piping systems covered by NFPA 58, the Liquefied Petroleum Gas Code
(4) LP-gas (including refrigerated storage) at utility gas plants *see* NFPA 59, the Utility LP-Gas Plant Code
(5) LNG facilities covered by NFPA 59A, Standard for the Production, Storage and Handling of Liquefied Natural Gas
(6) LP gas used with oxygen for cutting, welding, or other hot work
(7) Vehicle fuel dispensers
(8) Commissioning and maintenance of appliances or equipment
(9) Vent lines from pressure relief valves
(10) Systems regulated by the U.S. Department of Transportation, 49 CFR 191 and 192

3. Retroactivity

The standard clearly applies to plants built before the standard was published but in limited ways.

1.3.1 Unless otherwise specified, the provisions of this standard apply to facilities, equipment, structures, or installations that existed or were approved for construction or installation prior to the effective date of the standard.

1.3.2 The retroactive requirements of this standard may be modified if their application would clearly be impractical in the judgment of the authority having jurisdiction and only where it is clearly evident that a reasonable degree of safety is provided.

4. New Construction and Possible Retrofit Issues

Most EPC (engineer, procure, and construct) contractors will be following the codes for piping listed below when constructing flammable gas piping systems. However, there could be instances where the requirements for the documentation portions of these, such as pressure testing records, are not well conceived at this time.

4.1 Piping System Construction. Flammable gas piping systems must be constructed in accordance with ASME B31.1, the Power Piping Code, ASME B31.3, the Process Piping Code, or NFPA 54, the National Fuel Gas Code, as applicable.

5. Extensive Cleaning and Purge Procedures Will Be Required

Detailed procedures will need to be developed for handling cleaning and purging procedures for new and existing facilities. There is an extensive list of considerations and requirements for these procedures. There is also example material in the appendix.

4.3 Cleaning and Purging Procedures. Written cleaning and purging procedures must be developed and implemented by a competent person.

4.3.1 The written procedure for each cleaning and purging activity must address, as a minimum, the following items:

(1) Scope of work and site-specific purge procedure development
  (a) Cleaning and purging method
  (b) Piping and instrument diagrams
  (c) Chemical and physical properties of flammable gas, purge media, and discharge gas
  (d) Determination of purge endpoint introducing flammable gas, inert gas, or air
  (e) Assessment and control of purge inlet and discharge locations
  (f) Temporary piping system design
  (g) Personal protective equipment
  (h) Training and qualifications
  (i) Management review and approval
  (j) Restoration of service
  (k) Target design, launcher/receiver venting review for pigging operations
  (l) Regulatory permits
  (m) Evaluation of engineering controls to limit potential unintended ignition of gases (controlled oxidation, flaring)
  (n) Written stand-down instructions to stop activity in a controlled manner
  (o) Hazards

(2) Environmental conditions and work locations
  (a) Establish and clearly identify exclusion zones where flammable gas–air mixtures are likely to exist
  (b) Limit access for personnel not directly involved with purge operations
  (c) Assessment of potential for gas migration (building openings, adjacent structures)
  (d) Prohibit hot work within exclusion zones
  (e) Lockout/tagout
  (f) Impact of environmental conditions (wind speed and direction, temperature, barometric pressure) on purge operations
  (g) Vehicular and air traffic, if applicable
  (h) Topography
  (i) Noise control and monitoring

(3) Communication plans
  (a) Pre-job briefings
  (b) Work permits
  (c) Roles and responsibilities
  (d) Emergency response plan
  (e) Facility alarm, alert, and warning systems

- (f) General facility notification prior to the start of purge operations
- (g) General facility notification at the conclusion of purge operations
- (h) Notification of regulatory authorities as required (local emergency responders, utility operators, community officials, environmental authorities, etc.)

(4) Control of ignition sources
- (a) Bonding and grounding considerations
- (b) No smoking or spark-producing work within exclusion zones
- (c) Eliminate hot work within the exclusion zone
- (d) Static electricity ignition sources at discharge point

(5) Prepurge piping system assessment
- (a) Assessment of piping system for trapped liquids, pyrophoric solids, and other flammable or combustible deposits within the piping system
- (b) Ensuring that the piping system is isolated properly
- (c) Limiting site conditions that affect the safety of the activity

(6) Purge monitoring and instrumentation
- (a) Ensure that monitoring instruments are appropriate for gas being purged
- (b) Training
- (c) Calibration
- (d) Monitoring frequency and reporting
- (e) Selection of appropriate sample point(s)
- (f) General atmospheric checks in the vicinity of purge gas release

6. Development of Detailed Cleaning and Purging Procedures
Detailed plans and procedures for cleaning and purging will need to be written and well documented. These will need to be made available on the jobsite and include names of the primary developer and team members.

4.6 Documentation

4.6.1 Cleaning and purging procedures must be documented and available at the job site.

4.6.2 The safety validation documentation must include the following items:

(1) Names, company names, and addresses of the primary developer and other principal team members responsible for the safety validation.

7. Written Safety Validations of Procedures
Written safety validations of the plans referred to above, will be required each time the plans are used. The safety validation review will also need to document the persons who created it.

4.6.2 The safety validation documentation must include the following:

(1) Names, company names, and addresses of the primary developer and other principal team members responsible for the safety validation.
(2) Name, company name, and address of the principal operational personnel representing the plant owner or operator
(3) Date of preparation and any applicable modification dates
(4) The completed safety validation in accordance with Section 4.4
(5) Any procedures related to the safety validation and any limiting conditions identified in the management of change assessment required in Section 4.5

8. Record Retention Issues
Purging, cleaning, and safety validation plans must be kept for two years. Training documentation must be kept for five years.

9. Management of Change
Management of change issues needs to be integrated into gas and purging matters as indicated below.

4.5 Management of Change. Written procedures to manage change to process materials, technology, equipment, procedures, and facilities must be established and implemented.

4.5.1 The management-of-change procedures must ensure that the following issues are addressed prior to any change:

(1) The technical basis for the proposed change
(2) The safety and health implications
(3) Whether the change is permanent or temporary
(4) Modifications of cleaning and purging procedures
(5) Employee training requirements
(6) Authorization requirements for the proposed change

10. Training
There are considerable training requirements for all staff. Sites should have a process for providing training and certification of competent persons related to this topic and for personnel who would be participating in purge operations.

5.1 Training. Persons whose duties fall within the scope of this standard must be provided with training that is consistent with the scope of their job activities.

5.1.1 Such training must include the hazards of flammable gas, the hazards of compressed gas used for cleaning or purging, safe handling practices for flammable gas and compressed gas (as applicable), emergency response procedures and equipment, and company policy.

5.1.2 Personnel training must be conducted by a competent person knowledgeable in the subject matter and must be documented.

5.1.3 Training records must be maintained for a period of not less than five years from the date of completion of the activity.

NOTES AND REFERENCES

1. Sean P. George website, www.seanpgeorge.com.
2. Chemical and Process Technology, March 3, 2009, webwormcpt.blogspot.com/2009/03/pneumatic-test-explosion-in-shanghai.html.
3. U.S. Chemical Safety Board, Valero nitrogen asphyxiation incident, completed investigations, www.csb.gov.
4. www.airgas.com.
5. www.airproducts.com.
6. www.airliquide.com.
7. www.linde-gas.com.
8. www.mathesongas.com.
9. www.praxair.com.
10. U.S. Chemical Safety Board, completed investigations, Union Carbide asphyxiation incident, www.csb.gov.
11. U.S. Chemical Safety Board, Nitrogen Asphyxiation Safety Bulletin, and presentation, www.csb.gov.
12. Electric Power Research Institute, www.epri.org.
13. U.S. Chemical Safety Board, completed investigations, Kleen Energy, www.csb.gov.
14. National Fuel Gas Code Handbook, Theodore Lemoff, NFPA.

6

Understanding Fuel Trains and Combustion Equipment

Understanding Combustion Equipment

In previous chapters we have covered combustion and gas piping basics. In this chapter we discuss how combustion equipment is configured, how fuel trains function, the purpose and design of components, and how these components and systems work together (Exhibit 6.1). We also discuss how typical burner light-offs occur along with how fuel/air ratios are controlled.

6.1 FUEL TRAIN COMPONENTS AND THEIR PURPOSE

1. *Main fuel isolation valve.* This is a manual isolation valve that would be the primary means of isolating this fuel train from the main fuel delivery piping system. It is usually a lubricated plug-style valve or a ball valve. It should be capable of being locked out physically.
2. *Flanges to accommodate a blind.* You won't find having flanges to accommodate a blind to be a requirement in a code book. It's just a best practice that I have come to rely on and recommend. It's a set of flanges, possibly with a spacer, to accommodate the installation of a blind. Remember, all valves are only manufactured to a particular leakage specification, so they may all leak through in the closed position at some time, even when new, and still be considered to be within their original manufacturing tolerances. The only thing that completely isolates a piping system is a solid bulkhead of

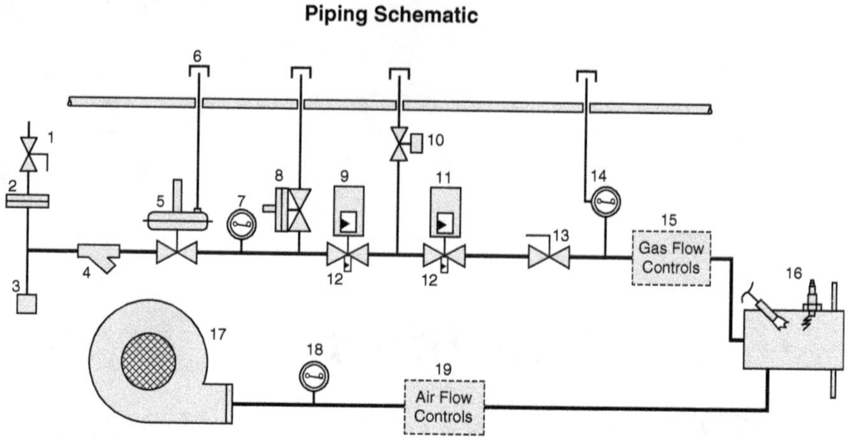

EXHIBIT 6.1 Typical main fuel train components and arrangement. Courtesy of Eclipse Combustion. Inc.

metal installed between a set of flanges (a blind). This accommodation is especially important where the firebox is large enough for, or intended for, entry by people. It also reduces the risks of leaked fuel into fireboxes for equipment that is intended to be down for extended periods of time or that has been abandoned in place.

3. *Fuel inlet drip leg or sediment trap.* Prior to World War II, most areas in United States did not have piped natural gas. In these areas, fuel gas was "manufactured" from coal or oil and called *manufactured gas*. The process of making and distributing manufactured gas incorporated small amounts of water and condensable hydrocarbon liquids in the gas. The drip leg was installed near the gas meter to have a place to remove any liquids that separated from the manufactured gas. Today in North America almost no manufactured gas is used. Drip legs are now commonly called *sediment traps* and are important for collecting particulate and debris in a piping system. In fact, they are required by NFPA 54 to be installed at the beginning of fuel trains for automatically operated appliances, which includes most industrial gas equipment.

   A sediment trap or drip leg (functionally the same thing) is a pipe tee, nipple, and pipe cap that provides a change of direction and reservoir so that contaminants that are being carried with the gas and can't change direction have a place to be deposited (in the pipe nipple), so they can't do any harm. These are normally installed at the gas service entrance, and as a part of each gas appliance fuel train as shown in Exhibit 6.1. In some cases, certain utility supplies will have hydrocarbon distillates or oil droplets in them that are either a natural part of the gas from the well, or lubricating oil carryover from pipeline compressors. Drip legs or sediment traps provide a place for this material to reside so that it won't do any harm. NFPA 54 calls for sediment traps to be the same pipe diameter as the fuel train inlet piping and the length of the capture nipple to be no less than three times the pipe diameter.

4. *Strainer.* Strainers are simple filters. They typically consist of a screen or perforated metal basket through which the gas flows. The holes in this material should correspond to the largest particulate size that can be tolerated by the components downstream. If they are too fine, they can more easily clog over time, restricting gas flow. If they are too large, they will pass harmful debris that can damage components such as valve and regulator seats. Many fuel train component manufacturers, for industrial fuel train applications, require that a size 100 mesh strainer be used (149-micrometer mesh size). Many European-style valves and valve proving systems use actual fiber mesh filters instead of strainers. Make sure you verify that the strainer installed meets the needs of your components. If you don't ask and don't specify, you are likely to get a steam pattern strainer with $\frac{1}{8}$-inch perforated holes inside that will provide you with very little value.
5. *Regulator.* A regulator is a variable-position control valve that is operated by a downstream pressure signal that causes the valve position to be reset to provide the desired flow conditions. Two types of regulators are commonly used for gas pressure control to industrial burners: self-operated and pilot-operated.

   Standard self-operated spring-loaded regulators have a variable-position valve that is operated by a diaphragm with a spring on top (Exhibit 6.2). When the system is started with the piping system at atmospheric pressure, the spring

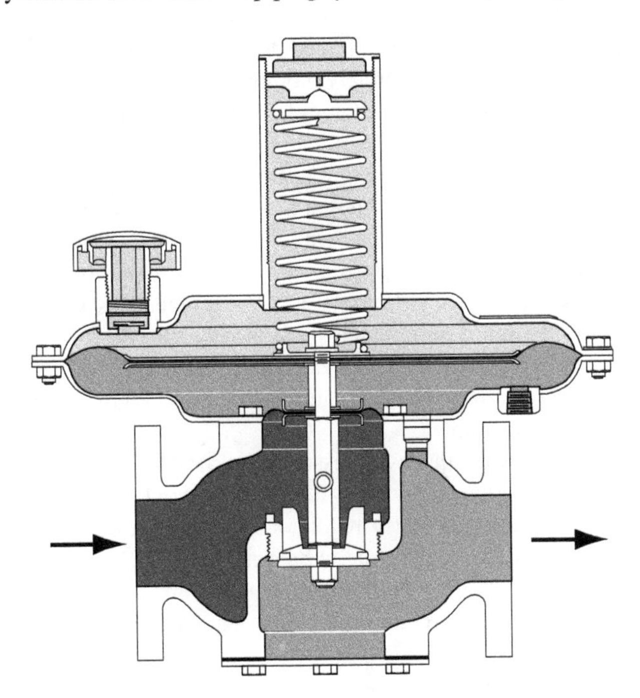

EXHIBIT 6.2 Schematic of standard spring-loaded regulator internals. Courtesy of Fisher, division of Emerson.[1]

located above the diaphragm pushes the diaphragm down, fully opening the regulator. As gas flows and fills the piping system, pressure increases against the diaphragm. This moves the diaphragm up against the spring to some balancing point. When this happens, the valve connected to the diaphragm moves closer to its seat, partially closing off flow. When gas is being used, the downstream pressure will be reduced (called *droop*). The lower pressure will cause the spring force to overtake the opposing diaphragm force temporarily and the valve will open, allowing more flow. This process occurs continuously to maintain a stable downstream pressure.

There are many variations of self-operated regulators, including service regulators that use a lever arm to provide a mechanical advantage to move the valve poppet open and closed. Another variation is the use of an air or gas spring. This is called a *pressure-loaded regulator*. In this case a pressure signal is supplied to the top of the diaphragm that comes through a special auxiliary regulator. These two variations of self-operated regulators have certain features and benefits that give them advantages over standard spring-loaded self-operated regulators for certain applications.

Pilot-operated regulators can provide the advantages of both spring-loaded and pressure-loaded regulators without some of the limitations associated with each.[2] The pilot, sometimes called the *relay*, *amplifier*, or *multiplier*, has the ability to multiply a small change in downstream pressure into a large change in pressure applied to the top of the regulator diaphragm. This gives pilot-operated regulators the ability to control downstream pressure very accurately compared to a standard spring-loaded regulator. Most pilot-operated regulators have a downstream sensing or control line, which may or may not have a valve or orifice in the control line to make it possible to create a lag to better tune the stability of the device.

In some cases a gas odor can be associated with regulators. This can be because the regulator is designed with a gas bleed-off of the air or gas spring, or it could be that the regulator diaphragm has become brittle and ruptured. Regulator diaphragms can become brittle with age. This is why when gas is turned on and off, it is important to do so slowly. One does not want to subject brittle rubber diaphragms to instantaneous shocks. This is also why you have to check the end terminations of regulator vents on a regular basis to determine if a diaphragm has failed.

Most codes require that overpressure protection be provided downstream of a regulator or that the components downstream can handle whatever pressure is immediately upstream of a regulator in case it fails. One form of overpressure protection that can usually be provided from the regulator manufacturer as an option is internal relief. This is in the form of a special built-in relief valve that directs overpressure out of the regulator housing to wherever the vent is routed.

Several other rules for piping regulator vents are identified in codes. These include never reducing the vent pipe size from the size of the fitting given. It is also important to consider the length of the vent discharge piping so that excessive backpressure is not imposed if the vent ever requires relief. Restrictions and excessive vent line length can reduce the vent flow capacity.

6. *Vents from components.* Vents are to be provided from all components that have a vent connection to release gas to a safe place in case of a failure. These are to be routed to a safe place away from building openings and ignition sources. They should be accessible, be turned down to avoid direct rain contamination, and have protection from insects on the termination end. There are rules in NFPA 54 for combining the piping from some vents. However, it is always preferable to run vents separately. Options for avoiding vent piping include the use of valve proving systems to eliminate vent valves on equipment other than boilers and ventless pressure switches where conditions allow.
7. *Low gas pressure switch.* A low gas pressure switch is there to ensure that you have enough fuel train gas pressure to light-off and operate a burner reliably, according to the requirements of the burner manufacturer. This switch is classified electrically as normally open. This means that unless the switch is seeing the set pressure no electrical continuity occurs across the switch. Getting the right set point for this switch is beyond the scope of this discussion. It needs to come from the equipment manufacturer, the burner manufacturer, or a skilled technician who has commissioned the equipment.

   There is no vent shown, to emphasize that ventless switches are available for the right environments. In other cases, where a vent connection is provided, a vent needs to be taken out to a safe place.

   There are many considerations for specifying and ordering a correct low gas pressure switch. If you're replacing an existing switch, you will need to look closely at the model number on the switch to understand all of the design features that were provided. As an example, here's the model number from an Ashcroft[3] switch: VB 54-643 789 HG 0001. Every one of these numbers and letters means something and some are for different options or capabilities that you can obtain for the switch. In most cases, depending on the device and the manufacturer, there is also a date-of-manufacture code so you tell the component's age. All automatic valves, switches, flame detectors, and burner management systems will have model and serial numbers that can be investigated to identify the component's age, intended capabilities, and features. You must make sure that components are compatible with your operating conditions.

   Another consideration when choosing control panels, switches, and other electrical devices is the NEMA (National Electrical Manufacturers Association) rating. These ratings indicate a device's relative resistance or suitability to the environment in which the device might be installed. A partial summary of the NEMA enclosure requirements is provided in Table 6.1 to introduce you to this concept.

   You may have options for the NEMA enclosure type associated with a component or cabinet enclosure. It is recommended that you consult NEMA to better understand the requirements for your equipment and planned installation environment. Each of the ratings has a definition such as the following:

TABLE 6.1 Comparison of Specific Applications of Enclosures for Indoor Nonhazardous Locations

	Type of Enclosure									
Provides a Degree of Protection Against the Following Conditions	1[a]	2[b]	4	4X	5	6	6P	12	12K	13
Access to hazardous parts	×	×	×	×	×	×	×	×	×	×
Ingress of solid foreign objects (falling dirt)	×	×	×	×	×	×	×	×	×	×
Ingress of water (dripping and light splashing)	—	×	×	×	×	×	×	×	×	×
Ingress of solid foreign objects (circulating dust, lint, fibers, and flyings[b])	—	—	×	×	—	×	×	×	×	×
Ingress of solid foreign objects (settling airborne dust, lint, fibers, and flyings[b])	—	—	×	×	×	×	×	×	×	×
Ingress of water (hose down and splashing water)	—	—	×	×	—	×	×			
Oil and coolant seepage	—	—	—	—	—	—	—	×	×	×
Oil or coolant spraying and splashing	—	—	—	—	—	—	—	—	—	×
Corrosive agents	—	—	—	×	—	—	×			
Ingress of water (occasional temporary submersion)	—	—	—	—	—	×	×			
Ingress of water (occasional prolonged submersion)	—	—	—	—	—	—	×			

Source: NEMA 250-2003, Table 1.

[a]These enclosures may be ventilated.

[b]These fibers and flyings are nonhazardous materials and are not considered class III ignitable fibers or combustible flyings. For class III ignitable fibers or combustible flyings, see the National Electrical Code, Article 500.

> *Type 4* Enclosures constructed for either indoor or outdoor use to provide a degree of protection to personnel against access to hazardous parts; to provide a degree of protection of the equipment inside the enclosure against ingress of solid foreign objects (falling dirt and windblown dust); to provide a degree of protection with respect to harmful effects on the equipment due to the ingress of water (rain, sleet, snow, splashing water, and hose directed water); and that will be undamaged by the external formation of ice on the enclosure.

It must also be noted that the ratings and protections provided apply only when the equipment is properly installed and maintained. This is another big issue. You must understand the manufacturer's instructions completely. In many cases, hazardous area installations require special accommodations (detailed in NFPA 70, National Electrical Code), such as fittings that must have sealants installed around the wires within the fitting. I have found many installations where upon temporarily removing the little fitting plug, no sealants were installed. This is one example of an easy thing to check for if you're doing a survey of an electrical system in a hazardous area.

8. *Gas pressure relief valve.* This device is designed to provide overpressure protection to downstream components in the case of a regulator failure by releasing gas through its vent. The use of this type of device is an option if internal relief in the main regulator, or other alternatives for overpressure protection, are not used. These devices are usually purchased for an intended range of field-adjustable set points and flow capacities. These parameters must be coordinated with the fuel train design.
9. *Safety shutoff valve.* This is the first automatic valve on the fuel train. It consists of an actuator driving some style of shutoff valve. This valve takes commands from the burner management system to tell it to open or close. Some valve styles can be ordered with different opening times to accommodate fuel train piping and burner needs. There are also many other possible options, including seat materials that can handle different fuel characteristics.
10. *Vent valve or bleed valve.* Some equipment fuel trains, like those for most boilers, are required to have a double block and bleed automatic fuel valve arrangement. Boiler systems under the jurisdiction of NFPA 85, those over 12.5 million Btu per hour, are also to have this valve arrangement on their pilot systems. Vent valves are normally open valves, meaning that they would be open when unpowered. A vent valve would be installed in a tee in between the safety shutoff and blocking valves. The discharge of the vent valve would run by itself (not tied into another vent line) to an outside safe place—away from building openings, air intakes, and possible ignition sources. The idea is that when the system is down and not operating, this arrangement gives any leakage past the safety shutoff valve a place to go instead of possibly through the blocking valve and into the firebox, where it could accumulate and make for a hazardous condition.

    When the equipment is operating, vent valves must provide tight closure to keep the fuel train pressure stable to the burner. If the vent valve leaks through, it could compromise the burner pressure and the flame stability. This is another reason why it's important to tightness-test all fuel train valves (including vent valves) on a regular basis according to code requirements.

11. *Blocking valve.* The second and final automatic shutoff valve in the fuel train is the blocking valve. This is a second redundant valve in series with the safety shutoff valve that's there to close in case there is no flame, or some upset condition occurs and fuel needs to be shut off immediately. In some designs, this may also have a proof-of-closure switch for verification of closure. This valve is usually of the same type and style as the safety shutoff valve.
12. *Proof-of-closure switches.* NFPA 86, Section 3.3.67.7, defines a proof-of-closure switch as, "a switch installed in a safety shutoff valve that activates only after the valve is closed."[4] These are usually not separate and distinct devices that you can see, but instead are mostly integral to a safety shutoff valve, a blocking valve, or both. These switches would provide a signal to the burner management system indicating that the valve or valves were in the closed position. This is a requirement prior to light-off being initiated by burner management systems.

    Proof-of-closure switches are an option purchased with valves. Not very many valve manufacturers have retrofit kits available to add these switches. When specifying or ordering valves, "proof of closure" is different from "position indication." Sometimes position indication switches indicate only what the actuator is doing and do not actually communicate what is happening to the valve stem itself. The best way to verify what you have is to investigate the model and serial number with the valve manufacturer.
13. *Burner firing valve.* The burner firing valve is a manual isolation valve located between the blocking valve and the burner. It can be before or after the modulating or control valve. This valve is necessary for leak-tightness testing of the safety shutoff, vent valve, and blocking valves. It also allows other fuel train testing since the main incoming fuel train isolation valve can be left open to make the low gas pressure switch and the unit dry-fired without actually introducing gas to the burner if the firing isolation valve before the burner is kept closed: (*Note*: In this case the pilot takeoff would also have to be isolated to keep fuel out of the burner during testing.)
14. *High gas pressure switch.* The high gas pressure switch is a normally closed pressure switch. This means that it has electrical continuity unless it is exposed to pressure above its set point. When that happens, the electrical contacts separate and the circuit opens. This would then trigger actions by the burner management system to close the safety shutoff and blocking valves and open the vent valve (if there is one).

    Excessive gas pressure can be present in the event of a regulator failure. Gas pressure that is too high or unstable can result in a flame-out. This should be detected by the flame detector, but the high gas pressure switch required provides one more layer of protection.

    Exhibit 6.1 shows a high gas pressure switch positioned before the modulating valve or gas controls. This is not always the case. In some cases, depending on the equipment design, it follows this valve. The exact configuration depends on what the equipment is and the overall design code that applies to the fuel train.

15. *Gas flow controls (firing rate valve or modulating valve).* The firing rate control valve is a flow control valve that adjusts the fuel flow to the burner to get the output that is expected. It can be a simple butterfly valve or a more traditional port or plug control valve. It usually has a position switch that feeds back its low-fire position for safe startups. You would never want to start your car or lawn mower with the gas pedal to the floor. Similarly, one would never want to start a piece of combustion equipment at high fire or full capacity.

    Simple single-function burner management systems (used in most packaged combustion equipment) usually watch the fuel control valve only during light-off to sense if it is at low fire. Once the unit is lit and operating, control is usually handed off to another control device, such as a thermostat, a programmable logic controller, and/or a steam pressure controller in the case of a boiler. These devices or systems then take control of the firing rate and work to achieve desired set points.

16. *Burner.* The firing rate control valve delivers gas to a burner. There are, of course, many types and styles of burners. Some of these were discussed in Chapter 2. Gas connections to burners are sometimes made with flexible connections to minimize the transmission of vibration and thermal stresses. In some cases metering and limiting orifices are added in the fuel lines connecting to burners. Limiting orifices are sometimes required by codes to limit to safe levels the maximum gas flow to burners. Metering orifices are used to accommodate measurement of gas flows and setting of fuel/air ratios. In some cases these are internal to burners, and test ports are used to measure fuel and air pressures to calculate flows.

17. *Combustion air fan.* Combustion air fans are usually centrifugal, radial, or plug fans that provide both purge air and combustion air to burner systems. If a combustion air fan is integral to the burner housing the burner is usually called a *packaged burner*. In this case the fan is shown to be separate, classifying it to be an external blower or fan system. Fans are usually connected with a flex connection to the air piping or duct system to minimize vibrations.

18. *Airflow proving switch.* The airflow proving switch can be a pressure switch or a differential pressure switch. It would be one of two forms of proving airflow that are required with many systems. This would be a normally open pressure switch, meaning that electrical continuity occurs only when the switch sees conditions above its set point.

19. *Airflow controls.* Like gas flow controls, airflow controls can take many forms. This is typically some type of damper that coordinates the airflow so that it is proportional to the fuel flow. It could be inlet dampers, outlet dampers, or even a variable-frequency drive to control fan speed, depending on the design. If there is a damper used, make sure that it cannot be fully closed and so compromise the purge airflow (NFPA 86 prohibits this).

Pilot burner fuel train (not shown) This part of the fuel train is applicable where there is a pilot for light-off. Remember, some fuel trains are direct spark ignited and have no pilot system. If one exists, the pilot fuel train system is a small parallel fuel train that regulates and delivers pilot gas for light-off. It usually consists of a manual shutoff valve, a regulator, and one or two automatic valves in series. NFPA 86 (applicable to ovens and furnaces) requires two automatic shutoff valves in series. NFPA 85 (applicable to boilers) requires a double block-and-bleed arrangement. To understand what you will need, you'll have to know your insurer's requirements and which codes apply to your situation.

6.2 BASIC OPERATIONS OF FUEL TRAINS

The development of safe fuel–air delivery and light-off systems (called *fuel trains*) and modern burner management systems (BMS's) was a somewhat treacherous path. It may be helpful to understand the history and mistakes made in early combustion system designs to better learn why systems today are configured as they are. In the early days of fuel trains, there was nothing more than a pipe feeding a burner with a manual fuel valve. Someone would partially open a fuel valve and light the burner at the end of the pipe manually. The lighting might have occurred with an oily rag or mop dipped in fuel that was blazing. The size of the burner flame would then be controlled by how much the fuel valve was opened manually. Of course, there were many safety hazards with, and deficiencies in, this type of fuel train system. For example, if the flame became extinguished, there was no automatic means to stop the fuel flow. This could allow an accumulation of unburned fuel and its sudden ignition—a very bad thing.

The modern development of fuel trains added the application of electromechanical actuators to the manual fuel shutoff valve. This would allow the automation of fuel introduction and ignition. This meant that fuel could be introduced only after some automatic sequence like a purge was complete or when a pilot flame was proven to be lit. It also meant that a flame detector could be used to close the fuel valve very quickly and reliably if there was a flame-out instead of relying on a person to be standing there watching.

Further refinement in fuel trains occurred as time went on because single fuel valves were found to leak fuel even in the closed position, and some failed to open or close mechanically when needed. It was quickly learned that two valves in series would serve the system much better. A further innovation in fuel trains was the creation of a double block-and-bleed system so that slightly leaking valves would not pose an immediate serious hazard.

6.2.1 Four Key Fuel Train Safety Functions

You've now learned about all of the components that usually exist on fuel trains. Now let's discuss how they work together to provide a useful combustion system. The four basic functions of all fuel trains are: (1) keeping fuel out of the firing chamber when the equipment is off, (2) initiating shutdown if an interlock trip or safety device is out of parameter, (3) helping to provide safe light-offs, and (4) controlling the combustion process.

Keeping Fuel Out of the Combustion Chamber Fuel trains help keep fuel out of the combustion chamber when no combustion is taking place through a series of specially designed shutoff valves that are usually spring-loaded to close. These are called the *safety shutoff and blocking valves*. These valves come in different variations and configurations. Some small systems have only one automatic or actuated valve for this function. In other cases, and especially with boilers, double block-and-bleed systems are required. The specific configuration that your site has depends on the age of the equipment and your insurance and local code requirements.

Initiating Burner Shutdown If an Interlock Trip Occurs Fuel trains also have a number of components designed to ensure that safe light-offs take place and that shutdowns occur immediately if anything goes wrong during operation of the equipment. These components are called *permissives* or *interlocks* and sometimes, *trips*. These are a series of devices, such as switches or transmitters, that are configured to help identify abnormal operating conditions, such as a loss of gas or air pressure or water-level problems in a boiler. Common trips for boilers and industrial ovens include flame detectors, high and low gas pressure switches, combustion air switches, and high-temperature-limit switches. Interlocks can be mechanical components, electromechanical devices, or software in a dedicated programmable logic controller (PLC) (see codes for special requirements of PLC's that are used in safety applications and as BMS's), or a more traditional single-purpose microprocessor-based BMS.

Helping to Provide Safe Light-offs There are three different types of light-off control or ignition systems: manual, semiautomatic, and fully automatic. In manual light-off systems, instead of a BMS system, an operator manages the light-off procedurally. This would still include a purge, to make sure that a vessel or combustion chamber is kept safe from fuel–air mixtures. The operator then carefully manages the insertion of some type of torch or electrical spark igniter to the pilot or main flame at low fire for light-off to occur while valves may be manually opened. These systems are generally used under very carefully controlled circumstances for large, complex equipment that might start only once every 5 to 10 years. This could be refinery equipment, glass-melting equipment, and/or equipment used in large paper mills that run for very long periods. Risks are controlled with extensive startup procedures and well-trained personnel.

In some cases this is done in conjunction with specialized contractors that have temporary heat-up equipment. This equipment has burners that get systems such as refractory and holding vessels near operating temperature before the equipment's integral built-in burners are started. Preheating using temporary equipment is common for coke ovens, some zinc-refining processes, steel ladles, and glass-making equipment.

In the case of semiautomatic systems, a start button is pressed. This starts a purge fan and a timed prefire purge sequence begins once airflow is proven. Once the purge is complete, manual intervention is required to actually begin the light-off. The system might then be configured for someone to hit an igniter button, which would open a pilot fuel valve and start a spark igniter. It might then be necessary for a main fuel valve to be opened manually.

In a fully automatic system a start button is hit and the operator has no further involvement; the BMS handles everything from start to finish. This sequence consists of verifying prestart interlocks (permissives), the preignition purge, the pilot trial for ignition, and finally, the main flame trial for ignition. These discrete steps are described below.

The Light-off Process

1. *Verifying prestart interlocks (permissives).* A number of basic prestart conditions must be met before the BMS can allow the light-off sequence to progress. These are equipment specific but could include having adequate fuel pressure, no flame in the firebox, verifying that the safety shut off valves and blocking valves are closed, and if the design is to meet European standards (i.e. seeing an air switch open before it is closed). Most systems also require the main fuel control valve to be at a low fire position prior to light-off unless the burner can be ignited safety at any firing rate (which is rare). This usually means that the fuel flow control valve is proven to be somewhere between 15 and 25% of high fire capacity. You'll have to check with the burner manufacturer to find the correct minimum firing rate for your particular equipment so that a qualified burner technician can make the necessary adjustments at commissioning. There may be additional items that relate to the specific equipment as well. For example, there has to be an adequate water level in a boiler.
2. *The preignition purge.* NFPA 86, the Standard for ovens and Furnaces, calls for a purge of at least four air changes of the firebox volume before an ignition source can be introduced. NFPA 86 is applicable to ovens and furnaces only. Boilers have a number of issues that must be considered to calculate the proper purge time. Different purge time requirements are identified in NFPA 85 for single-burner and multiple-burner boilers. For example, the purge requirement for some single-burner boilers is in terms of air changes, and can be four to eight changes. There is a minimum of 5 minutes or five volume changes, whichever is greater, for multiple-burner boilers (exceptions for simultaneous dual-burner systems do exist). Other requirements regarding minimum airflows and stipulations can increase this amount.

   In the case of refinery or other traditional chemical plant process furnaces, depending on their design, the prepurge may be done using mechanisms other than a traditional fan. This can involve a steam purge of the furnace or vessel for displacement of air volumes, or a steam jet eductor which acts as a fan to move air through the unit. American Petroleum Institute (API) standards are a good reference source for this equipment.

   It should also be noted that some classes of equipment, including some boiler systems, require a postfiring purge. Some BMS's are configured to remove residual firebox gases with a purge after the flame has been extinguished for any reason. This requirement exists in part because flame detectors and valve systems do not always act instantaneously when a loss of flame is detected. Hence, there can be some fuel flow, usually for up to 4 seconds, before a flame detector is required to send a signal for fuel shutoff. Then the safety shutoff and blocking valves have 1 second to close. The postpurge can remove the 5 seconds of residual flow if a loss of flame occurs. The duration and airflow for the postpurge is an equipment- and BMS-specific function.

3. *The pilot trial for ignition period.* Depending on the size and type of system, pilots are allowed 10 seconds to light for natural gas and 15 seconds for fuel oil. This is often a configurable setting on modern BMSs, although in the case of some equipment it's a permanent change. You can always decrease it but you cannot increase it when configuring on most BMS systems. It should also be noted that some European codes allow only 4 seconds. Actually, it should never really take 10 seconds. If you're not getting a pilot lit in 3 or 4 seconds, your system needs to be evaluated more closely. This is an important startup consideration. Once pilots start to approach 7 or 8 seconds to light, they can be audible and the condition can be an indicator of something going very wrong.
4. *The main flame trial for ignition period.* Once a pilot is lit and the main gas firing rate control valve is at a minimum position, the main automatic safety shutoff and blocking valves open while the vent valve, if it exists, closes. A timer allows a finite time for the main flame to light. This period of time is called the *main flame trial for ignition period.* Depending on the size and type of system, this may be 10 seconds for gas and 15 seconds for fuel oil. Once again, as in the case of the pilot trial for ignition period, it should never really take 10 or 15 seconds for the main flame to light. Again, how long this takes should be part of a regular observation and evaluation, just as in the case of the pilot trial for ignition period. It should never take more than 3 or 4 seconds to light a main flame.

IMPORTANT

Remember that although the equipment generally gives you 10 seconds to light a pilot or main flame, this should occur in 3 or 4 seconds. The longer it takes, the more you could be heading toward a dangerous condition. You can usually observe this and time it on the BMS display. Don't try to do it by watching the flame through the site port at light off.

Real-Life Story 21: BMS Evolution

Attempting to understand the evolution of burner management systems and their design and timing is like tracing the lineage of unicorns. One popular feature almost universal to all BMSs, and in fact in many codes, is the 10-second trial for ignition sequence timing. This means that fuel and air are allowed to flow into the combustion chamber for 10 seconds while attempts are made to light them for a pilot or a main flame. The ten-second pilot and main flame trial for ignition have always been a mystery to me. How and why was 10 seconds chosen? I asked a seasoned veteran of the burner control market this question, and his answer surprised me. He said that in the nineteenth century boiler operators on ships were deemed to be good operators if they could dip a mop in a bucket of oil, light it, shove it into the firebox of an oceangoing vessel's boiler, and get the main burner lit in 1 minute or less. He said the time just moved down from there as people and systems got better. This was not the well-developed, path-of-science answer I was looking for, but it kind of explains the trial-and-error evolutions of other components and systems.

Lessons Learned The state of fuel system and combustion equipment technologies is ever evolving. Codes and standards change on about a three- to five-year cycle. Pay attention to what the changes are with relevant standards organizations such as NFPA and ASME for your industry and equipment. Assign someone the task of subscribing to these standards and monitoring changes, including tentative interim amendments and interpretations. Many people also don't know that anyone can attend code committee meetings, provide comments, and ask questions. The only thing you cannot do is vote. I highly recommend attending such meetings and staying in touch with the evolution of safety related to fuels and combustion equipment.

Pilot Systems To better understand the light-off process and flame safety, you also need to know about pilot control systems and how they work. Remember also that not all burner systems have pilots. Some have main burners with the capability to be directly ignited by a spark plug. Pilot flames are small flames that are lit near or integral to the main burner. They are designed to provide the ignition energy for the main burner flame to light-off quickly and effectively. On larger industrial boilers pilots are also called *igniters* and are classified by their Btu capacity. NFPA 85, Section 3.3.85 provides definitions of the igniter classes.

There are three basic types of pilot control systems: standing, interruptible, and intermittent.

1. *Standing pilots.* Standing pilots are always lit, just as are hot water heaters in many homes. They are generally used in small-capacity equipment and appliances, mostly residential and commercial. They are usually monitored by thermocouples or flame rod–based flame detectors.
2. *Interruptible pilots.* Interruptible pilots light and stay on for only a short period of time in the main flame burner operating cycle. They are then shut off by the burner management system. If the flame detector continues to see a signal, it is probably from the main flame and the system can proceed to operate safely.
3. *Intermittent pilots.* Intermittent pilots are lighted and stay on for the entire call-for-heat cycle. These are not as safe as interruptible pilots because the flame detector might be fooled into thinking that the main flame is lit when it is only getting a signal from the pilot. This can lead to an explosion since fuel from the main burner that has not ignited could be accumulating and then ignite all at once. Codes describe whether a pilot system needs to be intermittent or

interruptible. This hazard is sometimes mitigated by the use of two detectors, one for the pilot system and one for the main flame and very careful sighting or positioning of whatever detectors exist.

Real-Life Story 22: Failed Light-off Attempts Result in Death

A devastating industrial paint oven explosion resulted in one death and caused tremendous property damage at a manufacturing facility.[5] A maintenance crew was troubleshooting a large oven in which painted products were dried and cured. The burner had shut down and three technicians were attempting to relight the pilot with the electronic igniter in a semiautomatic light-off system. After several attempts, an explosion occurred in the oven. The force of the explosion broke loose the 800-pound fan housing, propelling it across the aisle and toward one of the workers. He was crushed and killed by the impact.

How did this happen, and how could it have been prevented? Fuel-fired equipment has the potential of accumulating gas in the combustion chamber every time there is a failed light-off attempt. If there is no source of ignition or inadequate purge airflow, fuel can continue to increase in concentration with every failed light off attempt. Ultimately, the fuel released can explode when a source of ignition is found if it has accumulated in concentration to a flammable mixture (minimum of about 5% by volume for natural gas). To prevent this from happening, all fuel-fired equipment, such as paint ovens, is required to have purge cycles that sweep fresh air through the firing chamber for every light-off attempt.

NFPA 86 requires at least four volume changes of a firebox or combustion chamber for ovens and furnaces before an ignition source is introduced. After completing the purge, the flame safety system or burner management system will continue with the firing sequence. Next, the igniter will attempt to light the pilot. If the pilot has not been proven in the time allotted, the flow of gas will be stopped. The purge cycle must be repeated before the next attempt to relight so that any accumulated flammables can be removed. In this case, there were repeated attempts to relight the pilot without proper purging taking place. Purging deficiencies can include things that are not known or obvious to users.

One of the defects identified in the investigation was that a pneumatically operated butterfly damper located on the discharge side of the purge fan was somehow closed during the purge cycle. This blocked the purge airflow. It was later concluded that the butterfly damper was neither calibrated nor properly aligned. This prevented the combustion air fan from delivering the purge airflow required. NFPA 86 prohibits the use of dampers in combustion air systems that could be closed and thus impair the possibility for a proper purge to take place. If dampers exist, they must be cut away or sized so that even in the closed position a proper purge can occur.

The purge time was also a problem. The oven manufacturer recommended a purge time of 6 minutes. The purge timer used was equipped with an adjustable dial. Today, BMSs are available with tamper-resistant computer chips or cards that dictate the timing. The setting of the timer at the time of the incident was only 1.5 minutes.

Additionally, the exhaust duct, specified by the manufacturer, was supposed to be routed directly outside. In this case, the exhaust duct was restricted by two 90° elbows, which further limited purge airflow.

When the pilot failed to light, natural gas was allowed to enter the combustion chamber with each successive startup attempt. The natural gas built up to higher and higher concentrations throughout this process. Ultimately, the natural gas concentration in the combustion chamber increased to the lower explosive limit. At this point, a source of ignition, probably the spark igniter, triggered the explosion.

The force of the explosion was not able to relieve as originally designed, due to modifications made to the oven. When an explosion did occur, the pressure front found relief locations to be either restricted or reinforced. This caused the pressure to relieve itself at the structurally weakest location, by blowing out the fan and housing. Following a previous incident, the plenum of the heat exchanger was reinforced with a heavier gage of sheet metal. The original, thinner sheet metal was designed to deform in the case of an explosion. This might have provided some pressure relief (management-of-change issue).

There were several warnings which if heeded could have prevented this tragedy. Several years earlier, there was a smaller explosion at another of the manufacturer's facilities. The furnace involved was the same type as the one under discussion. In the earlier incident, a contract employee was troubleshooting the oven and was injured when the explosion occurred. The details of this incident were not communicated to others involved with similar equipment. Additionally, the oven with the fatal explosion had been repaired at least twice in recent history, due to smaller explosions. The cause of the explosions was never fully investigated or widely reported.

Lessons Learned All safety devices need to be inspected at their recommended intervals for proper operation and to ensure that they have not been defeated. It is likely that combustion airflow switches had not been set properly for some time, along with the purge timer. Properly functioning combustion air proving switches would not have let the permissive be made to the BMS without proper air flow. This would have stopped successive light off attempts.

There is no replacement for policies that require detailed investigation of every incident and widespread communications to all who may be involved with any combustion equipment. When natural gas–fired equipment malfunctions, it is necessary that the repair do more than merely get the unit running again. It is important to consider the history of the unit and to conduct a systematic root cause analysis.

Operators and service personnel need to be given the proper training to help them fully understand the combustion sequence and its faults and to understand hazards such as multiple starts and audible starts. Personnel need to be aware of safe operational procedures, including daily and periodic testing and maintenance requirements.

Controlling the Combustion Process Controlling the combustion process is what makes combustion useful. This means controlling the fuel/air ratio and firing rate. There have been many incidents related to improper fuel/air ratio control. Setting the right ratio and properly having control systems that maintain that ratio accurately are still challenges today.

Before we move on, there's a basic burner term related to firing rate capacity that you should understand: *turndown*. If a burner has a maximum capacity of 2 million Btu's per hour and a minimum capacity of 500,000 Btu's per hour, we would say that this burner has a 4 : 1 turndown capacity. This is simply the overall firing rate or capacity adjustability range of that burner.

Burner capacity controls come in three basic types, with several variations within each. There is on/off control, high-fire/low-fire control, and full modulation control. On/off and high-fire/low-fire control are pretty much as their names imply. Full modulation control means that the burner can fire at an infinite number of capacities anywhere between the minimum low fire start setting and the maximum (high fire) setting.

We discuss capacity control and fuel/air ratio control together since these systems are interrelated with most burner controls. The four most common fuel/air ratio and capacity control systems are cross-connected single-point parallel-positioning mechanical linkage, independent actuators, and mass flow control.

Cross-Connected Systems The most common form of fuel/air ratio control for industrial ovens and furnaces is the cross-connected system (Exhibit 6.3). These have very few moving parts and are almost never found on boilers. In these systems an actuator operates a control valve or damper on the incoming combustion air. A signal line will be taken after the control damper to the top side of the diaphragm of a control regulator (often called a ratio regulator) on the fuel system. This is not the main fuel regulator. These systems still have a main fuel regulator upstream and prior to the ratio regulator to maintain pressure stability. As the airflow changes, the pressure to the diaphragm of the regulator controlling the fuel also changes. This keeps the air and fuel flow synchronized.

One of the nice features of this type of system is that if there is some sort of failure on the air supply fan, actuator, or linkage, there would be no airflow or signal to the fuel supply regulator diaphragm. This means that it's unlikely that you would have fuel flow without airflow. One variation with these systems is the availability of

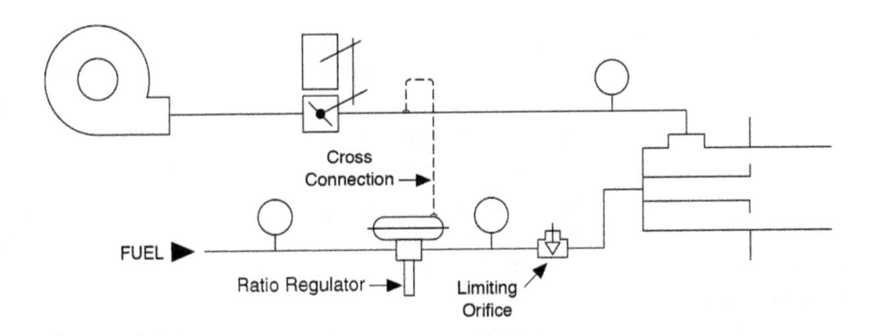

EXHIBIT 6.3 Cross-connected firing rate fuel/air ratio control. Courtesy of Eclipse Combustion.

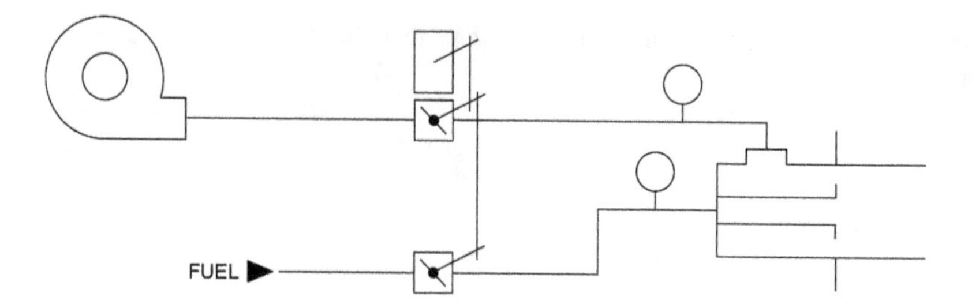

EXHIBIT 6.4 Single-point parallel-positioning linkage firing rate fuel/air ratio control system. Courtesy of Eclipse Combustion.

nonlinear response ratio regulators. This means that you can, for example, make the fuel delivery richer or leaner at high-fire if that would be desired.

Single-point parallel-positioning Mechanical Linkage Systems In a single-point parallel-positioning mechanical linkage or jack shaft type of system, firing-rate control is accomplished by moving a gas valve and an air valve simultaneously with the aid of linkages (Exhibit 6.4). The linkage positions are set by a trained technician during the initial installation and commissioning of the equipment. One of the limitations of this type of system includes differences in fuel/air ratios that can occur because of the different flow characteristics that can be expected from partially open air and gas valves (i.e., the gas and air valves would usually have different flow curves).

Unlike cross-connected systems, if the linkage falls off the air damper or valve and there is no airflow, there can still be fuel flow. This type of failure has caused numerous incidents over the years. Linkages need to receive periodic lubrication and a review to see if excessive stress or binding is occurring. In some cases, bearing blocks, damper blades, or valves could be binding and making for problems that appear to be only linkage related. This is why frequent checks of these linkages are important, along with the practice of match marking. Match marking can easily be done at no cost by using a permanent marker to denote the proper orientation and position of linkages. A quick visual check can then be made to see if slippage is occurring on a regular basis by comparing the marks to see if anything has moved or shifted.

Individual Actuator Systems Fuel/air ratio control in individual actuator systems is accomplished with separate actuators on the fuel and air dampers or valves. These can have linkages or be coupled directly. A temperature or pressure controller sends a signal to increment both the fuel and the air settings simultaneously to meet some desired load (firing rate) condition. There may or may not be position feedback on these valves or actuators. Feedback control for the position of these valves is a safety enhancement that can be added. It can often be obtained as feedback of the actuator or valve movement. Valve movement is, of course, a more reliable means of feedback than actuator movement.

These systems provide much more accurate combustion control than linkage-based systems since each valve can be stroked independently at something other than a linear positioning. This can allow flow characteristics to be better matched for fuel and air under all load conditions.

I have seen several accidents involving these types of systems due to improper installation and commissioning. These systems are usually driven from a proprietary programmable logic controller (PLC), which means that only an authorized technician, usually the installer, can make adjustments or repairs. Spare parts are typically not available except through the original manufacturer. If you put one of these into an "old school" maintenance environment where people like to tinker, it could be trouble. Once you own one of these, you also just formed a close relationship with the vendor's service team.

IMPORTANT

Be aware of proprietary burner control systems. Review your site to see if you have them. If so, how are they being serviced? Are your personnel trained, and do they have the right software? Are you vulnerable to limited service and replacement options for this equipment? If the equipment goes down, how long might operations be down?

Mass Flow Control Systems The most sophisticated of all firing rate control methods is mass flow control. This involves individual actuator systems as described above except that they have one very important refinement. These systems employ flowmeters for both the gas and the air that is delivered to a burner. Instead of relying on valve flow characterizations and a one-time setup of the fuel and air systems and hoping for repeatability, these systems monitor the actual mass flow of fuel and air and make adjustments continuously. The flowmeters in use are usually thermal mass flowmeters. The flow data are presented to a PLC along with other important parameters (possibly flue gas compositions, combustion air temperatures, and/or draft conditions) so that combustion efficiency and other process parameters can be optimized continuously. These types of systems are expensive and complex. They are generally used in very large process heater and boiler systems where the incremental fuel savings justify the incremental costs.

The old saying "garbage in, garbage out" applies very accurately to these systems. I have seen issues with these systems related to inaccuracies in the flow-measuring elements. I have seen process heating applications where thermal elements have become coated with dirt or process contaminants, causing them to read inaccurately. This has made for garbage into the PLC and some treacherous combustion conditions. This is usually dealt with in the PLC program code by adding safety features such as cross-limiting[6] (also referred to as cross-tie limiting). The overall concept is a continuous monitoring of fuel valve position versus air valve or damper position and/or comparisons of flows and feedback of valve positions. If the gas and air ever get beyond a specified range apart from each other, the condition is alarmed or an automatic shutdown occurs. In some cases a built-in minimum exists for specific air or fuel flows. This is a safety feature put into many PLC-based fuel/air ratio control systems, and it makes sense especially where fuel and air are being metered and there are separate actuators.

Real-Life Story 23: A Failed Tuning Job Destroyed a Boiler

This case involved an 800-horsepower fire tube boiler at an agricultural processing plant that had exploded a few days before my arrival. The incident was suspiciously in alignment with a contractor service visit to "tune" the burner. It's not clear why the contractor was there for this work. I believe that it was either a regular seasonal review for improving efficiency or because the site had been having burner issues and a problem was suspected. Many facilities tune their burners twice a year, in winter and summer, to optimize efficiency, since the characteristics of the combustion air may change seasonally. However, sometimes when there is a problem we would get calls to come "tune" a burner.

The end result of this contractor's well-intentioned service visit was a very serious explosion. The rear of this boiler had bolt-on inspection plates. These were blown off, as were $\frac{1}{2}$-inch bolts and nuts, which went flying around the boiler room like shrapnel from a grenade. If anyone had been standing near the back of this boiler when it exploded, they would certainly have been severely injured or killed.

A good first step in any explosion investigation is to ask who was recently involved in servicing the equipment for any reason, be it in-house or contract work and why the person was there. It's very rare that equipment is just running one minute and blows up the next. In the majority of situations I have found that explosions are associated with some type of change or service event.

In this case I was made aware of the "tuning visit," and when I asked for the documentation associated with the event, I was given the data shown in Table 6.2, left by the service technician. I mentioned that the technician should have left

TABLE 6.2 Boiler tuning flue gas composition results

Boiler A	Flue Gas	
Percent Fire	CO (ppm)	O_2 (%)
10	27	3.5
20	29	3.2
30	125	2.1
40	430	0.4
50	>2000	0
60	>2000	0
70	380	0.5
80	122	2.0
90	15	3.9
100	4	4.1

documentation of the before and after conditions. The maintenance manager insisted that all he got was this "as left" information.

The results were surprising. As a general rule, readings of oxygen and carbon monoxide should be somewhat consistent through the firing range. In this case, the numbers were bouncing up and down the firing range. When the firing rate was around 30%, the CO levels started to climb, and then spiked at 50 and 60% as oxygen decreased. This all started to clear up at about 70% load. Carbon monoxide is often a surrogate indicator of other, more volatile unburned hydrocarbons that might be in the flue gas and create conditions for an explosion.

These boilers had a small adjustable cam that moves the gas linkage up and down proportionately to the air linkage. It was obvious when looking at this gas cam that it had a bulge in it after a little over 30% of its travel. I don't understand how someone with any competence did not see this and associate it with the jump in CO above the 40% firing rate.

The operator said that the process runs on low fire (less than 20% firing rate) for a number of hours while the product slowly gets up to temperature. Then the steam load is suddenly ramped up (over 50% firing rate). It was during this sudden ramp-up that the explosion occurred. It's likely that an unburned slug of fuel or unburned fuel derivative accumulated in one of the rear passes of the boiler near the back door when it finally found the air it needed to ignite. Pressure pulses from explosions are usually higher near where they are ignited; they dissipate from there. Hence, the higher forces developed in the rear pass at the back door by the access plates were enough to shear the bolts and send shrapnel everywhere.

Lessons Learned Setting the fuel/air ratio is a high-risk activity and there are many ways that fuel/air ratio problems can cause explosions. If you understand some basics, you will know that you never want to see anything that indicates fuel-rich conditions like this in a boiler. Most packaged firetube boilers are set up with less than 100 ppm carbon monoxide under all firing-rate conditions. Specific flue gas composition requirements for your site may depend on the fuel/air ratio controls that are installed and on air permit requirements.

It's important to retain original commissioning information and refer to it during subsequent tuning and setup work. This should include fuel and air pressures at the burner for various firing rates, along with corresponding flue gas compositions.

It is also important to have protocols established for anyone conducting fuel/air ratio adjustments so that specific procedures are followed consistently every time. These might be requirements for the equipment being through an adequate warm-up period, waiting in between completing settings at different firing rates, and even for verifying the repeatability of results going from low-fire to high-fire and then back down to low-fire. Employees need to be trained to understand basics regarding proper fuel/air ratios and flame observation techniques so that out-of-adjustment flames can be readily identified. Remember too that just because someone is a contractor and claims to have expertise there is no guarantee of competence.

6.3 OIL FIRING SYSTEMS

Before discussing fuel oil supply systems it is important to understand a few things about fuel oils themselves. Fuel oils are designated by a number, such as 2, 4, or 6, or as Bunker C. These describe the thickness and grade of a fuel oil. No. 2 is very much the consistency of diesel fuel, whereas No. 6 is very thick and viscous. You could throw a rock at a pool of No. 6 during the winter and the rock would skip over the top of it. No. 6 needs to be heated to flow very well and be put into a condition for it to be burned. Tanks for No. 4, No. 6, and Bunker C are usually heated or have suction heaters. Tank and suction heaters are often steam or electrically heated. Suction heaters heat a localized pool somewhat near the tank discharge rather than the entire tank. In these cases the fuel oil tanks and much of the piping will usually be insulated and possibly heat-traced.

In the case of fuel oil supply systems, it all starts with the site's tankage and pumping equipment. Let's start our discussion of fuel oil systems with the storage of the fuel oil at the site. This is usually an aboveground or underground tank. Potential tank problems to be aware of include inadequate diking, leaks, and temperature issues. In the United States, the requirements for the installation of fuel oil–burning equipment and liquid fuel storage tank installation and diking are covered in NFPA 30, the Flammable and Combustible Liquids Code, and NFPA 31, the Standard for the Installation of Oil Burning Equipment. NFPA 30 describes the use of different types of tanks and has safety features to prevent overfilling, overpressure, and vacuum collapse. Spill control is required for all liquid fuel storage tanks. If diking is used, its net free volume must be at least the capacity of all tanks in a diked area. Refer to NFPA 30 for more specific requirements. In other countries NFPA 30 may be accepted as a standard for liquid fuel storage areas. Be aware that local environmental regulations may exceed the NFPA 30 requirements. Also, countries other than the United States may have their own requirements for the installation of tanks and the storage of flammable liquids, or may allow internationally accepted standards to be used.

Pumps move fuel out of tank systems and move it to either smaller day tanks local to the fired equipment or directly to the fired equipment itself. If local day tanks are part of the system, additional pumps will exist to move the fuel oil to the equipment fuel train. The pumps are typically positive-displacement fixed-volume flow gear pumps. Gear pumps have intermeshing gears or rotors that send a fixed volume of oil into the discharge line with each rotation. Gear pumps make for very little slip or bypass and if you close off the discharge line with the pump running, considerable heat and pressure builds up. Pressure relief valves are usually installed internal to the pump or externally downstream with a return line to the tank in case valves are closed off. Excessive operation in recycle mode can heat the fuel oil significantly and cause problems.

The pumping systems will be fed from or through strainers. Every fuel oil pumping system should have strainers somewhere to clean up residual solids and other contaminants that probably exist in the oil. Strainers are usually of duplex type: that is, two strainers in parallel. Duplex strainers are usually designed so that they can be switched between each other without a fuel shutdown. It's important never to shut off or interrupt the fuel supply to an operating system. The system should have a low oil pressure switch that trips if all works well and the pressure dips below the set point. However, a drop in oil pressure, even if its reestablished quickly, can also cause flame instability and a flame-out, and put the system at risk of an explosion.

Strainers should also have gages so that the pressure drop associated with normal operation can be compared to dirty or clogged operation. Operators at every site that burns fuel oil should understand when strainer performance is degrading. Strainer systems sometimes have a scraper that allows online cleaning of the filter mechanism. These scrapers rout foreign materials to a cleanout line for removal. Make sure you understand what you have installed and how it should be operated.

The overall intent of most fuel oil systems is to allow a pump to send a constant flow to the burner firing rate control valve for it either to be burned or returned to the tank. If a fire occurs in the building, fuel flow in both of the systems (delivery to the day tank and fuel train recirculation) must be stopped so as not to continue to feed fuel to the fire. Some codes call for the installation of fusible-link fire shutoff valves onto the piping systems to mitigate this risk.

Liquid oil does not burn. It must first be converted into a mist and then the fine droplets evaporated and the vapor burned. The entire process of conditioning the fuel and converting it to mist is called *atomization*. Atomization takes place using one of three methods: pressure, steam, or air. Pressure atomization is simply the process of forcing high-pressure oil through small holes and making a very fine spray. Atomization is also done with steam or compressed air. Where compressed air is used, it is usually supplied by a local compressor or from a dedicated compressor on the equipment. In these cases the steam or compressed air is forced into the burner with the fuel to help make for finer sprays. The adjustment of fuel pressure and atomizing medium pressure is vital to safe and proper operations.

In the case of heavy oils (above No. 2) proper temperature is also vital. It is not typical to heat No. 2. Heavy oils are usually heated to the range 150 to 250°F, or they will not flow properly or be atomized effectively. In fact, pumps can be damaged if the temperature is not correct. The concept of tank heating or suction heaters has already been discussed. However, once the oil nears the boiler or furnace there are usually temperature interlocks and further temperature trim systems (could be an electric or steam heat exchanger) to produce specific and carefully controlled temperatures so that atomization and combustion are consistent.

Oil system fuel trains serving combustion equipment are constructed with similar pressure regulation, high and low oil pressure switches, safety shutoff and blocking valves, and firing-rate control valves as exist with natural gas systems. There are also relief valves for recirculation of what is not burned; refer to NFPA 31 and specific equipment codes for detailed guidance on fuel train safety configurations. There is usually no difference in the combustion air safety devices provided from the devices used for gas-fired equipment. Additional safety controls and interlocks for oil systems that do not exist on gas-fired equipment include oil gun position interlocks and interlocks that may be associated with fuel heating and atomization media for pressure, differential pressure, and flow.

6.3.1 Oil Ignition Systems and Burners

Lighting-off oil systems means first going through a purge cycle and lighting a pilot. Oil system pilots are often configured for a fuel that is something other than oil. Oil pilot systems often use liquefied petroleum gas (LPG), and often from a small storage tank located near the equipment. An LPG pilot system can usually provide better flame stability and reliability than a fuel oil pilot.

If oil pilots are used, ignition transformers are usually higher in voltage than for natural gas, since igniting oil takes more ignition energy than does igniting natural gas. If you have a dual-fuel piece of equipment, remember that the ignition transformer and igniter device may be not compatible with those for natural gas alone.

Usually, oil firing involves the use of an *oil gun*, usually a long pipe that has a special nozzle or tip where the oil comes out at a precise location within the burner. In the case of multiple fuel burners, you will usually find this in a rack sitting somewhere in the boiler room. It is typical that these are not installed while the burner is firing gas. If oil guns are left in during periods of natural gas firing, there can be heat damage to the tip and coking of residual fuels left in the gun.

Inserting the oil gun occurs with the unit not firing. The oil gun is picked up, inspected, partially inserted, atomization and oil feed hoses connected, and then the gun is inserted completely until interlock switches are made telling the BMS that the gun is properly positioned. This interlock has to be satisfied or the BMS will not allow the light-off sequence to occur.

Proper nozzle tips are vital to the safe and effective operation of oil systems and need to be chosen specifically for the type of oil, pressure available, the atomizing medium, and the flow capacity desired. Oil nozzle tips need to be cleaned routinely and checked for wear to be sure that the holes have not become enlarged from erosion over time.

Once the oil gun is deemed properly inserted, a precise sequence of operation needs to be in place for both light-off and shutdown of the system and the atomizing mediums (if they exist). Improper shut down of oil firing systems where oil gun residuals have been cleared into fireboxes has caused a number of explosions that I have been familiar with over the years.

6.4 OVEN AND FURNACE TYPES

Now that you understand fuel trains, it's important that you know a little about how and where this heat energy would be applied to real-world equipment. There are six major categories of fuel-fired heat processing equipment in which water is not heated: (1) ovens, (2) furnaces, (3) thermal oxidizers/incinerators, (4) fluid heaters, (5) space heating equipment or dryers, and (6) flares.

I'll start with the most popular styles of equipment: ovens and furnaces. Generally speaking, ovens are lower-temperature furnaces. These terms are used somewhat interchangeably for ovens and furnaces that operate under 800°F. At design operating temperatures over about 800°F, the equipment is usually referred to as a furnace. The oven/furnace dividing line is somewhat vague.

Another designation with all of this equipment is whether or not it is direct- or indirect-fired. Direct-fired implies that flue gases are coming into direct contact with the material that's being heated and performing the work of the oven or furnace. This would be the case in a metal-melting furnace where hot combustion products are directed at metal that is to be melted. Indirect applications are those such as radiant tube furnaces where combustion products are fired into a special tube so that they don't come into contact with something like a flammable atmosphere that is actually performing the work of the oven or furnace. Another example of an indirect unit is a baking oven with a heat exchanger. In this case it is intended that the combustion products stay contained within the heat exchanger and only recirculated air (not combustion products) would be moving past the heat exchanger and coming into contact with the food being baked.

6.4.1 Oven Types

NFPA 86, the Standard for Ovens and Furnaces, is the Bible when it comes to ovens and furnaces used in industrial applications. This document is very detailed and directly applicable to a wide range of ovens and furnaces. The designations in Table 6.3 are from the definition section of NFPA 86, which describes ovens and furnaces of being one of four types: A, B, C, or D.

Class A ovens are those where flammable materials are present from processing and not provided purposefully as in the case of class C ovens. A good example is a paint-drying oven, where solvents are liberated intentionally. A more subtle example is that of melting recycled aluminum cans, where residual organics, coatings and paints can make for a flammable smoke inside the oven. Owners sometimes don't understand that the liberation of things such as residual oils or other surface contamination can create unforeseen flammability hazards inside an oven. NFPA 86 defines many special safety design considerations to keep these ovens safely operating.

TABLE 6.3 NFPA Oven Types by Classification

Furnace Type	Description
Class A	A heated oven or furnace where there is a potential explosion or fire hazard that is due to the presence of flammable volatiles or combustible materials processed or heated in the furnace.
Class B	A heated oven or furnace where there are no flammable volatiles or combustible materials being heated.
Class C	A heated oven or furnace that has a potential hazard due to a flammable or other special atmosphere (such as hydrogen or ammonia) being used for treatment of materials in process.
Class D	An oven or furnace that is a pressure vessel that operates under vacuum for all or part of the process cycle.
Thermal oxidizer	An independently controlled, enclosed combustion system whose purpose is to destroy volatile organic compounds, hydrocarbon gases, or vapors, or both, using elevated temperature, residence time, mixing, excess oxygen, and in some cases, catalysts.

Class B ovens are the largest single category of ovens. They operate at or near atmospheric pressure levels where no flammable volatiles are produced or combustible materials are heated. Examples include metal-melting or metal-holding furnaces, aging ovens, crucible heaters, some annealing furnaces, thermal fluid heaters, and some heat-treating ovens.

Class C ovens have a special atmosphere inside to change or sometimes to protect the properties of the material being heated. An example is a carbon-rich environment to do surface hardening of tool steels or gears. There are many specialized atmosphere ovens that have a myriad of safety controls for obtaining the proper environment and keeping it safely in the oven when doors are open and closed.

Class D ovens are vacuum ovens or furnaces. In these cases, processing is done in a special pressure vessel. The materials to be processed are placed inside, a vacuum pulled, and then heat is applied. Class D ovens are typically used in batch-type metal heat-treating applications. Some of the unique hazards they pose are issues related to the pressure vessel shell. Shells can be overpressurized or can implode from excessive vacuum. Failures of seals can allow leaks of atmospheres into surrounding work areas. Air can also leak into a flammable gas atmosphere that is already at ignition temperatures.

6.4.2 Furnace Types

Furnaces are also categorized similarly to ovens in the NFPA world (A, B, C, and D). However, the most popular and common furnaces either anneal, change surface properties, or melt something. Annealing furnaces are used to heat metal to a certain temperature to change the grain structure and metallurgical properties. Special atmospheres are sometimes used to control surface finish characteristics in furnaces. For example, certain carbon steels can readily absorb carbon from a surrounding atmosphere when they reach a certain temperature. In these cases the atmosphere may be a carbon-rich gas such as natural gas or a modified version of it. Some materials are surface-hardened this way. Metal-melting furnaces can span the range of small foundry units to large iron-making blast furnaces at steel plants.

6.4.3 Thermal Oxidizers and Incinerators

Thermal oxidizers and incinerators use high temperatures, around 1400 to 1500°F, to break down unwanted materials into carbon dioxide and water and then release them safely into the atmosphere. Incinerators are usually discussed in the context of the disposal of solid materials (not gases or liquids). Thermal oxidizers are commonly discussed in terms of the disposal of undesirable liquids or gases.

These devices are somewhat unique among other combustion systems since the design has to consider if the materials being disposed of actually have Btu value. The Btu value can often also change during the disposal process. Consider the case of drying paint on a car. The first part of the drying cycle inside a heated oven often includes a considerable amount of solvent that flashes off. This solvent-laden air is in many cases taken to an on-site thermal oxidizer for disposal. This puts a substantial loading of volatile organic compounds (VOCs) into the thermal oxidizer's combustion chamber but it can be a variable loading. These volatile organics have fuel value and contribute to keeping the temperature of the firebox at the level required for complete combustion; however, these organics require a considerable amount of oxygen for complete combustion to take place. Hence, you cannot set up a burner on a thermal oxidizer the way you would any other piece of equipment. Most other types of combustion equipment need to consider only air that is coming through the main burner. In most cases the goal is to minimize the excess air to this main burner. In this case, providing the right amount of excess air is much more complex, since it requires some estimate of VOC loadings. If in doubt about proper settings for airflows or burners for thermal oxidizers or incinerators, refer to the equipment manufacturer for guidance.

Thermal oxidizer designs also include catalytic units and systems with natural gas injection. Catalytic thermal oxidizers are built with a catalyst bed through which the airstream has to pass. Catalysts do not themselves react, but contribute to and enhance other chemical reactions. Catalytic thermal oxidizers usually operate at lower temperatures and rely on combustion to be completed in the catalyst bed.

There are special safety considerations in NFPA 86 for catalytic thermal oxidizers. Catalyst beds can have a tendency to become fouled and/or plugged and may require periodic replacement or regeneration of the catalyst material. Catalysts can also be poisoned by the wrong materials coming through, which renders them ineffective.

Natural gas injection systems are sometimes used in thermal oxidizer systems when there is little or no fuel value to the material being introduced. In these cases the introduction of natural gas with the process stream to be destroyed provides fuel value. Consider the case of a fragrance or spice manufacturer. In some cases the manufacturing process requires odors to be destroyed so that they do not contaminate the neighborhood on a continual basis. In some of these applications, natural gas is carefully injected in specific measured quantities into the disposal airstream immediately prior to it being submitted for burning.

Thermal oxidizers can also come as regenerative thermal oxidizers (RTOs). These are devices that make for more energy-efficient operations. They involve heat recovery beds or chambers that take turns heating or absorbing heat and then being cooled (rejecting this recovered heat) by incoming process air. This swapping of heat dramatically reduces the overall energy input required and makes for substantial energy savings. This is why a combustion chamber temperature can be 1500°F and yet the flue discharge temperature from the unit can be only 300 to 400°F.

Hazards surrounding thermal oxidizers have included fuel/air mixtures not being set properly, catalytic and heat recovery beds being clogged, and natural gas injection systems putting in too much or too little natural gas and at the wrong times.

6.4.4 Process Furnaces and Fluid Heaters

Process furnaces and fluid heaters come in many shapes and sizes somewhat specific to their function. However, many are of cabin style, looking somewhat like a log cabin with four walls and a stack coming out the top. The walls are usually lined with tubes containing process fluids, with special burners in the floor firing up.

The lower section, in this case nearest the burners, would be a radiant section with the highest temperatures available and the primary mode of heat transfer being radiant energy. Additional tubes might be located higher and in the combustion product stream to represent a convection section that transfers heat to colder incoming materials.

The process fluid to be heated could be a liquid or a gas. There are many American Petroleum Institute (API) standards for process furnaces or heaters serving the petrochemical and refining industry. The International Society for Automation (ISA) also has information on this equipment as it relates to process and petrochemical applications. Another common type of fluid heater used for industrial heating is a called a "hot oil" heater. There is a new NFPA recommended practice, NFPA 87, which covers these fluid heaters. These systems involve some type of thermal fluid transfer medium, such as DowTherm™.[7] In these systems the fluid is circulated through tubes in a firebox, where it is heated to some temperature, usually below 800°F, and then circulated to the process through heat exchangers or coils.

6.4.5 Space Heaters, Makeup Air Units, and Dryers

Space heaters, makeup air units, and dryers are somewhat similar, in that they are all usually heating air. They can do this directly with airflow-style burners or indirectly, by first passing the air to be heated through a heat exchanger so that the flue products do not come in contact with the air that's being heated.

6.4.6 Flares

The last category of equipment that does not heat water is that of flares. Flares are used in process applications such as wastewater treatment plants generating methane, landfill gas operations, steel blast furnace operations, petroleum refineries, and chemical plants to burn off flammable materials that are a normal by-product. This prevents the release of an unburned gas cloud that can mix with air and create a much larger flammable mixture cloud that could be ignited and make for a large explosion. The idea of a flare is to keep a small fire burning (pilot system) at a vent release point to ignite combustible materials as they pass by. Some designs also include a knockout pot to remove liquids before they get to the discharge point. Still others include a mixing system at the end of the gas discharge tip or release point to encourage more complete combustion. The discharge mixing system can be such that steam is injected as a release occurs to encourage turbulence and mixing.

Although flare systems seem relatively simple in concept, there are many critical design considerations such as the use and application of flame arrestors to keep flames from back-feeding, keeping pilots lit and stable, and addressing radiation heat transfer from releases that can heat things in the immediate vicinity of a large release. There are many very competent and experienced long-time flare design and installation companies, such as John Zink,[8] that can be of tremendous help to you if you're having problems with a flare system.

NOTES AND REFERENCES

1. Fisher regulators, www2.emersonprocess.com/en-US/brands/fisher/Pages/FisherValvesIn-struments.aspx.
2. Steve Berry, Fundamentals of Pilot Operated Regulators, Class 6030, Fisher Controls International, Inc.
3. Ashcroft Instruments, www.ashcroft.com.
4. NFPA 86, Section 3.3.67.7 Definitions: Proof-of-closure switch.
5. Lessons from a Fatal Paint Oven Explosion, www.uaw.org/hs/00/02/hs01.html.
6. Air/Fuel Control, Getting There and Staying There, Chap. 4, www.cleanboiler.org/resources/primer/primer_chap4.pdf.
7. DowTherm, www.dow.com/heattrans/products/synthetic/dowtherm.htm.
8. John Zink, www.johnzink.com/products/flare-systems.

7

Understanding Boilers and Their Special Risks

Why Boilers Are a Special Risk

Before I describe the world of boilers as I know it, I want to emphasize the fact that the potential for catastrophes is much greater for them than for any other category of combustion equipment, because there is a twofold risk, fuels and saturated water/steam. Besides this, heating water, in boilers or hot water heaters, is by far the single biggest application of heat energy and fuel trains on the planet. In the United States alone, a 2005 study[1] indicated that there are over 163,000 commercial and industrial boilers. There are millions of residential boilers and hot water heaters as well. In this chapter we describe different boiler types and also provide insights into some of the hazards associated with steam systems, including safety relief valves and steam piping.

Real-Life Story 24: Ten Sailors Die in a Boiler Steam Accident

You would never think that a small steam leak can turn into a tragedy that ends up taking 10 lives. This story is the ultimate warning about how careful one has to be around high-pressure steam systems and pressure vessels. The following is a direct account from a long-time friend and colleague, David Peterson, and information contained in the official U.S. Navy JAG report.[2] He was personally at this incident, which changed his life. I have found Dave Peterson to be one of the most passionate, competent, and dedicated boiler inspectors and safety professionals I have ever known. It is a privilege to be able to provide his story in this book.

Fuel and Combustion Systems Safety: What You Don't Know Can Kill You!, First Edition. John R. Puskar.

Today, David Peterson is a machinery and equipment field representative with the Cincinnati Insurance Company. In 1990 he was a master chief boiler technician assigned as the boiler inspector for Amphibious Squadron Twelve. The *Iwo Jima* was one of the ships in this squadron. A few days after the accident, David was on an airplane bound for Bahrain to board the *Iwo Jima.* The following is his account of the incident.

Let's begin the story by reviewing some of the facts of the events that led to this accident. On August 2, 1990, Iraq invaded Kuwait. A few days later, Iraq forces control Kuwait and declare Kuwait to be a providence of Iraq. Thus begins an armed forces buildup that would later become Operation Desert Storm. Toward the end of August 1990, a large number of U.S. naval ships began deploying to the Persian Gulf. Included in the ship deployment was the USS *Iwo Jima* (LPH-2). The *Iwo Jima*, commissioned on August 26, 1961, was an amphibious assault ship designed as a helicopter carrier that carried approximately 2000 Marine ground force troops. During mid-October 1990, the *Iwo Jima* was experiencing a number of difficulties with her engineering plant. To effect the necessary repairs, a decision was made to send the ship to Bahrain, where she could go pier side and shut down her boilers. Among the various pieces of equipment to be repaired were main steam valves 1MS-7 and 2MS-7. The ship's engineers suspected that steam leaked by one or both of them, thereby preventing two-valve protection to be achieved. Two-valve protection is required if routine maintenance is to be accomplished on either boiler while the other boiler is in operation. The *Iwo Jima* arrived pier side in Bahrain on October 25.

The ship had two D-type water tube boilers, each boiler's normal operating pressure was 600 psi and 850°F at the superheater outlet. Each boiler was rated at approximately 250,000 pounds per hour. Just past the main steam stop valve for each boiler was the pipe branch providing steam to the No. 1 turbine–driven generator. These were stop valves 1MS-7 and 2MS-7, respectively. These valves were steel gate valves of the outside screw and yoke design. The valves were approximately 6 inches in pipe size and were butt-welded to the piping system, with the valve stem being in a horizontal position. The valve bonnets were attached to the valve bodies by threaded fasteners with a spiral-wound compression gasket sealing the two pieces. The valve bodies were drilled and tapped. Studs were screwed into the valve bodies, and nuts were used to secure the bonnets. Ten fasteners were installed to make this flange joint, and I believe that each stud was $\frac{3}{4}$ inch in diameter. Each valve also had a $\frac{3}{4}$-inch bypass valve and piping installed for system warm-up and pressure equalization.

On October 28, a pipe fitter from the Bahrain Shipbuilding and Engineering Company arrived and began work on these valves. Both of the 6-inch valves and their bypass valves were disassembled by removing all the fasteners that held the valve bonnet to the main valve body. After disassembly the valves were inspected by the *Iwo Jima*'s engineering officer. The bypass valve for 2MS-7 was found with defects in the seating surfaces, and the other valves were found without any visual steam cuts, cracks, or other flaws. The pipe fitter was directed to repair the seating surfaces on the 2MS-7 bypass valve by lapping the seating surfaces with a grinding compound and to reassemble all the valves. The official investigation report notes that the pipe fitter completed reassembly of these valves and departed the ship at approximately 7:00 P.M. on October 28.

At 2:00 A.M. on October 30, 1990, fires were lighted in No. 1 boiler in preparation for getting underway at 8:00 A.M. that morning. Everything appeared to be just a routine light-off watch. The No. 2 boiler was lit-off at approximately 5:30 A.M., and by 7:30 A.M. both boilers were on line. The engineering officer had reported that the engineering department was ready to get under way, or in navy jargon, ready to answer all bells.

The ship set the special sea and anchor detail and prepared to go to sea. When a navy ship is maneuvering under restrictive conditions such as entering or leaving a port, or in general quarters ready for a battle situation, all standby equipment is placed in operation as a ready contingency. On the *Iwo Jima*, the No. 1 turbine–driven generator, which was located in the boiler room, was designated as the standby generator. Main steam valves 1MS-7 and 2MS-7 were boiler stop valves that routed steam to this generator. Therefore, just prior to the ship getting under way, these valves were opened and the No. 1 generator was placed in operation.

The ship got under way at 7:56 A.M. that morning. Very shortly thereafter, valve 2MS-7 started leaking steam. Within moments the valve was leaking very badly. The boiler room personnel reported to the engineering officer in the engine room, which would be referred to as main control, that there was a major steam leak behind the No. 2 boiler and they requested permission to shut it down. At 8:11 A.M. main control reported a major steam leak to the ship's commanding officer on the ship's bridge. Before the commanding officer had a chance to take any action, the bonnet completely blew off valve 2MS-7, dumping steam at 600 psi and 850°F into the boiler room.

The ship lost propulsion power and electrical power immediately and was dead in the water. The diesel emergency generators started, and the ship regained some electrical power but could not maneuver. The ship went to general quarters and at approximately 8:35 A.M. a repair party entered the boiler room to investigate damage and search for survivors. Ultimately, all 10 crew members in the boiler room died as a result of this accident. Four crew members managed to escape the boiler room, where they were transported by helicopter to a navy hospital ship. The other six personnel were found dead in the boiler room. At 11:30 P.M. that night, the last survivor died on board the hospital ship *Comfort*.

One can only imagine what was going on behind the scenes in the minds of those who perished. But we do know some of the details of the communications that occurred from their perspective. Shortly after getting under way, the boiler room operators knew they had a major leak and that it was getting worse. The boiler room operators were very frightened and their instincts were telling them to shut down the boilers and get out of the boiler room. Under normal circumstances, that is exactly what they would have done. However, the ship was at sea detail, and their training was to continue to operate until they received permission to shut down. In a panicked voice they reported to main control that they had a major steam leak behind the No. 2 boiler and requested permission to secure the boiler. The phone operator in main control replied "main control aye," indicating that the message was understood and followed with the reply "wait one," which meant that he was waiting for a reply. That would be the last communication. Seconds later the bonnet blew off the valve completely. When this occurred, a loud "boom" was heard in the engine room. From the loud boom and with the steam pressure gages dropping suddenly, the engineering officer knew immediately what had happened. He ordered the main engine throttle man to open the throttle to try to draw as much steam out of the piping as possible. Four of the boiler room crew members immediately ran to try to escape the boiler room. The other six crew members stayed and tried to shut down the boilers. It was reported that the main propulsion assistant (MPA) had positioned himself on the stairs to enter or exit the boiler room. I would assume his primary motive was to prevent anybody else from entering the boiler room. The MPA was a young officer in excellent health and physical condition, and he was already on the stairway when the valve failed. Yet he could not escape the boiler room without sustaining fatal injuries. That fact illustrates how quickly the boiler room filled with intense heat. It demonstrated exceptional bravery for these men to remain at their post as the steam leak worsened increasingly.

The No. 1 SSTG on the *Iwo Jima* was considered to be a backup generator. This is because it required two additional watch standers to run this generator than it did to run the No. 2 SSTG, which was located in the engine room. That is why these valves were not opened until shortly before the ship was ready to get under way. As a normal part of setting the special sea and anchor detail, the ship would run standby equipment such as the No. 1 SSTG as an added precaution. Once the ship had safely left the harbor, they would resume normal operations.

Although it has been 23 years since this accident, I can still vividly remember some aspects of my time aboard the *Iwo Jima*. Shortly after I arrived I was having coffee in the chief's mess and talking with senior chief Johnson, who was the engineering officer of the watch during that morning's light-off watch. Shortly before the ship set the sea and anchor detail, and about 30 to 40 minutes before the valve failed, he had taken a tour in the fire room. He became emotional as he described a conversation that he had with one of the young fire room watch standers. This person was nearing the end of his enlistment and as the senior chief walked by him, he called out jokingly, "only 40 more days to go, senior chief." I also remember a conversation with the lone survivor of this accident. BT1 Hamilton had started his way into the fire room to draw boiler water samples. As he was climbing down the ladder into the fire room, the main propulsion assistant screamed at him to get out of the room. He had no more than exited the fire room when the valve blew. I remember him saying that if he had just been distracted by a 30-second conversation on his way to the boiler room, he would have been walking into the disaster instead of running away from it.

Of the people who perished in this accident, I knew only one personally. BT1 Robert Volden was the boiler technician of the watch: in navy jargon, the topwatch. We had served together on board a destroyer, the USS *Coontz*, from about 1982 until 1984. In 1983 the *Coontz* underwent a one-year overhaul at the Philadelphia Naval Shipyard. Bob and I drove from Philly to Norfolk many times that year to visit our families over weekends. Bob was a devoted husband and the father of two daughters. He was also a very good boiler technician. He died bravely following orders and trying to operate the boiler plant in the manner in which he had been trained.

It is virtually impossible to describe the way I felt when I entered the boiler room for the first time. Normally, a boiler room has many contrasting colors, but the *Iwo Jima* boiler room was a very eerie white color. This was because the velocity of the steam had sandblasted all of the pipe insulation from nearby pipes and the insulation was evenly distributed on virtually every surface in the boiler room. I could picture the terror the operators felt as they tried desperately to shut off the boilers and escape the heat. Quick-closing valves were installed on the fuel oil pumps and on the fuel supply headers for each boiler. These valves could be tripped by pulling a remote cable attached to the valves. The operators pulled these cables so violently that they were broken on each valve. The main steam stop valves for the boilers were opened manually, but pneumatic motors were installed so that the valves could quickly be shut down remotely in an emergency. However, when the bonnet blew off 2MS-7, it sheared the hoses providing air to the pneumatic motors, so the main steam stop valves did not close.

Beyond the impact of feeling what the operators must have felt in their final minutes, I began assessing what a mammoth undertaking it was going to be to get this boiler room operational again. We had to disassemble and clean all of the electrical switchgear and many pieces of machinery. Thousands of hours were expended on this project. The navy brought a repair tender to Bahrain to do the majority of the repair work.

Once the repair work began winding down, our duties shifted to training replacement crew members and building confidence in the crew that the boiler room was safe to operate. This was no minor undertaking. Many of the young boiler room crew members had seen a friend's dead or injured body and were scared to death to return to the boiler room. One of the hats that a navy chief petty officer wears is that of a counselor. I had many long heart-to-heart conversations with these young men, asking them to trust that I will make sure that the boiler room was safe before it was restarted. Everything imaginable was inspected and tested, and then tested again.

Lessons Learned Ultimately, the cause of this catastrophic accident was that improper nuts were installed to hold the valve bonnet to the repaired valve body. The nuts used were of an improper material. The pipe fitter from the Bahrain Shipbuilding and Engineering Company that performed the work on the valve was Pakistani and didn't speak much English. As he was preparing to reassemble the valve, he thought the existing nuts were corroded and he wanted to replace them. It was stated during the JAG investigation that he asked one of the ship's crew members for new nuts and was told to look through the spare parts bins in the boiler room. The nuts he chose to install were made of brass. The nuts chosen and used by the pipe fitter were not visibly distinguishable as brass because the manufacturer had applied a black oxide coating to the nuts, which gave them the appearance of ferrous metal. This was a problem long before this accident. When I was a boiler room supervisor, one of my subordinates tried to use such black brass nuts on a steam piping job. Like Gibbs in the TV show *NCIS*, I would slap the back of his head and explain that these nuts were not suitable.

During normal operations, valve 2MS-7 would be closed because its purpose was to supply steam to the No. 1 SSTG, which is the standby generator. As steam at 600 psi and 850°F began flowing through this valve, the brass nuts were expanding at a greater rate than the steel studs, thus quickly losing the strength to secure the bonnet to the valve body. After less than 30 minutes of operation, the valve failed catastrophically.

So, what types of studs and nuts should have been used? I remember from my navy days that anytime I was working on superheated steam components, any threaded fasteners had to be ASME/ASTM grade B-16, a heat-treated alloy steel designed for high-pressure, high-temperature service. The proper studs and nuts to use were marked as B-16. If I am remembering correctly, the studs had "B-16" etched on the flat surface, and grade B-16 nuts had a raised letter "H" or possibly the letter "B" and the number "4" on the face of the nut. The letter "B" identified the nut as being manufactured from bar stock. The letter "H" identified a nut as being heat-treated. The number identified the grade of steel used in the nut. A nut marking of "2H" indicated a heat-treated nut made of carbon steel. A nut marking of "4H" indicated a heat-treated nut made of carbon alloy steel. The brass nuts used during the repair to 2MS-7, while black in color from the oxide coating, had no identification markings on them whatsoever.

In the case of the repair to valve 2MS-7 onboard the USS *Iwo Jima*, the use of any type of steel nuts would have been preferable to the use of brass nuts. Use of an incorrect grade of steel might have shortened the time period before another repair to the valve was required, but the valve would not have failed catastrophically after about 30 minutes of service.

The JAG investigation determined that the required quality assurance procedures had not been followed. Specifically, the type commander's quality assurance manual, in this case COMNAVSURFLANT, designated main steam components as quality assurance level 1. This meant that all materials, including threaded fasteners, had to be positively identified as meeting the required specifications. The initial charges considered by the JAG investigators was manslaughter. However, these charges were never filed, as willful negligence was not proved. The ship's commanding officer received a letter of reprimand. The chief engineer was relieved of his duties and permitted to retire from naval service. The main propulsion assistant died in the accident. The boiler technician chief petty officer was relieved of his duties and was discharged from active duty.

7.1 BOILER INCIDENT STATISTICS

According to the National Board of Boiler and Pressure Vessel Inspectors (NBBI),[3] from 1999 through 2003 there were 1477 reported power boiler accidents, resulting in 143 injuries and 26 deaths (Table 7.1). The NBBI stopped collecting this data in 2003.

TABLE 7.1 Power Boiler Incidents

Year	Accidents	Injuries	Deaths
1999	335	41	8
2000	459	20	8
2001	296	56	7
2002	282	14	3
2003	105	12	0
Total	1477	143	26

TABLE 7.2 Power Boiler Incidents (1999–2003) by Cause

Cause of Incident	Accidents	Injuries	Deaths
Burner failure	95	26	15
Faulty design or fabrication	46	8	1
Improper installation	40	2	0
Improper repair	63	1	0
Limit controls	65	10	0
Low-water condition	579	13	1
Operator error or poor maintenance	535	74	9
Safety valve	15	1	0
Unknown/under investigation	39	8	0
Total	1477	143	26

TABLE 7.3 Boiler Incidents (1999–2003) by Type of Boiler

Type of Boiler	Accidents	Injuries	Deaths
Power boilers	1477	143	26
Steam heating boilers	3615	83	6
Water heating boilers	4297	20	2
Totals	9389	246	34

Power boilers include utility boilers as well as boilers used by other industries for cogeneration and on-site power production.

As Table 7.2 illustrates, there are several causes of power boiler accidents. While a low-water condition was the leading cause during the period identified, operator error or poor maintenance was a close second. In addition, operator error or poor maintenance was the leading cause of injuries and second only to burner failure in the number of deaths. It is also important to note that when all boiler types—power boilers, steam heating boilers, and water heating boilers—are considered (Table 7.3), power boilers were involved in only about 16% of the total number of accidents, yet they were responsible for more than 76% of boiler-related deaths and approximately 58% of all injuries.

From my years of experience, I believe that boiler accidents are seriously under-reported. Much information regarding boiler and fuel system–related fires and explosions never gets out to the public. Even if the numbers reported were completely accurate, the NBBI data clearly show just how dangerous power boiler accidents can be. These numbers emphasize how important it is for all boiler owners, including utilities and other power generators, to do everything possible to mitigate power boiler accidents.

7.2 BOILER TYPES

Before we explore safety issues related to boilers, let's better understand some things about basic types of boilers. The dictionary defines a boiler as a steam generator consisting of a metal shell and tubes in which water is converted under pressure into steam. Of course, there are many different styles and types of boilers.

Boilers don't always make steam; in some cases they simply recirculate heated water. One of the key differences between a hot water boiler and a hot water heater is that boilers retain the hot water within the system and keep recirculating it. Hot water heaters are open devices in which water leaves the system and has to be replaced. If the circulating water never leaves the system, it's a hot water boiler. Hot water boilers are rated as low or high pressure. Most hot water boiler systems operate at 30 psig or less for heating homes and commercial buildings. High-pressure, high-temperature hot water systems exist are used principally in large central plants.

To understand steam boilers, it's important first to understand a basic steam loop. In a basic steam loop, heat, is added to water, which turns it into steam (water vapor). Steam is transported to the load by the steam distribution piping, where it gives up its latent heat by condensing or turning back to water (liquid phase). *Latent heat* is the heat of vaporization as opposed to *sensible heat*, which is the heat that can be sensed from a temperature loss. When steam gives up its latent energy, it condenses and turns back into water (condensate). The condensate is returned back to a holding tank or a dearator, where it is processed and conditioned for reuse.

At this point, the (condensate) water's pressure is increased substantially by a feedwater pump and returned to the boiler to be heated and converted again to steam. Within the steam distribution piping system are special devices called *steam traps*, designed to allow the flow of condensate only, and not steam, back to the return holding tank system.

Steam is a much more effective means than hot water of transferring heat. Every pound of steam condensed provides about 1000 Btu of energy. In hot water systems, 1 pound of water provides only about 20 Btu in a recirculated system with a 20°F design temperature drop. This is why you don't see hot water heating systems for many process applications: It's just not efficient to transfer large amounts of energy in hot water systems.

Boilers are also designated as packaged or field-erected. As one might guess, packaged boilers can be hauled around on trucks or railcars as complete units. Field-erected boilers are actually fabricated, piece by piece and tube by tube, at a plant site.

Steam boilers are also classified as being of either high or low pressure. Steam boilers designed for operating at or over 15 psig are considered high-pressure. There are also other classifications used for boilers, including fire tube, water tube, cast iron, steam generators, and low-water-volume hydronic types.

7.2.1 Fire Tube Boilers

In fire tube boilers, sometimes called Scotch marine boilers because they are similar to the boilers used on ships in the nineteenth century, combustion products pass through tubes submerged continuously in a drum of water (Exhibit 7.1). They are generally sized up to about 2500 boiler horsepower, or about 70 million Btu per hour input, which is about 65,000 pounds of steam per hour. They typically have maximum allowable working pressures of about 600 psig. Fire tube boilers are termed single-drum vessels because they are represented by one large pressure vessel containing a significant amount of water. This large reservoir of water, under pressure and at a

EXHIBIT 7.1 Cutaway view of Cleaver Brooks fire tube packaged boiler.[4]

saturated temperature, creates an element of danger for all boilers of fire tube design compared to some other boiler types.

In normal installations, fire tube boilers will respond well to the steam load demanded of them. They are not as fast as water tube boilers to respond to large load swings. The materials used in the construction of a fire tube boiler are normally very resistant to corrosion when proper water treatment practices are maintained. They are also relatively simple to operate. These are just a few of the reasons that fire tube boilers are by far the most popular style of boiler for commercial and medium-sized industrial applications throughout the world.

Fire tube boiler burners' first direct flames and combustion products through a central, relatively large-diameter tube called a Morrison tube (named after an early boiler manufacturer). This is often called the *first pass*. The hot flue gases next enter a chamber in the rear of the boiler, where they are redirected to pass back through the shell toward the burner through a series of tubes, called the *second pass*. These tubes are generally less than 4 inches in diameter and are sometimes dimpled to create more surface area. The gases usually make three or four back-and-forth passes before they are discharged out a flue. Generally speaking, more passes means more surface area, more capacity, and greater efficiency.

The tubes are arranged horizontally and are attached at both ends with sheets of steel called *tube sheets*. These tubes may contain twisted sheet metal inserts to cause more flue gas turbulence to increase heat transfer and efficiency. The turnaround zones at the rear of the boiler can be of either dry-back or water-back design. In dry-back designs, the turnaround area is refractory-lined. In water-back designs, this turnaround zone is water-cooled, eliminating the need for a refractory lining.

In most cases these boilers have a bolted-on front and rear circular door. These gasketed doors are held on by a number of bolts. These can generally be removed and the doors swung out on specially designed hinges to allow access to the tube sheets and tubes for inspection and repair. In some cases there is no front or rear door, only access plates that can be removed.

Tubes are held to the tube sheets mechanically using a special tool that rolls the ends of them so that they hold tightly against the tube sheets. If leaking occurs, ends can be rerolled a limited number of times. Tubes that are failed have to be cut out and removed from one end or the other. Hence, it's important to consider future access for tube removal whenever these types of boilers are installed.

Access to view the water side of tubes for corrosion or deposits is usually through hand holes or manholes. These are gasketed ports that are bolted in place. These are usually removed on an annual basis for a jurisdictional inspector to assess the exterior condition of the tubes.

7.2.2 Water Tube Boilers

Water tube boilers function with water inside the tubes or pipes. Hot flue gases heat the water from the outside of the tubes. The water in the system circulates between two reservoirs, one above the other, connected by the tubes. These reservoirs are called drums. The higher of the two drums is the steam drum; the lower drum is the mud drum (Exhibit 7.2).

EXHIBIT 7.2 Water tube boiler drums and tubes. Courtesy of Babcock & Wilcox.[5]

As flue gases and radiant energy heat the water in the tubes, small steam bubbles are formed. These rise in the tubes to the steam drum, where they find their way through the water level and eventually break the surface. The steam is sometimes sent through a mechanical cyclone or mesh pads to minimize water droplets that may be carried out of the drum with the steam in its pure gaseous state. Water droplets entrained in the steam, called *carryover*, are undesirable and make for a condition known as *wet steam*, which can overload steam traps, damage turbines, and carry minerals into the steam system that over time can form deposits. There can be many reasons for carryover, and some of which can be associated with other styles of boiler systems as well. These can include too-high water levels, too-rapid steam load demand on a boiler, and even problems with water treatment chemicals that cause foaming in the steam drum.

If there is a need for very dry steam and better performance in turbines, superheater sections can be incorporated in water tube boiler designs. Superheaters simply provide sensible heat to steam vapor. Superheaters comprise a series of tubes that pass steam only to increase its temperature further so that condensation is avoided in its transport and conversion to a lower pressure through a turbine or steam engine. Superheater tubes are vulnerable to damage, due to the significant amounts of heat to which they are exposed and because they only have passing through them steam vapor to remove the heat applied, not water that changes phase. Starting up boilers that have superheaters requires great care and experience if damage, is to be avoided.

The arrangement of tubes in a water tube boiler is usually in a series of banks or groupings called *risers* or *generating tubes* and *downcomers*. These tubes work in parallel to create natural circulation of water between the steam drum and the mud drum. The risers, often called the *radiant section*, are located closer to the firebox and get direct radiant heat exposure from the burner.

The downcomers contain denser (cooler) water and concentrated minerals. The minerals accumulate in the mud drum, giving it its name. To avoid boiler water quality-related problems, water must be discharged periodically or "blown down" from the boiler mud drum to control the concentrations of suspended and total dissolved solids. For the same reasons the blowdown process needs to occur as an operational maintenance task in fire tube and all other steam boilers. The importance of proper boiler blowdown is often overlooked. Improper blowdown can cause increased fuel consumption, additional chemical treatment requirements, tube damage, and heat loss.

Water tube boilers are also classified according to the arrangement of their tubes; some of the popular styles are the A, D, and O types. Water tube boilers can be constructed to produce millions of pounds of steam per hour with pressures of nearly 4000 psig. Water tube boilers are sometimes referred to as *power boilers* because they are used to make steam for electrical power. Water tube boilers are the predominant style of boiler used by electric generating utilities.

7.2.3 Steam Generators

Steam generators, sometimes called instantaneous coil boilers, have gained popularity over the past 50 years (Exhibit 7.3). These boilers have very little water storage and

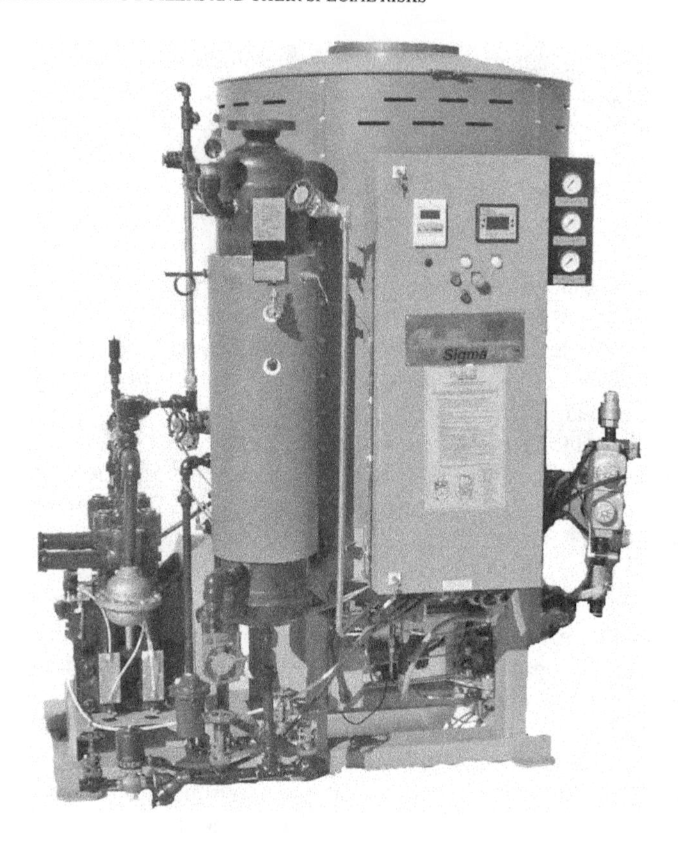

EXHIBIT 7.3 Steam generator boiler. Courtesy of Clayton Industries.[6]

usually no discernible drum level. Hence, from a water-side explosion perspective, they are much less risky than, for example, a fire tube boiler.

Some steam generators force water under pressure through heated coils, where steam is produced. Others operate by forcing water under pressure through coils to superheat the water. This superheated water is then pushed through a lower-pressure separation tank, where part of the circulated water flashes to steam. These boilers can be started very rapidly, have little or no refractory, and deliver steam faster than many other styles of boilers. They are usually available with a capacity of up to about 20 million Btu per hour. They are among the highest-efficiency boilers available for their size.

These boilers also require meticulous attention to their water treatment. They are much less forgiving on this one issue than almost any other boiler style. Some of these boilers also have special proprietary pumping systems and controls that need careful and regular maintenance to function properly.

Boiler Tube Failures The categories of boilers discussed so far all utilize tubes that can fail from overheating. There are several ways that this has been known to happen. The simplest is overheating when water levels are low. The water in the tubes removes heat and keeps the surfaces from getting hot enough to change metallurgicaly. Tube damage can also happen when tubes are coated with minerals from poor feedwater treatment. When this happens the minerals make for an insulating effect and heat then cannot transfer effectively through the tube surface to the water. This means that heat is not adequately transferred from the tubes to the water, causing overheating and damage. Burner problems can also cause flames to impinge directly on tube surfaces. This makes for concentrated high temperatures that are beyond what can be removed effectively by the water in the tube.

Tube overheating can make tubes twist and distort and can cause tubes to be pulled out of drums. You may recall from Chapter 2 that a phenomenon called *creep* can start to occur at temperatures above about 800°F.[7] Creep is the tendency of a solid material to move slowly or deform at elevated temperatures. The heat-affected area loses strength. The surface may blister because the internal tube water pressure is pushing out the weakened steel tube surface. The blister gets larger and larger and eventually bursts. This sometimes causes steam to enter the firebox directly at a very high velocity. This can quench flames and can also erode refractory and other tubes if it is at a high enough pressure.

If you ever look into a firebox sight port and see what appear to be droplets of water vapor condensing on the port from the inside, it's very possible that you have experienced a tube failure. Other evidence of a tube failure may be a considerable amount of water vapor coming out of the flue. Excessive sooting from flames being quenched can also be an indicator. Inside firebox inspections for water tube boilers are one of the primary mechanisms for identifying pending tube failures. Remember, these may require special confined-space entry protocols.

7.2.4 Cast Iron Sectional Boilers

Cast iron sectional boilers[8] are a popular choice with residential and commercial applications and for low-pressure steam and hot water applications. These boilers are modular, can be fit in pieces through small doorways, and can be assembled on site. These boilers, which are available with a capacity of up to about 10 million Btu per hour, are very durable and forgiving. When they do fail, a common failure mode is to have a cracked cast iron section. This can happen from rapid temperature changes or because poor feedwater quality has caused sludge to accumulate that has insulated sections and made for localized overheating. In many cases, a new section can be installed where an old section is cracked.

7.2.5 Low-Water-Volume Hydronic Boilers

Low-water-volume hydronic boilers[9] are very popular for low-pressure, low-temperature residential and commercial space heating applications but are rarely used for process heating applications. In most cases they are ganged together (modules) to create staged heat applications in a very efficient manner. This very modular technology is, at present, one of the most efficient means of providing hot water heat.

Hot water boilers require very little in the way of chemical water treatment since no water is ever supposed to leave the system. In a device called a *shot feeder*, an initial charge of chemicals, usually an oxygen scavenger and a corrosion inhibitor, is added when the system is started. The amount of these chemicals is proportional to the hydronic systems volume. The water treatment chemicals level is usually checked annually to see if the residual has diminished. There is really no reason to expect to have to add anything else unless water has left the system because of a safety relief valve leak, safety relief valve lift tests, or some type of repair that involved opening the system. Many sites install water meters to identify if and when fresh water has entered a system. In any case, where fresh water has entered a system, the chemical residuals should be checked.

I have been party to very few explosion- or reliability-related incidents related to hot water boilers or heaters. When these incidents have occurred, it is usually because circulation or flow has been stopped and a safety relief valve has failed. Safety controls for hot water boilers commonly include water-level sensing devices and flow switches.

I have seen fireside issues with some styles of burners that are located directly below coils in these boilers. In these cases when units were started cold, or where the system circulation design is done incorrectly, cold water in the boiler tubes condensed combustion products that were firing up. This condensation then dripped down back onto the burner, sometimes extinguishing flames and causing incomplete combustion and sooting. This has been known to clog and corrode the heat-exchange surfaces, restrict the flow of flue gases, and in some cases make for flame roll-out.

7.3 BOILER-WATER-LEVEL SAFETY DEVICES

Ask any old school boiler inspector and he or she will tell you that most common failure mode for boilers throughout the history of their existence is water-level issues. In this section we discuss ways to reduce risks related to the two most important categories of equipment related to water level in boilers: water gage glasses and low-water cutoffs. Many of the issues presented here come from ASME Section 1, Rules for Construction of Power Boilers; ASME Section 7, Recommended Guidelines for the Care of Power Boilers; ASME Section 4, Low Pressure Steam and Hot Water Boilers; and ASME CSD-1, Controls and Safety Devices for Automatically Fired Boilers. It is recommended that readers obtain the relevant sections of these standards and understand how they apply to specific equipment and conditions that may exist at their sites.

7.3.1 Water Columns and Gage Glasses

Let's talk first about how water levels are monitored on boilers. Drums are connected with pipe to water columns (Exhibit 7.4). This is a cylindrical vertical pipe manifold that acts to reduce the turbulence and fluctuation of water level on a steam boiler so that a more accurate assessment of water level can be made. Water gage glasses are installed on water columns as are low-water cutoffs. There is conflict among some codes regarding the placement of valves in the piping that interconnects boilers and

EXHIBIT 7.4 Boiler water column with level gage and control. Courtesy of Clark Reliance.[10]

water columns. On steam boilers firing more than 12.5 million Btu per hour, it is common to have valves in the interconnecting piping from the steam drum to the water column. If these exist, they must be locked or sealed open. ASME CSD-1, which deals with boilers rated at less than 12.5 million Btu per hour, prohibits valves in this interconnecting piping. In the case of hot water boilers, codes have recently been changed to require that new boilers have special valves installed so that low-water cutoffs can be tested more easily. These can be purchased and retrofit into existing hot water boilers to make testing more accommodative. Make sure that you discuss with the jurisdictional inspector the issues of valves in this piping on your systems and the possible need to lock or seal them open.

There are many different types and styles of gage glasses. Direct-reading gage glasses are those that make for direct viewing of water in the boiler. Indirect-reading units involve a transducer of some type that converts the water-level signal to some electronic-level indication, which can, for example, manifest itself as lights on a display in a control room. All boilers are required to have at least one direct-reading gage glass. Two direct-reading gage glasses, or one direct and two indirect gage glasses, one in the area where control occurs, are required for boilers over 400 psig.

Water gage glass types and technologies need to be compatible with boilers' maximum allowable working pressure. There are four basic types of water gage glasses: tubular glass (for up to 150 psi), reflex glass (for up to 350 psi), flat glass with mica protection (for up to 2000 psi), and ported type (for up to 3000 psi).

7.3.2 Gage Glass Safety/Risk Reduction Review Considerations

Being able to see and assess the water level accurately is a critical part of all boiler operations. The following are things that you can review to help to make sure that your gage glass systems are functioning properly.

- Verify that the water level is readily discernible. In some cases, illumination behind the gage glass can be provided or is installed later. This is a requirement on ported gage glasses. If existing gage glass illumination has incandescent lamps, consider converting to LEDs. If the gage glass is rust stained or obscured with mineral deposits, make sure that it is repaired promptly if a blowdown attempt does not clean the glass internally. Use caution to ensure that reflex-type glasses are installed with the prismatic surface facing toward the inside of the gage glass assembly. Otherwise, the level will not be viewed correctly.
- Verify that the gage glass is installed properly. Every boiler has a lowest permissible water-level mark or line provided by the manufacturer. ASME code for Section 1, high-pressure boilers, requires that the lowest visible water level in the gage glass must be at least 2 inches above the lowest permissible water level, also sometimes called the danger mark. In the case of Section 4, low-pressure boilers, the requirement is 1 inch above the lowest permissible water level.
- Verify that a means to blow down (clean out) the gage glass exists. This should be a piped drain with a valve. The drain piping must be a minimum of $\frac{1}{4}$-inch-diameter bore. A drain allows the gage glass water level to be tested daily by operators. A common test is to open and close the gage glass blowdown rapidly. This should move the water out and then it should reestablish very quickly. If it does not, or is sluggish, it could be evidence of sludge accumulating in parts of the water column assembly or within the mating piping.
- Consider replacing tubular glasses on a regular basis (at least annually). These can become eroded and thinned, making them susceptible to breakage and blowing out. If you find one leaking near the insert points, it could be evidence of thinning and an impending failure. Make sure that these are repaired immediately. You should always have spare gage glasses and correct packing rings at every boiler plant.
- Verify that protective rods are in place if the design calls for them. These can help the alignment of the gage glass with the valves and provide some protection.

7.3.3 Low-Water Cutoffs

Water-level controls are the single most important safety device installed on boilers and why most of them include a redundant form of the device. Low-water cutoffs serve to shut down firing if the water level drops to an unsafe level. There are many different types and styles of low-water cutoffs, from the simple float type to the more complex conductance probe type. These devices are usually installed in, or as a part of, water columns.

Two low-water cutoffs are usually installed in parallel so that there is always a backup. The main cutoff is called the primary low-water cutoff and the backup is called the auxiliary low-water cutoff. The primary low-water cutoff must be installed such that its tripping is visible within the gage glass. This means that its trip point is likely to be about 3 to 4 inches above the lowest permissible water level. The auxiliary is usually mounted about 1 inch in water level below the primary. Its trip should also be visible within the gage glass.

The primary will shut the boiler down when it senses a low-water condition and then will usually allow the boiler to restart once the condition is corrected. The auxiliary puts the boiler into a safety lockout condition where a manual reset has to occur before the unit can be relighted. These devices must be installed on separate water connections from the drum but may share the same steam connection, so one piping malfunction cannot take them both out of service. The two devices used are sometimes of two different technologies to enhance their overall reliability as a layer of protection (e.g., a float and a conductance probe instead of two float-type low-water cutoffs). However, the recent trend is to rely more on conductivity probes, thereby eliminating the risks associated with the wear factors common with float connections.

Water flow switches are commonly used as an alternative means of proving water flow in hot water boilers. An interlock on the water circulation pump motor starter is also sometimes used but is not as reliable as measuring water flow directly. Usually, these boilers will have a flow switch and a probe type of low-water cutoff device.

7.3.4 Testing Considerations for Low-Water Cutoffs

There are many test procedures for low-water cutoffs (depending primarily on the style and type) and many do's and don'ts. Please be aware that it's up to each facility to determine what their needs are for water-level device testing and to be in compliance with applicable codes and insurer requirements. Guidance is also available from jurisdictional inspectors. The frequency of these tests must be coordinated with risks to processes in industrial plants and utilities that could happen during testing if the boiler is shut down by accident during the test. For example, there are some pharmaceutical plants that run carefully controlled batches where the entire batch can be lost if the temperature or steam pressure fluctuates out of specification for even a moment, and months of work can be lost. In these cases testing should be coordinated so that additional boilers are online during testing.

7.3.5 Standard Low-Water Cutoff Tests

A standard water column test is a check of functionality of the float or conductance rod condition in the chamber. It is done by opening the valve below the water column chamber to drain water from the water column while the boiler is firing. If no alarm or shutdown occurs as water drops below the minimum water-level indication in the gage glass, the low water cutoff has failed and the boiler should be shut down immediately. Failures like this could be due to float or conductance rods that are contaminated, damaged, or stuck.

In some cases, where outages can be a nuisance, shutdown bypass switches are installed for testing. These bypass switches allow for this test to be done without bringing the unit down. If one of these is installed, it should always be a momentary contact type of switch. This means that it can bypass the safety device only when it is being pressed. These are usually installed within arm's reach of the water column blowdown valve. The idea is that the button is held in while the column is drained for this test. Check with your jurisdictional inspector for approval before adding this feature. In a number of field safety inspections, these blowdown bypass switches have been found with paper clips and other devices jammed into them, so that the alarm would not go off and the shutdown would not occur. This rendered these low-water cutoffs inoperable and unsafe. Operators caught in the act said things like: "Well, the drum levels were swinging because of a process upset and I would be up and down all day if the thing was not bypassed; besides, the auxiliary low-water cutoff is still in service." This is completely unacceptable behavior and very dangerous. Make sure that this kind of thing is not going on at your facility.

Auxiliary low-water cutoffs also need to be checked. This means, however, that the primary low-water cutoff needs to be bypassed temporarily. This test should be done in the same way as the primary test at the water column drain valve. Temporary bypassing of the low-water cutoff for this test should only be carried out by an experienced technician and then double-checked to make sure that it is put back into service properly before normal boiler operations continue.

7.3.6 Slow Drain Test (For Low Water Cutoff Function)

A slow drain test is required by some codes on an annual basis for steam boilers (ask your jurisdictional inspector about your site). During a slow drain test, the feedwater is shut off temporarily and a drum blowdown line is opened a small amount so that the water level comes down very slowly. This can be dangerous since one is removing water with the burner still firing. It must be done very carefully and with much oversight by trained personnel.

One might ask, why a slow drain test? How can this make a difference? Believe it or not, it does. I have seen many failures in low-water cutoffs being found in this way. I have taken some of these defective low-water cutoffs apart and have seen strange things like little nicks in the slide devices that catch on one another, and bent linkages that bind only when they move very slowly. Obviously, the need for this test has been validated throughout history and for good reason. Again, I must emphasize that this test must be done under very carefully controlled and monitored conditions by trained personnel.

IMPORTANT

Water-level issues have always been the single biggest cause of boiler catastrophes. Make sure that your operations include regular attention to blowing down gage glasses and water columns. Make sure that regular operational testing of low-water cutoffs occurs and include slow drain tests where applicable.

7.4 BOILER PRESSURE SAFETY CONTROLS

Steam boilers all have at least two different types of pressure controls. One is an operating controller, which maintains a desired process pressure. This controller brings the burner on and off to maintain this pressure. The other is a high-limit control, which shuts down firing in case the pressure exceeds the pressure setting of the operating controller by a predetermined amount. Safety codes require high-limit pressure controllers to be manual reset devices. This means that the device will shut the burner down and lock out the system when tripped until someone resets the device (presses a button or moves a small lever). The set point of high limit pressure controls cannot exceed the maximum allowable working pressure of the boiler. Many codes require the careful placement of these devices so that valves cannot render them ineffective.

In many cases, on older boilers these high-pressure limit switches are mercury switches. Care must be taken to install these levels so that they function accurately. These can be installed mistakenly with the mercury bulb tilted (not level), which makes their set point not what is indicated. Through the years I have also observed many high-steam-pressure limit switches that failed due to diaphragm failures within the switches. Many manufacturers of these switches recommend a pigtail loop or a dead leg of piping to be installed in the sensing lines of these switches so they do not experience direct steam impingement. Direct steam impingement to the switch diaphragm can shorten the useful life of these switches. Make sure that these switches are part of your regular preventive maintenance safety device testing program. Manufacturers provide such test methods as set-point adjustment tests. Taking the boiler pressure up to the switch set point is a very high risk method for testing and is not recommended.

7.5 SAFETY RELIEF VALVES

Safety relief valves (SRVs) provide the final layer of mechanical protection from overpressure and sometimes over-temperature conditions. These valves are unique in that they open at a preset pressure to allow steam to escape and lower the pressure in the boiler or pressure vessel before reseating at a lower pressure. This is one of the differences between a safety relief valve and a relief valve. In some cases, relief valves open and close at the same set pressure.

Safety relief valves are in service on devices as simple as residential hot water heaters and as complex as utility generating station boilers. Safety relief valves have a number of major functional components, including the stem (spindle), the seat, the blowdown ring, and the tri-lever (handle). An important thing to remember about safety relief valves is not to try and repair them or adjust them yourself. They must also be handled with great respect and care. They should be stored carefully and serviced only by ASME-certified repair facilities.

7.5.1 Reviewing Safety Relief Valves for Hazards

Several chapters could be written specifically about safety relief valves. Many of the fine points are beyond the intent of this book. Instead, the material presented here is meant to acquaint you with some of the common failures and issues that have been witnessed inspecting boilers and steam systems over the past 30 years. The following are the five most common safety relief valve defect issues that I have witnessed along with an explanation of each.

1. *Slight discharge or weeping from the device.* If you see a safety relief valve that is constantly discharging a small amount of water or steam, you have a problem. A slight release or weeping from the safety relief valve indicates that fluids are getting past the valve seat. When this occurs, these fluids could be at a high enough pressure to cause erosion of the metallic surfaces, called *wire drawing*. This condition can make a small leak bigger in a short amount of time. Also, the discharge will tend to accumulate and puddle at the exterior surface of the valve seat. Minerals in these fluids, along with oxidation of the metal, could also cause the seat to seize. This would render the valve useless.
2. *Lead seals removed (applies mostly to high-pressure SRVs).* There are usually small wires wrapped around and sometimes through critical calibration adjustment devices on safety relief valves. These usually have a factory-applied lead seal which indicates that the device is as set by the manufacturer or a certified calibration and testing facility. If the seal no longer exists, most boiler inspectors identify this as a problem and ask that you have the valve removed and recertified, since the integrity of the valve settings is not known.
3. *Installation issues.* Safety relief valves must always be installed so that there is no weight from the discharge piping transferred to the valve itself. Always review the installation of hangers to make sure that the valve itself is not supporting any piping load like the discharge piping. If the valve body is loaded with a force or bending moment, it may not function properly and perhaps even shear off the boiler, causing a catastrophe.
4. *Drains.* Larger safety relief valves are usually installed with drip pan elbows that include drains. You will usually see $\frac{3}{4}$- or 1-inch piping leading away from these. Make sure that these drains are open and can flow. If they can't, testing discharges or possibly even rainwater will result in water lying on the outside of the seat, making for the possibility of corrosion and seizure of the valve stem to the seat.

5. *Studs and bolts.* For larger safety relief valves, be sure to examine flange bolts to see that they are not short-studded. The threads should be fully engaged. Verify that high-strength fasteners, including studs, are being used. Bolts are easy to check since there are standard-grade markings on the heads. For studs, you must make sure that they are truly high-strength studs, not just threaded rod. High-strength studs are supposed to have a marking stamped into the ends that is visible upon examination. See ASME B16 for more information on fastener criteria.

7.5.2 Safety Relief Valve Testing and Service

It is important that you understand the nature and type of safety relief valves located at your site. We have, up until now, been focusing on safety relief valves associated with boilers. However, safety relief valves should also be installed on all pressure vessels and downstream of pressure-reducing stations on steam systems. You should catalog them all and review the set points and relieving capacities. You should also review how they're installed to make sure that they have proper discharge piping and that this piping relieves to a safe place and that regular testing is occurring. Most of what has been presented applies to all safety relief valve installations.

All safety relief valves, whether or not they are on boilers, have to be tested regularly. Three tests are required with some frequency, which is jurisdictionally dependent: (1) that the valve actually functions and that the seat can be lifted, (2) that the valve lifts or relieves at the proper set point, and (3) the valve reseats at the proper set point after a blowdown period.

There are lots of different jurisdictional requirements for the frequency of testing. In some cases certain tests, such as lift tests, are required several times a year, depending on pressures; in other cases it may be yearly. The National Board Inspection Code (NBIC; available at www.nationalboard.org), Section 2.5.8, contains a table identifying test requirements. If you are in doubt or cannot access this document, ask your jurisdictional inspector what your schedule should be.

The most common form of regular operational testing, usually done by boiler operators or plant maintenance staffs, is lift testing, which needs to be carried out very carefully. Often, valves do not reseat. Hence, it's best to do this type of testing right before a planned shutdown in case the valve needs to be replaced. It's also important that every site have spare valves prior to doing lift testing. Lift testing for high-pressure boilers should always be done from a remote location without exposing anyone to potential harm. High-pressure relief valve testing should never be done by an employee pulling directly on the tri-lever (handle). Many sites employ a system of cables and pulleys so that this can be done remotely from the ground with no risk to staff.

It is recommended by most high-pressure safety relief valve manufacturers that there be at least 75% of the set pressure on them before a lift test is attempted.[11] Otherwise, there could be too much force on the spindle. Excessive spindle force can mean bending the spindle and rendering the valve useless. There are many other safety issues involved in lift testing valves, including making sure that discharge piping and vent discharge area are clear. Lift testing should be done only by properly trained personnel and according to the valve manufacturer's instructions.

Verification of a set pressure test is not something that it is recommended be done with an operating steam boiler. In my opinion this presents a tremendous risk to sites. If the relief valves or the high-pressure burner cutout device should fail, a catastrophic explosion could result. Instead, many sites rotate freshly calibrated valves into place when pressure verification is required.

Another lower-risk way to deal with this issue is to use an electronic valve testing (EVT) service.[12] In some jurisdictions this service can also validate or substitute for the lift test. When EVT is done a technician attaches special equipment to the safety relief valve and then loads the valve with force hydraulically to counteract the spring. The device measures the force required at the precise moment the valve was supposed to lift.

The only practical way to validate relieving capacities, what's called the *blowdown rate*, is by sending the valve out to a certified rebuilder or test shop. These facilities will have to carry the ASME stamp and certification for servicing safety relief valves. All maintenance work on these valves, including set point adjustments, should be done only by an ASME code stamp holding facility. Most sites have two sets of valves. One set in service and one set fresh, calibrated, and waiting in reserve.

7.6 STEAM SYSTEM PIPING SPECIAL ISSUES

The piping, valves, and components for the steam, feedwater, and even blowdown systems for any high-pressure (over 15 psig) steam piping system have to be rated for the maximum allowable working pressure (MAWP) of the boiler. The MAWP can be found on a tag usually affixed to the steam drum. The piping attached to the steam discharge flange of the boiler must also be designed to meet or exceed the MAWP, usually up-to-the-second manual stop valve from the discharge flange. After the discharge flange, the piping must usually be rated for at least the set point of the highest safety relief valve protecting the system. There are very good diagrams in ASME B31.1, the Power Piping Code, that will show you how and where the ratings apply. If in doubt, the jurisdictional or insurance inspector assigned to your site will no doubt be of great assistance in this matter.

When servicing existing steam piping, be aware that it may be insulated with hazardous asbestos-containing materials. It's also possible that older boilers can be insulated with asbestos-containing materials. Sometimes this is prevalent on elbows in a plaster-looking form. If you are in any doubt, have the insulation tested by a qualified asbestos contractor before proceeding. Although most facilities should have had surveys and have these areas abated or marked, there's still plenty of it around.

When removing old flange gaskets, be aware that many that have been in service are asbestos based. These will require special abatement techniques that are beyond the scope of this book. The last thing you want to do is to handle these improperly and/ or discharge these materials into the air with a grinder. Consult your facilities environmental team or an asbestos-abatement contractor for guidance on removing these. For the same reasons, be careful of handling any old gaskets that are in storage or boiler seals and door gaskets.

7.6.1 Piping System Water Hammer

If water cannot be removed from the piping system adequately because of excessive carryover, steam trap failures, poor piping design, or a number of other factors, accumulated slugs of water can be accelerated by the speed of the steam traveling in the pipe and cause tremendous damage. Steam piping systems can sometimes be designed for velocities of 100 feet per second. Water piping systems are usually designed for speeds of 3 to 8 feet per second.

When water gets accelerated at the velocity of steam, it can create tremendous impact forces if it has to change direction or if it hits something like a control valve. Think about a 15-gallon slug of water weighing about 100 pounds suddenly getting accelerated to 68 miles per hour (100 feet per second) and hitting an elbow. It's easy to understand what can happen. When this occurs, entire piping systems will shake or can be moved. It can cause a banging or clanging noise in the piping system called *water hammer*. In extreme cases piping systems can be broken and torn apart, creating dangerous conditions.

Real-Life Story 25: Water Hammer Causes Extreme Damage

How much damage can water really do? Consider the case of a manufacturing facility that had undertaken a steam line asbestos removal project. Specialized demolition crews removed hundreds of feet of asbestos insulation in a particular section of the plant. This was all done outside the heating season. Unfortunately, the piping was not reinsulated prior to starting up the steam system. When the steam system was started for the first time in the fall, considerable amounts of condensate were generated in the piping system because of the lack of insulation. This flow far exceeded the capacity of the steam trap's ability to remove it. The excess water (condensate) inside the pipe was picked up and accelerated until it hit an elbow in the piping. After 3 or 4 hours of clanging, an 8-inch elbow was shattered. This occurred on a section of steam line where the asbestos had not yet been removed.

A shower of debris and fine white flakes began to fall from above inside the plant. Much of the plant had to be evacuated and shut down until abatement crews could get in for the next six days and do a complete cleanup. Needless to say, this cost the facility hundreds of thousands of dollars in lost production.

Lessons Learned Don't ignore water hammer or pipes that are moving around. This situation can only get worse as hangers loosen and forces increase over time. Verify that an annual steam trap maintenance check occurs at your facility. Be especially cautious of steam system startups. There are two types of steam system startup designs, attended and automatic. In attended designs drains are supposed to be opened manually while the steam system is slowly, and gently brought up to temperature over time. In automatic startups there is still a slow and gentle warming of the system, but traps are more frequent and large and sometimes there are double-trapped legs, designed with the capacity to remove the condensate being generated with no human intervention or monitoring. Steam systems and equipment need to warm up slowly. Warm-ups can take many hours and even days. In all cases remember to open any fluid system valve (steam, water, or gas) very slowly. Similarly, close valves slowly so that there are not rapid decelerations of the mass in the pipe.

7.6.2 Certifications for Welding and Repairs

Welding on pipe, fittings, flanges, and pressure vessels must only be done by welders with the proper credentials. Similarly, repairs to boilers and safety relief valves should only be done at shops or by vendors that have the proper NBBI or ASME certifications. In many jurisdictions, pressure vessels and piping can be worked on or welded legally only by someone with an "R" stamp certification from the National Board of Boiler and Pressure Vessel Inspectors (NBBI; www.nationalboard.org). This indicates successful completion of special education, testing, and evaluation of repair processes. There are different welding certifications from many groups. A resource for understanding the world of welding certifications is available from the National Certified Pipe Welding Bureau.[14]

ASME has a number of stamps and symbols that indicate special equipment validations (Exhibit 7.5). This stamp could be present but possibly obscured by insulation. Sometimes it is printed on a stainless steel tag attached to the piping. Similarly, whenever pressure vessel repairs are made, a tag is to be attached identifying the nature of the repair, date, and credentials of the responsible party. When in doubt about what a stamp means or if something should be code stamped, contact your jurisdictional inspector.

EXHIBIT 7.5 ASME pressure piping stamp.[13]

NOTES AND REFERENCES

1. 2005 study published by Energy and Environmental Analysis for Oak Ridge National Laboratory, www.cibo.org/pubs/industrialboilerpopulationanalysis.pdf.
2. U.S. Navy JAG report, www.jag.navy.mil/library/investigations/IWO%20JIMA%2090.pdf.
3. National Board of Boiler and Pressure Vessel Inspectors, Journal Bulletin Incident Reports from 1999 through 2003, www.nationalboard.org.
4. Cleaver Brooks, www.Cleaver-Brooks.com.
5. Babcock & Wilcox, www.Babcock.com.
6. Clayton Industries, www.claytonindustries.com.
7. Creep formation, NBBI, www.nationalboard.org/Index.aspx?pageID=181.
8. Weil-McLain, www.Weil-McLain.com.
9. Lochinvar, www.Lochinvar.com.
10. Clark Reliance, www.ClarkReliance.com.
11. NBIC, Safety Valve Inspection and Testing Guidance, www.nationalboard.org/SiteDocuments/NBIC/NBIC%20Pressure%20Relief%20Device%20Inspection%20Guide%201-19-10.pdf.
12. Electronic valve testing, www.ge-energy.com/products_and_services/services/valve_services/consolidated_evt_pro.jsp.
13. American Society of Mechanical Engineers, boiler and pressure vessel code stamps and certifications, partial list, and their meanings, www.asme.org.
14. National Certified Pipe Welding Bureau, www.mcaa.org/ncpwb/.

8

Controlling Combustion Risks: People

The Three Keys to Risk Mitigation and Protection: People, Policies, and Equipment

In the next three chapters we focus on three key concepts that I have found form the basis of long-term sustainable fuel and combustion system safety. These are the three legs of a three-legged safety and risk management stool. Any successful program must contain elements of each to be successful. This "people" piece involving controlling human error is among the most important. Human error has been the leading cause of fuel and combustion system accidents that I have witnessed.

Real-Life Story 26: Culture and Design Cost a Person's Life

When I arrived at a site where a worker had died less than a week before, I stood next to the furnace and one of the workers said: "Come here, right here, this is where we found him." I teared up and at that instant this was about something much bigger than me. I had always felt passionate about my work, but from time to time my focus became very sterile and technical. Visualizing the man laying there and thinking about his family gave me a special emphasis.

It was likely that he didn't speak English at this site. For many of the men I met those days, English was clearly a second language. I understood that he and many of his co-workers were immigrants like my father and the rest of my family, who had emigrated from Czechoslovakia. This man, like my family, came to America to fulfill hopes and dreams for a better life. My father worked in the steel mills of Youngstown, Ohio, with others who did menial, sometimes dangerous tasks in conditions

Fuel and Combustion Systems Safety: What You Don't Know Can Kill You!, First Edition. John R. Puskar.

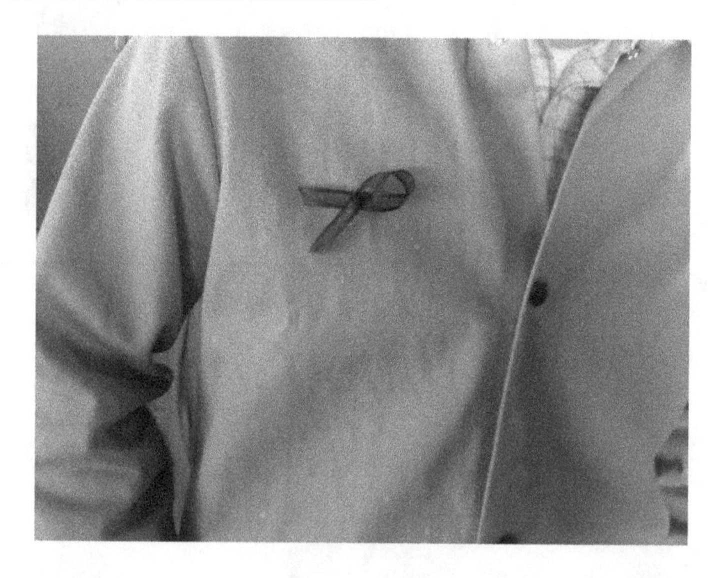

EXHIBIT 8.1 Black ribbon worn by workers after a fatal furnace explosion.

unthinkable by today's standards. I wanted to make this right for the man who died, all who came before him, and all who would come after him.

There was rubble everywhere. The top of the furnace that took the man's life was peeled open like a sardine can. Part of the furnace had ejected some 40 feet in the air and punctured the light sheet steel roof of the structure where it was housed. Everyone was nervous and tense. Management wondered what I would find and say. The plants workers didn't want anyone else to die, and they wanted justice for their fallen co-worker if that was anywhere to be found. When I came to the site, the president of the local union came over and asked me to wear a special black ribbon (Exhibit 8.1), in sympathetic memory of the man who died. All of the plant workers wore one of these ribbons. I was honored that they asked me to wear one.

The root cause of the incident was improper application of a flame safety system. This kept the main fuel valves open even though the main flame on one or more of the burners did not light or was extinguished. The flame safety system was one called an *intermittent system*. This style of system kept the main fuel valves open as long as any kind of flame signal was being received. Today's codes for this type of equipment call for interrupted pilot systems. These shut the pilot flame off after some period of time when the main flame should have started. This means that the flame safety system can't be fooled by only the pilot being ignited while gas from the main burner pours into the combustion chamber and is not ignited.

The furnace that exploded had been having trouble lighting-off during the previous two months. The pilot flame was weak and unstable and would frequently go out during the light-off process. It became such a problem that plant personnel positioned a chair next to the control panel so they'd have a place to sit while trying to get the unit to light. Other issues that were found that did not help flame stability included considerable dirt, rust, and particulate fouling the fuel lines and burners. This meant that some of the burner zones might not have had stable main flames. It was clear that very little in the way of preventive maintenance was going on.

In my investigation I found many things that spoke negatively about the fuel and combustion system safety culture of this facility. For example, I opened a series of control panels and was horrified to see old-style burner management systems with open relays that had cardboard stuffed into some of them to hold them open. When I left, I knew that whether or not it was this particular incident, something was bound to happen here sooner or later—it was a just a matter of time.

Lessons Learned Make sure that a review of equipment for code compliance and "gap analysis" take place and include the type of flame safety system that is installed. NFPA 86 defines these requirements very clearly. It is never okay to bypass safety devices. Annual safety interlock and valve tightness testing would have found many problems here, including the cardboard relay bypasses. If annual fuel/air ratio setting had been done, the fuel line fouling and main flame stability problems might have been discovered and corrected.

Eventually, all of these issues contributed to the explosion and the severity of the incident. I was involved with the plant's corporate leadership for a fix along with the union representing the workers. The fixes included retrofits of the equipment to bring it up to today's code requirements and extensive training of the plants staff in their native tongue. This was a challenge. It meant translations and finding presenters who could speak the native language and identify with the workers.

The knowledge and skills that were now in the hands of the workers meant that they could better protect themselves in the future. I was confident that the circumstances I found would never be repeated at this site. When I left I saw a team effort, management and labor, working together toward a much needed cultural change.

8.1 PERSONNEL ISSUES

Perhaps you're in charge of part of a plant that has combustion equipment installed. The project for installing the boilers or ovens might have been conceived of and directed by someone on your corporate staff. The plans were then probably reviewed by a number of people, including the local building department, the local fire department, an architect, and maybe even an insurance company representative. A licensed contractor probably did the equipment installation. You may expect to feel comfortable knowing that a half-a-dozen skilled professionals have reviewed every aspect of the installation and have given their blessing.

The problem is that the day after all the specialists leave, one person and a misplaced screwdriver can reduce your entire facility and everything in it to rubble. Operations and maintenance are where the real human error issues tend to lead to disasters. These issues usually present themselves through a lack of training and operating procedures.

Most people come to work each day and want to do a good job but just don't always know how. In tough economic times, where fewer employees are being asked to do more, training isn't always the priority that it should be. Accumulated knowledge sometimes retires with a worker when he or she leaves. And the pace of technology today, especially related to industrial combustion systems, is far outstripping the ability of operational and maintenance staffs to assimilate it.

In the past, combustion equipment tools were pipe wrenches and screwdrivers. Today, laptops and software are just as important. Systems that include complex components and PLCs are often purchased and then bundled into a piece of equipment that has minimal local support. Unsuspecting buyers don't understand that in some cases, they are getting married to one vendor or contractor, regardless of the cost, availability, or quality of their service.

Companies must make a conscious decision about how much technology is correct for their particular organization and its employees. In some cases, customers later take out sophisticated systems only purposefully to regress to more simple controls and components, just so that their people can actually keep the equipment running and reduce downtime caused by waiting for special parts or expertise. This can be especially important for multinational companies that install plants in less developed countries with fewer technological resources.

Avoiding human error is a subject that has been studied for many years. It is an important consideration for space flight, air travel, and even operating nuclear power plants. More needs to be done regarding human factors related to combustion equipment operations. I have found that regardless of the industry involved, improving combustion equipment human factor issues needs to begin by addressing the organization's culture. The culture needs to be one of respect for the power of the equipment and the energy that can be released from the fuels involved. If I could take every person involved with combustion equipment on a time-travel trip to some of the scenes of devastation and death that I have witnessed personally, cultures could be fixed very quickly. Since this is not possible, it's up to you to change behaviors on an incremental daily basis through much reinforcement and training.

Real-Life Story 27: A Death Caused by Human Error When Troubleshooting a Boiler

This incident began as a service call by in-house maintenance personnel regarding a boiler that would not light. It ended with one of the operators dead. The boiler was used for space heating at a water treatment plant in a major midwestern city. It was a Scotch marine type of fire tube boiler that had round-the-clock operators. One of these boiler operators was taking a break while the service technician was trying to figure out why the boiler would not light. The operator was standing about 25 feet away and during a lull in the action started to read his Bible a favorite way of his to pass the time.

The technician's troubleshooting efforts focused on the controls and the burner management system. This particular style of boiler was manufactured with a control panel near ground level. Because it was so low, the service technician had to duck down and squat to reach it. During the course of the troubleshooting the boiler suddenly exploded. One of the firebox doors (about 3 feet by 3 feet and about $\frac{1}{2}$ inch thick), was ripped off its hinges, sailing right over the head of the crouching service technician and right at the boiler operator. The door struck him in the neck, killing him instantly.

It's not clear exactly why this explosion happened. Some claimed that the technician installed jumper wires, defeating the safety controls temporarily. When someone mentions "jumpers," they refer to an electrician or service person installing a wire to render a safety circuit inoperable or in a bypassed state. This is done by technicians sometimes to diagnose control issues. Very experienced technicians sometimes do this with great caution and with manual gas valves closed to eliminate explosion risks. It's clear that somehow during the troubleshooting process, a gas valve was opened, allowing gas to fill the boiler.

Lessons Learned Marginally skilled or inadequately trained service persons or electricians troubleshooting combustion controls are a recipe for disaster. I have seen it many times. People who just work casually on gas-fired equipment can be more deadly than a toddler with a loaded sawed-off shotgun.

Troubleshooting protocols should include keeping all manual gas isolation valves closed when electrical panel troubleshooting is taking place. Some sites have protocols for using jumper wires. They make these different in color and sometimes they attach tags that record the installation date. Some sites require that two people be present to witness whenever this is being done, so it can be discussed and verified that fuels are isolated.

I have told hundreds, if not thousands, of people over the past 30 years that none of this work is a spectator sport. It's always better to have only essential personnel around whenever fuel or combustion system work is going on.

8.2 TRAINING

A lot of plants and other facilities assume that training is something that happens on the job in an informal sense. To them, training is information that gets passed on from person to person over coffee or in between baseball scores. Training related to fuels and combustion systems must be more formal and planned to be effective. It must begin when an employee starts a job; it must then be regular and on a schedule. Training shouldn't end after a couple days of orientation to a new job.

Many companies in the nuclear industry spend over 200 hours a year training operators.[1] I'm not suggesting that this is needed for operators of all industrial combustion equipment. However, my experience is that the overall average for combustion equipment training is no more than 4 hours a year at most industrial sites. This is obviously too little. You need to look at everyone who comes into contact with the equipment (operators, maintenance personnel, managers, etc.) and think through everyone's different needs. In some cases it will be communicating simple awareness issues. In other cases there will be a need to transfer detailed knowledge and skills in depth.

In the course of developing and implementing training and certification programs, the issue of validating that knowledge transfer took place always comes up. Tests or assessments can be a touchy subject in corporate cultures, and for good reason. You can get in a lot of trouble with assessments if they are not conducted properly. Your company's human relations department should be part of the discussion about any assessments that may be used. If assessments will influence job positions, overtime assignments, or anything else that has to do with money, they can and will be challenged.

In a 2009 case known as *Ricci* v. *DeStefano*,[2] a group of white firefighters from New Haven, Connecticut presented the U.S. Supreme Court with a case that challenged conventional notions of bigotry. They argued that the city of New Haven discriminated against them in 2003 when it threw out a test that white firefighters passed at a 50% greater rate than blacks. Because performance on the test was the basis for promotion, none of the blacks in the department would have advanced had the city accepted the results. Your assessments could be facing editing from specialists who will normalize for education levels and other issues that might be considered discriminatory.

8.2.1 Mock Drills for Practice and Validating Knowledge Transfer

Another tool (besides assessments) to consider for validating knowledge transfer is the establishment of mock drills that facilitate students practicing what they have learned. The following is an example of a mock hazard drill that boiler operators could use for knowledge transfer validation.

> Upon entering a boiler room to take over your shift, you smell smoke that has an odor similar to that of burned paint. You walk toward one of the boilers where the smell is strongest, and you notice that it is running. However, the water gage glass has no discernible level. There's more heat coming off the boiler than normal and parts of it are smoking. What do you do?

Select one person in the training group to provide a response, explaining that there is not necessarily just one right or one wrong answer. Instead, you want to discuss the general principles behind each person's approach.

In this case the last thing that anyone should want to do is add water. If the boiler is being dry-fired (i.e., if it has exposed hot tubes), any water that enters could be vaporized immediately to 1600 times its volume in milliseconds. This is likely an evaporation rate faster than the safety relief valves can remove the steam. Water entering such conditions would result in an immediate catastrophic explosion.

Although an operator's initial thought may be to figure out how to solve the low-water problem, that's not always the best first step. In this situation, operators should talk about things such as notifying emergency personnel about a possible hazard and evacuating the area, as well as shutting down the burner, shutting down feedwater pumps, and closing feedwater lines. Yes, you do want to solve the problem. But before doing so, we must play it safe and get people cleared.

Now that you have the concept of a mock training drill, you can develop site-specific scenarios for many things surrounding your operations. I have found mock scenarios tailored to specific facilities and equipment to be among the greatest knowledge transfer validation and learning tools available.

8.2.2 Licenses and Certifications

Few universally recognized credentials exist for operating or maintaining combustion equipment in the United States. In many places, the greatest number of requirements exist for boilers. Many states at least require licenses for boiler operators and jurisdictional inspection certifications for some classes of equipment. Some areas require that only licensed contractors with specializations perform work on fuel trains and combustion equipment; however, in many cases, industrial maintenance work is exempt. You need to find out which credentials are required in your area. You can start by examining websites related to your state's licensing requirements for contractors. Read the state or province laws surrounding these to better understand who needs what licenses and certifications.

In Ontario, Canada, rigorous standards and qualifications exist for those who work on gas-fired equipment. The Technical Standards and Safety Authority[3] offers programs and enforces the standards. The information they offer is good reference material if you want to consider in-house training and certification.

8.3 CULTURE CHANGES

You've probably heard the phrase "There's nothing new under the sun." What you may not realize is that its origins go back to the christian Bible. In Ecclesiastes 1 : 9, it is written: "There is nothing new under the sun, what has been done will be done again." Although this is certainly true in Hollywood, where every television show or movie these days seems to be a sequel or reboot, it is also quite true in terms of corporate disasters involving fuels and combustion systems. When a serious explosion or incident occurs at a corporate facility, the typical response falls into a well-known cycle that I view as a conventional incident response cycle. I have seen it many times, as I am sure you have as well.

The company or organization vows that something like this will never happen again, makes some big public changes, and increases training and safety protocols. But then over time, the pain of that incident begins to subside, people begin to question why they're spending all this time and money on training or special procedures, and safety starts to slip. That cycle, shown in Exhibit 8.2, is why many incidents recur within large organizations every seven to 10 years. It's not about people being evil or not very smart. It's about sustainability and culture changes.

There's another well-worn cliché at work here, too: "If you don't learn from history, you are bound to repeat it." In many organizations, most of the staff turns over at least every 10 years. The business world, especially today, is so dynamic that things get lost and champions of certain programs move on. This cycle is exactly why the policy part of the recipe "people, policies, and equipment" is so important. If new policies can get implemented and made a part of the culture, you have a fighting chance for long-term sustainable success.

The path to revisiting disaster by companies was studied and well documented by a very experienced and knowledgeable engineer from England named Trevor Kletz. In his book *Lessons from Disaster*,[4] Kletz illustrates this point by providing

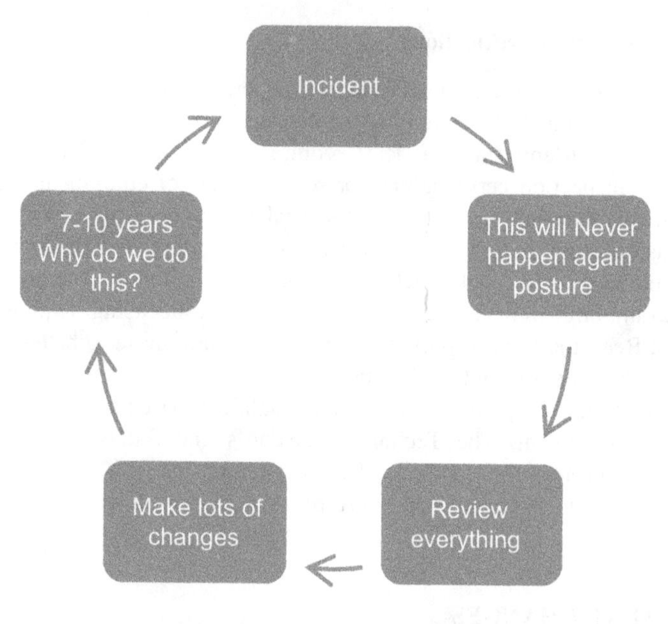

EXHIBIT 8.2 Incident response cycle.

three real-life stories (28, 29 and 30) relaying incidents that happened to the same company over time.

Real-Life Story 28: Explosion Caused by a Leaking Valve During Repair

In 1928, a 36-inch-diameter low-pressure gas line was being modified and a number of joints had been broken (opened up). Before work started, the line was isolated by a closed isolation valve from a gasholder (tank) containing hydrogen, swept out with nitrogen, and tested to confirm that no flammable gas was present. Unknown to the workers on the job, the isolation valve was leaking. Eight hours after the job started, the leaking gas ignited. There was a loud explosion and flames appeared at a number of the flanged joints on the line. One man was killed by the pressure (not the flames), but equipment damage was slight.

The source of ignition was a match, struck by one of the workers near an open end of the pipe, so that he could see what he was doing. He thought it was safe to strike a match, as he had been assured that all flammable gas had been removed from the plant. Once a flammable mixture is formed, a source of ignition is always liable to turn up, so the real cause of the explosion was not the match but the leaking valve.

The accident report included three recommendations:

1. Never trust an open gas main attached to a system containing gas, and keep all naked lights (non-intrinsically safe light sources) clear.
2. Make sure that adequate lighting is available when working on pipe bridges at night.

3. Never place absolute reliance on a gasholder valve, or any other valve for that matter. A slip plate (blind) is easy to insert and absolutely reliable.

These recommendations were repeated in a safety handbook given to every employee the following year. But over the years, this sound advice was forgotten and the use of slip plates for isolating equipment under repair was allowed to lapse. No one knows why. Perhaps it was difficult to insert slip plates because the plant's piping systems were not originally designed to take them. Perhaps, with changes in staff, the accident was forgotten and slip-plating was abandoned to save time and trouble. Meanwhile, the company expanded, the site became full, and a new site was developed about 6 miles away.

Real-Life Story 29: Another Explosion from a Leaking Valve During a Repair

In 1967 on the new site, a large pump was being dismantled for repair. When a fitter removed a cover, hot oil came out and caught fire, as the 14-inch-diameter pump suction valve had been left open. The temperature of the oil was about 280°C (536°F), above its autoignition temperature, so ignition was quick and inevitable as soon as oxygen was available from air to complete the fire triangle. Three men were killed and the unit, only a year old, was destroyed.

We do not know why the suction valve had been left open. The foreman on duty said that he inspected the pump before issuing the permit and found the valve already closed. Either his recollection was incorrect or, after he inspected the pump, someone opened the valve. There was no lock, tag, or slip plate to prevent any one from doing so.

After the fire, instructions were issued that before any equipment is handed over to the maintenance organization:

1. The equipment must be isolated by slip plates or by physical disconnection and blanking unless the job to be done is so quick that fitting (or disconnecting) slip plates would take as long as the main job and be as hazardous.
2. Valves used to isolate equipment for maintenance, including isolation for slip-plating (or disconnection), must be locked shut with a padlock and chain or an equally effective device.
3. If there is a change in intention—for example, if it is decided to dismantle a pump and not just work on the bearings—the permit to work must be returned and a new one issued.

Real-Life Story 30: Yet Another Explosion from a Leaking Valve During a Repair

The second incident was so traumatic that it was not forgotten on the site where it occurred. But it was forgotten on the main site, the site where the first incident had occurred. In 1987 a hydrogen line about 12 inches in diameter had to be repaired by welding. The hydrogen supply was isolated by closing three valves in parallel (and a fourth valve in series with two of them). The line was then purged with nitrogen and tested at a drain point before welding started to confirm that no hydrogen was present.

When the welder struck his arc, an explosion occurred and he was injured. The investigation showed that two of the isolation valves were leaking. It also showed why hydrogen was not detected at the drain point: It was at a low level and air was drawn through it into the plant to replace gas leaving through a vent. The source of ignition was sparking because the welding return lead was not securely connected to the plant.

The accident report recommended that slip plates be used in the future for the isolation of hazardous materials. There was no reference to the earlier incidents, which were probably not known to the author of the report.

Lessons Learned All three incidents occurred because workers failed to isolate the gas piping properly during maintenance. The use of blinds had been recommended following each incident but had not yet been put into practice. Each accident killed or injured at least one worker. Again, I emphasize that all three disasters happened to the same company. But this is not an isolated case; it happens again and again in many companies. Please think about this and try to implement long-term sustainable solutions. Sustainability often requires long-term top management commitment.

Few plant management teams, organizations, or industries can implement true cultural changes that will persist over time and prevent painful incidents from recurring. These incidents are precisely why there is a paradox in preventing incidents. It's well recognized that human error is the leading cause of incidents. Knowing this, and that organizations have a poor track record of changing behavior, is why upgrades of safety systems and keeping up with code changes over time are important. The more that one can do to design in safety, the less risk there is associated with inevitable human fallibilities.

For example, changes were made in NFPA 86 to require the use of two automatic blocking valves in series on pilot gas lines. This requirement was added due to repeated incidents where one valve leaked through when in the closed position, resulting in incidents. If these valves are checked regularly, the chances of an incident occurring are less. If two valves are installed in series, a lapse in the regular valve tightness program brings less risk.

IMPORTANT

Cultural changes start at the top. You must have a company's top leadership fully engaged in supporting fuel and combustion system cultural changes in visible ways. Start with a letter that goes to everyone, signed by the most senior top executive, announcing the effort and providing full support. Otherwise, many may question the authority and legitimacy of every aspect of your efforts. There is no time for that when there are lives to be saved and new procedures need to be implanted.

Even though designing out risk is important, it's still astonishing how little attention is paid to fixing the human element in fuel and combustion system risks. Everyone likes to change equipment or controls, but the more difficult and sometimes more meaningful thing to do is to change the human and cultural parts of an organization. If you really want to open your eyes to some innovative safety culture concepts, consider reviewing the material that Sidney Dekker has put out involving "Just Cultures." There's a great YouTube video in which Dekker[5] engages in some very thought-provoking discussion about cultures in the workplace. Organizations such as American Electric Power have benefited from "Zero Harm/Just Culture" implementation. In such an environment there is little emphasis on blame and punishment for what appears to be a mistake, and more emphasis on learning and digging deep into the basic causes of certain behavior. Good fuel and combustion safety culture starts with open communications and nonthreatening environments. There is too much at stake with this equipment and its safe operation to be hampered by cultures where no one wants to cooperate.

8.4 HUMAN LAYERS OF PROTECTION ANALYSIS

Mitigating risks related to human error is as complex as human beings themselves. The problem gets easier to solve if it's broken into a number of pieces or layers. The concept of layers of protection analysis (LOPA) is a classical means of providing protection from hazards. In his paper "LOPA and Human Reliability,"[6] William Bridges of the Process Improvement Institute explains that independent layers of protection are at the heart of LOPA. Each independent layer of protection must be:

- Independent of the initiating event and other independent protection layers
- Capable (big enough, fast enough, strong enough, etc.)
- Maintained or kept in practice/service
- Validated/proven to provide the probability of failure that has been chosen

All of the layers must be documented and audited periodically to ensure compliance with these definitions.

So how does this apply to fuel systems and combustion equipment? Let's consider the layers of protection afforded to steam boilers to avoid catastrophic overpressure incidents:

1. Pressure vessel codes and standards for construction are applied, including the choice of materials selected and their fabrication requirements.
2. Pressure operating controls are provided which decrease or shut off firing when specific pressures are achieved.
3. High-pressure burner cutout switches are provided to shut down burner firing if an overpressure condition is detected from the failure of an operating pressure control.
4. Safety relief valves are provided that will release steam if overpressure occurs.

5. Explosion relief walls may be included at the boilerhouse that are meant to release pressure excursions from the building in a predetermined and desired way.
6. The central boiler plant itself may be located in a less populated area of a campus or part of a site with minimal traffic.

The same sorts of layers of protection can be afforded to people and how they are related to fuel and combustion systems. Following are LOPA factors to consider that can minimize the chances of human error related to fuels and combustion equipment.

1. *Procedures exist and are accurate and clear.* Concise written and validated procedures must exist for starting and stopping combustion equipment, maintaining it, and for fuel system isolation, venting, purging, and reintroduction processes. Studies have shown that without checklists, error rates go up dramatically.[7] This is why they are commonly in place for airline and military pilots. But in my own experience, written procedures exist less than 50% of the time when it comes to fuel systems or combustion equipment startup and maintenance practices. In the aftermath of incidents, OSHA frequently identifies a lack of written procedures as a contributory factor in accidents. It is often the subject of citations and fines. Once there are OSHA citations and fines, civil litigation by plaintiffs becomes much easier.

   In most cases, procedures include general warnings, PPE requirements, and conditions that must also be reviewed prior to even starting to get to the parts that involve the machine controls. Other best practices for procedures include having troubleshooting guides or similar contingency procedures. These are often best done with flowcharts.
2. *There is continuing training of operators regarding knowledge and skills.* Combustion equipment safety requires both knowledge and skills. Remember that knowledge is book learning, whereas skills are things people do with their hands. You can pass on knowledge with Web-based learning and classroom instruction but you cannot pass on skills this way. Skills have to be passed on person to person on an individual basis or in small hands-on groups.

   Most codes include some requirement for training. For example, NFPA 86 contains a section regarding training (Section 7.2) which requires that all personnel involved in the operation, maintenance, and supervision of equipment be trained by a qualified person. Operating and maintenance personnel also must be able to demonstrate that they understand the equipment's function and safe operation. Refresher training is also required, but the time between training is not called out in NFPA 86, so each plant should establish a reasonable period for refresher training.

   The American Society of Mechanical Engineers (ASME) boiler code in Section VII, Subsection C2.110, states: "Safe and reliable operation [of boilers] is dependent . . . upon the skill and attentiveness of the operator and the maintenance personnel. Operating skill implies knowledge of fundamentals, familiarity of equipment, and a suitable background of training and experience. Regularly scheduled auto-manual control systems changeover, manual operation during upsets or abnormal conditions, and emergency drills to prevent loss of these skills is recommended."

Some labor unions, including the United Auto Workers[8] and the United Steel Workers,[9] have programs to provide training and pathways to developing skills for its skilled trades such as electricians & pipe fitters. Some of the information in these programs relates to topics important to fuels and combustion systems. One of the things I have found very effective with some of these programs in particular is the use of witness documents or books. This is a system by which a student demonstrates skills to others over time. Once the skills are witnessed, someone signs off on a skills tracking document. When someone has filled his or her skills book with signed sheets, he or she may be eligible for advancement through the program. This same type of process can be used to validate the transfer of fuel system- and combustion-related skills.

Skill training steps must include walk-throughs of procedures under real life conditions and shadowing to ensure that the reliability of actions can occur under the increased stress caused by alarms and "Real life" conditions. The time it takes to complete each step should be measured during each practice/drill to ensure that response-time limits are not exceeded. Each person who is expected to implement the action on demand must be capable of performing the required actions in the amount of time allotted.

3. *Operators and staff are fit for duty. This includes many sub factors, such as fatigue, stress, illness, medications, and substance abuse.* Company policy and enforcement actions to limit fatigue, including limiting work hours per day and per week, should be considered. Hours worked, including working outside the workplace, should be factored in. A detailed study by three prominent medical doctors indicated that medical error rates increase by 25% after missing one night of sleep.[10] I have known people to have worked 20 to 30 hours in a row in sensitive roles and have seen this poor judgment after the fatigue sets in. In fact, the matter of fatigue was identified by the U.S. Chemical Safety Board as a significant factor in the 2005 BP Texas City disaster. A new recommended practice, API RP 755, Fatigue Risk Management System, has been published in an attempt to provide guidance on this subject.

   Other issues related to fitness for duty can include drug and alcohol addiction and chronic medical conditions such as diabetes or sinus allergies. Life issues such as recent deaths in families or divorces can also be a distraction.

4. *The workload is managed well.* Too small a workload and the mind is bored and looks for distraction, but too many tasks per hour can increase human error and create fatigue. The advances in data collection and monitoring systems have been incredible. There are now control rooms with virtually everything available at a quick glance. However, there's no substitute for being able to assess sounds, vibrations, and smells that no displays or data collection systems can replace. There are always trade-offs between using remote monitoring and having someone walk around and actually see the equipment. I have found that there is no substitute for the walking-around part. There should be a roaming hands-on walking-around aspect to any operator or monitoring function.

Whoever does these tasks should always be in radio contact with others and their position monitored in case of an injury or incapacitation caused by something unexpected.

Where considerable combustion equipment or gases are in use, it's not a bad idea to have the employees who do the actual equipment inspections wear gas monitors. The steel industry does this extensively and in my business I required technicians to wear CO monitors when servicing combustion equipment. In a number of cases, monitors went off in completely unexpected circumstances, and dangerous conditions such as flue gas leaks were discovered.

5. *Communications need to be clear and well managed.* Miscommunication of an instruction, set of instructions, or the status of a process or piece of equipment is a leading cause of human error in the workplace. An example of how to improve this can be borrowed from some of the excellent boiler operations I have seen. In some central plants white boards are used to communicate the status of equipment and repairs. Boiler log books are also a conventional, well-understood means of communicating among operators on different shifts. Boiler log books, journals in which operators write down significant events and maintenance issues daily, can be extended to any combustion equipment operations. A neat innovation is to use different-colored inks for different indications. For example, any abnormal operating problems can be recorded in red ink. This helps these concerns to stand out in an otherwise black-and-white log.

   Some electric utilities have morning lineup or hand-off meetings. When there is an overlap of shift personnel, those leaving explain important operational issues that occurred during their shift to the staff taking over. Supervisors leading a hand-off meeting must follow a rigid and defined agenda, which often involves both visual and spoken communication policies. Before breaking up the meeting, the supervisor goes around the room to each person and asks for input and comments. In some organizations these types of meetings take place periodically for common operations on one large conference call even across enterprises.

   Some sites practice mandatory repetition of instructions and policies and enforce the use of common jargon. Such meetings are a place where safety emphasis is reinforced and near-misses are discussed. Also, don't forget three point communication—you tell me, I repeat it back to you, and you validate what I repeated back. It's simple, basic, and very effective.

Real-Life Story 31: Poor Communications and Human Factor Issues Lead to a Major Explosion

Consider the case of a major North American specialty chemical manufacturer where a large process heater was found to be experiencing a fuel-rich condition. An operator went into the field to make adjustments to a manual air register (a type of air damper on a burner). He was in communication with the control room, where they were able to change fuel flow rates. He was able to make air adjustments to the manual dampers on the burners.

It was not clear to the operator in the field exactly what type of change on a manual damper setting would produce a specific required airflow difference. It was sort of a trial-and-error effort for him in the field. He went the wrong way with an adjustment and instead of increasing airflow, decreased it. The opposite error was made in the control room with fuel flow. The operators increased the fuel flow instead of decreasing it. This compounding of errors made for way too rich a fuel condition.

Before the combination of these two errors could be identified, it was too late. A large slug of too rich fuel–air entered the unit and made its way to an upper pass, where it leaned out and became flammable. This led to a horrible explosion. The result was over $10 million in damage. No one got killed, but the heater stack came down on a pipe rack and interrupted a significant amount of processing capability at this petrochemical site.

Lessons Learned It is important that there be clear and concise protocols and procedures for communicating things between those in the field and those in remote control room locations. Better visual indications for changing airflow in the field might have helped to identify proper directions for damper positions to increase or decrease airflow. A good rule of thumb to remember for fuel/air ratio adjustments is never to add air but to pull fuel slowly when trying to correct a fuel-rich condition.

6. *The work environment can be an important layer*. If the work environment is not lit well, is too noisy, too hot, too cold, or has many distractions, the chances for an operator or person performing tasks to do things correctly obviously diminishes. Many issues related to work environments can be vital to safety. I have observed boiler rooms that were chronically over 110°F. There could be no escaping the heat unless you left your post and sat outside the building, and that's where these operators could usually be found. This did not put them in a position to respond effectively to issues related directly to the boilers. Also, remember that even such things as ambient noise levels can be fatiguing and distracting. Having to wear earplugs all the time to meet OSHA requirements means that the likelihood of hearing alarms and pending equipment failures such as bad pump bearings or water hammer also diminishes.
7. *Human factor interfaces, such as displays, instrumentation, layout, size, color, shape, and location on a wall, can all have an important impact*. The design of displays and their layout for instrumentation has been a much-studied subject. If you're not aware of standards and guidelines for the design of control rooms and information displays, you can find great information in a document such as NUREG 0700.[11] This is a document published by the U.S. Nuclear Regulatory Commission. Although the design of nuclear facilities seems off the topic, the consequences of mistakes in both the fuels and combustion industry and the nuclear industry can be tragic. This document can give you the ability to

EXHIBIT 8.3 Match-marked linkages.

conduct a human factors review of displays and control elements. This can focus on such things as the arrangement of gages and information displays for users, including the colors used, the size of displays, and many other factors.

Outside the control room there are a number of things that can be done in the field to improve the likelihood that equipment or operational errors will be detected. Simple things such as match-marking of linkages and gages can have an immediate impact on reducing human error and can be done at little or no cost.

Match-marking linkages (Exhibit 8.3) is the simple practice of marking linkage clamps and rods with paint or an ink marker. The lines and spots on linkages and holding clamps will be displaced if there is relative movement and slippage and will be seen more easily. It's the same with gages. They can be marked with information on normal operating conditions. During every prestart walk down, emphasis can then be placed on looking at these lines and spots to verify that all is well.

IMPORTANT

Match-marking of linkages is a simple "no cost" but effective procedure that can be carried out across an enterprise in one afternoon via an e-mail. Mark linkages in place with a marker or paint. Lubricate the linkages while you're at it. Watch them operate to make sure that they move smoothly. Make the review of linkages part of a daily startup routine.

Another simple but effective gage readability issue is making sure that the units are consistent with all devices or a piece of equipment. If you're measuring pressure on a single system fuel train in several places for example, everything should be in common units such as psig or inches w.c.—not one gage in ounces, another in psig, and yet another in inches w.c. or kilopascal. Make sure that everything is in either metric or English units. Don't make people continually have to convert in their heads from kilopascal to psig, for example.

8. *Task complexity can also be important.* Things that can make tasks more complex can be a real problem. Sometimes the design of the process can make a number of choices available that are wrong but appear similar. Consider the case of similar switches or valves (shape and size) on a control panel or operator board. Do many of the buttons on the panel look the same? Are they small and thus difficult to read for someone over 50 who does not have his or her glasses on? The number of parallel tasks that may distract a worker from the task at hand can also be an issue, as can the number of people involved in the task.

   Fuel switching is an example of something that has often gone wrong and caused many disasters. I am talking about a boiler or other piece of equipment capable of firing multiple fuels, such as oil, natural gas, or some other process-related fuel. Many incidents have occurred during fuel switching while operation of the equipment is continuing (i.e., when the burner is not shut off and people try to do this on the fly). This is not recommended. It's always safer to shut down and restart on the alternative fuel if there is a choice. Depending on the fuels and their relative heating values, complex adjustments are required to do it on the fly. In the case of those who have very complex procedures for changing fuels on the fly, if everything is not done exactly as required in some of these procedures, a disastrous incident can occur, as demonstrated in the following real-life story, which demonstrates a series of tasks that in my opinion were not operator friendly and exhibited poor procedure design. To demonstrate this point I could have chosen the story that follows among three or four different fuel switching explosions I have seen.

Real-Life Story 32: A Tricky Switch

An employee at a steel plant was trying to change fuels on the fly within a large boiler that burned both blast furnace and natural gas. Blast furnace gas has a heating value of about 90 to 100 Btu per cubic foot; natural gas has a heating value of about 1000 Btu per cubic foot. The combustion air requirements for these fuels are dramatically different (by about a factor of 10). The controls for this boiler were somewhat old and unsophisticated. A lot of the operation was made dependent on the art of watching cameras, monitoring gages, and doing the right things at precisely the right times. This was a lot to ask of a relatively new operator—still, at some point every pilot gets his or her first solo flight, and in this case every operator at some time gets his or her first "solo light." If you don't have really good training and validation processes in place, it's hard to know if the person is really ready. Unless, of course, an explosion occurs, and then it is very positive validation that the person was not ready.

In this case the procedures called for attempting to stay on low fire while natural gas was switched out and the blast furnace gas was brought in. Although this can be done and had been done for years, it's never as safe as a complete shutdown of the boiler and a re-light with a fresh air purge of the firebox before the alternate fuel light off. In most cases, if this can be planned, the small reduction in steam output for the purge and light-off period does not affect operations significantly.

For this procedure to work properly, the operator had to monitor draft gages, fuel flow instruments, and even watch the flame on a monitor screen switching among all of these and constantly thinking through what was happening and what to do about it. This process had gone right for years, until one operator, who had not been on the job long, made an error. All was going well for him and the unit had been transferred successfully from natural gas to blast furnace gas. Then within seconds of what seemed like success, the unit suddenly flamed out. The burner management system closed the blast furnace gas valve. The operator reopened the gas valve for only seconds to try to get it re-lit within the hot boiler. This put a lot of blast furnace gas into the firebox. Unfortunately, it did not light immediately. The unlit gas accumulated, and finally lit, in the boiler's convection section. The resulting explosion blew out a flue gas heat recovery device on an upper floor of the boiler. The cost to repair the damage was well over $200,000. The disruption to operations cost considerably more. Fortunately, no one was injured.

Lessons Learned Review the human factors related to complex tasks where a mistake could create a tremendous disaster. Are there just too many steps? Is there very little time for the right call to be made, and are the timing and the tasks design risky? What are the consequences of incorrect decisions along the way?

Make sure that written fuel switching procedures exist. If it all possible, make sure that these call for a complete shutdown and relighting by the alternative fuel. Training must include hands-on skill transfers and methods for evaluating proficiency. In many cases, simulators can be an effective means of practicing and validating skills. Simulators are commonplace among pilots in the air travel industry and in the nuclear and process industries. I have participated in designing and fabricating small-scale furnaces and ovens and firing systems for use by clients so that their staff can practice and experience things in a safe environment instead of in "real-life" situations where lives and millions of dollars are at stake. These can be designed and fabricated in some form for almost any process.

9. *Time must be available to perform any action.* This could be the time required to react to a high-pressure alarm on a boiler before a safety relief valve opens. It could be the time required to respond to a steam load change or pressure change. In many cases the time required for human interaction is driven by set points on interrelated systems (i.e., the sizing of equipment). For example, if safety relief valve set points are at the maximum allowable operating pressure of a boiler, there might be little time to intervene before a drum rupture. If the set points are somewhat below that pressure, there will be more time before a catastrophe.

   You just have to count on people not reacting perfectly for the myriad of reasons that make us human. Make sure that there is considerable slack left in the system for things to get to the human response and action step before really bad things happen. It can be important to run mock drills on these types of things with many staff members, to understand the reality of what timing might be in different situations.

10. *Physical capability is required to perform an action*. There are many ways to mitigate human error related to physical capabilities. It might be possible to minimize the strength needed to operate a large valve many turns by installing a geared wheel operator. Think about how this might help for older operators or perhaps someone recovering from an injury that might have to climb many steps and then be expected to close quickly a valve. Strength, dexterity, and range of motion are just a few of the ergonomic considerations that must be given to what we ask people to do for safe operations to happen.

Real-Life Story 33: A Fruit Cup Saves the Day

I was at a combined cycle power plant control room and noticed that the operators had chairs with long backs on wheels that if moved backward could contact a series of buttons on the wall. Over one big red button was a plastic cover Scotch-taped™ over a red switch that stuck out of the cabinet. It did not look like something that had been provided from a professional panel shop. I said, "What's that?" as I pointed to it. Suddenly the control room erupted with laughter. The operator on duty said, "Oh, you don't want to know about that." A voice called out, "Go ahead, Bill, tell 'em; I know you will anyway." Bill said, "That's a little plastic snack cup from peaches that Jerry over there loves. One day he leaned back in his chair and shut the entire plant down. Now we found a use for those cups when he's done with lunch." Jerry had leaned back in his chair one day and shut the entire plant down. The big red poorly placed button sticking out of the panel was the master plant trip emergency shutdown button.

Lessons Learned Yes, really stupid things happen. Review the critical panel switches and where and how they are located. Are they tamper resistant? Are they vulnerable to human error? You can buy pushbuttons that have protective guards so that it's difficult to hit them by mistake. In some cases you might want to relocate some of these mission-critical control devices because where they are doesn't make sense anymore.

You may want to conduct a review of how operators interact with the combustion systems. What are the location and size of critical displays? Is the information presented adequate and clear? Are the control elements on the panel large enough, uncluttered, and well labeled? All of these factors can reduce the risk of human error.

8.4.1 Other Factors That Influence Human Error

The location of some equipment and controls can also have a lot to do with human error. In several facilities, the location of fuel trains has been observed to be more than 6 feet above the floor level. When this is the case, one cannot readily review fuel pressures, switch set points, or the condition of valves and actuators.

In some circumstances, abnormal conditions or upsets are not easily detectable. It's always important in any inspection process to think through whether the defect can be detected. Consider safety relief valves that are located such that their discharge cannot be seen easily. These can be weeping or leaking through for days or even months without being detected.

Specialty equipment such as cameras can achieve faster detection rates and may need to be a consideration in expanding the convenient reach of your operating staff. Operating staffs in some industries install cameras to observe things such as stacks, flames inside equipment, and even safety relief valve discharges. These days, camera technologies are inexpensive and simple to install and configure. Remember, if it takes extraordinary effort at all times to monitor things properly, it's just not going to happen.

8.5 CONTRACTOR ISSUES

So, perhaps you've decided that it's better to call out experts when something goes wrong, since you understand what it might take to maintain certain disciplines in-house. Nevertheless, you will still need to find a way to ensure that you're receiving the service you know you should be getting. Contractors for fuels and combustion equipment need to be qualified, assessed, and validated, not just their company's reputation but that of the service personnel they send as well. The number of contractor-related accidents involving combustion equipment that I have witnessed is staggering. Just because someone has a company name on his shirt and drives a nice van with graphics does not mean that he knows everything or even anything.

It should be simple enough, right? I need a new burner or controls for my boiler, furnace, or oven, and you have these to sell. You give me a price, do the work, and then we're all happy. Wrong! Unless you do some very careful planning and communicating, you could be in for trouble. I have often been called in to serve the role of arbitrator, mediator, and neutral third-party expert on projects that have gone wrong. The end result of such projects can range from damaged equipment that is out of service for months to long and ugly lawsuits. What's even worse is the wasted money and time on the part of the equipment owner.

Even competent contractors must be familiarized with your equipment, controls, and maintenance procedures. Isolation, startup, and shutdown procedures must be reviewed by contractors before they begin work, as illustrated in the following real-life story.

Real-Life Story 34: Know Who You Are Dealing With

I arrived at the site of a primary metal manufacturer expecting to do a regular combustion-related safety inspection and testing of fuel train components. Instead, I was whisked away to a meetingroom full of management personnel. "So," they asked, "are you going to be able to test the equipment without causing an explosion like the last guy did?" I looked around the room and was in awe at the number of people assembled and their roles: all levels of management and union representation.

They certainly had a reason to be cautious. I was going to be looking at the fuel train for a piece of equipment that was at the heart of their operations. The unit was about 10 feet in diameter and 20 feet long. It had a burner rated at million 8 Btu per hour. Inside the circular vessel were coils of 4-inch pipe circulating hot oil used to heat pitch, which in turn was used to make solid blocks of carbon from graphite. Without the heater, the plant could not operate, and most of the hundreds of workers at the site would have nothing to do all day.

A close examination of the unit indicated that the outer shell was repaired. I could see where a fresh patch was welded on representing about a third of the unit's outer skin. Now I started to understand their concern. I found out that about a year earlier, a technician who had serviced the site's combustion equipment for many years was testing safety interlocks. I was told that he was manipulating wires in the control panel to open the fuel train's automatic gas valves to get gas onto the high gas pressure switch to test it with the burner off (a really dumb and dangerous way to do this test). On one occasion (and it only takes one), he forgot to close the downstream burner firing valve. Fuel flowed into the hot firebox for about 30 seconds before it exploded. The unit peeled open like a tin can and sprayed refractory bricks and debris all over the room where it was located. Although none of the workers in the room were hurt badly, the plant nearly had a disaster on its hands. There could easily have been three deaths. If it weren't for a massive response to get the unit running again, the two 10,000-gallon pitch tanks and all the pipes full of pitch might have solidified. This could have kept the plant down for weeks and cost the company millions of dollars.

One of the men who was there when it happened refused to come anywhere near the building until I had finished my work and left the site. I then clearly understood why.

Lessons Learned Just because someone comes to your site as a contractor, it does not mean that he or she knows what he or she is doing. Ask for the training records and experience of the technician being sent to your facility. Discuss the need for safe practices and isolating fuels.

Simple validated procedures for testing interlock devices and fuel train valves must be in place. It's also a must that someone in your organization be knowledgeable enough and trained to know that what is being proposed is correct and makes sense. Also, remember, as has been presented before, none of this is a spectator sport.

8.5.1 High-Risk Activities

There are five specific combustion system tasks that require very special consideration and validation of skill levels and competence. These should be considered high-risk activities that need to be monitored and planned more carefully than others: (1) changing or adjusting fuel/air ratios (burner tuning); (2) changing fuel modulating or control valves; (3) burner startups where new systems have to be commissioned; (4) changing over to or using alternative fuels (i.e., propane systems, landfill or digester gas, or a different type of fuel oil); (5) venting, purging, and reintroducing fuel gas to piping and distribution systems.

This means that folks conducting these activities need to display evidence of competence, be carefully validated, and demonstrate knowledge of codes and standards related to these activities. One of my large manufacturing clients had developed special guidelines for plants when these types of high-risk activities took place anywhere in the world. It was a set of screening criteria that would help to acquire some of the relevant information required for proper contractor selection. This is a great practice that everyone should consider to provide guidance to purchasing departments and staff engineers who do not regularly do fuel system- or combustion-related projects. This is another place, like brain surgery, where you don't necessarily want the lowest bidder.

Real-Life Story 35: Selecting the Right Contractor and Defining the Rules Saves Money

I was on site to conduct a safety review of a plant's boilers. The scope and focus shifted a little, as it often does, as more was uncovered about who had recently worked on the equipment and why. This particular investigation turned out to be more about saving the plant money than about saving lives.

One of the boilers was a two-year-old Scotch marine fire tube that came with a proprietary fuel/air ratio control system. This system was basically a small PLC custom programmed to work with the air dampers and gas valve provided. When I asked to see fuel/air ratio records and burner performance data, I was referred to the great data logging features that come with the system. In reviewing the information, I saw that six months earlier the response to changes in load made for rapid responses and corrections to deviations from set points. Since the latest tuning (by a different service technician and organization), the boiler was responding much more sluggishly to changes in load. This different response was costing the site money because poor response meant that the boiler operated at conditions that were not optimal for longer periods of time than before.

I found out later that a new technician made changes to the control system's proportional–integral–derivative (PID) loops. A PID loop is an automatic feedback control sequence that is like the cruise control for your car. It means that as a control disturbance occurs, the computer receives feedback of actual conditions and then moves to restore some desired condition after realizing what results the change actually produced. The speed at which it does this is a complicated series of mathematical equations that ultimately results in driving the fuel and air valves to specific locations and then looking for the resulting changes.

This boiler ran at over 50% loaded for basically 24 hours a day to serve process loads. This meant that very small changes in efficiency made for considerable cost changes over time. Changes to the PID parameters were made without the boiler plant management's approval or awareness. This changed PID loop tuning issue was costing the site nearly $30,000 a year in additional fuel costs—in fact the plant also paid $800 for the service technicians visit to unknowingly harm them. It's sometimes very difficult to understand what a technician is doing and what the impact is if you've got no awareness of the equipment or how it works. In cases like this, it's important to document the complete before and after performance information and settings within the PLC and all PID information so that at least things can be restored. You can usually get into these controllers and record what all the PID settings are without changing anything. This is good information to have along with basic commissioning information like fuel and air flows or pressures at different firing rates and valve/damper positions.

Lessons Learned This site did not have clear communications about protocols for setting fuel/air ratios. There should have been rules communicated such as: "You do no changing of PID set points unless you discuss it with us." In many cases it is also a good idea to assign an experienced operator or technician to shadow someone setting fuel/air ratios so that all changes can be understood and witnessed.

Make sure that all "as found" conditions are well documented to include all PID settings whenever fuel/air ratios are changed. If a hard copy of the PLC code is available, make sure that it's archived before anyone starts work. This can help things to be restored if someone messes things up.

This case is one more example of how setting the fuel/air ratio, if not done properly, can become problematic. Again, it's important to have procedures for this and expectations spelled out very clearly, rather than to be at the mercy of whomever is sent to your site to do whatever he or she believes to be involved in fuel/air ratio setting. Remember, just because someone is a contractor does not mean that the person knows what he or she is doing.

NOTES AND REFERENCES

1. Nuclear industry training, www.exeloncorp.com/assets/energy/powerplants/docs/Limerick/fact_LimerickTraining.pdf.
2. Ricci v. *DeStefano*, U.S. Supreme Court, Case 07-1428, April 22, 2009, www.law.cornell.edu/supct/html/07-1428.ZS.html.
3. Ontario Technical Safety Standards Authority, www.tssa.org.
4. Trevor Kletz, Lessons from Disaster: How Organisations Have No Memory and Accidents Recur, Institution of Chemical Engineers, Warwickshire, UK, 1993.
5. Sidney Dekker, www.sidneydekker.com, Just Culture video, www.youtube.com/watch?v=t81sDiYjKUk.
6. B., Bridges, "LOPA and Human Reliability: Human Errors and Human IPLs," Process Improvement Institute, presented at AIChE Spring 2010 Meeting, www.aicheproceedings.org/2010/Spring/data/papers/P177345.html.
7. NASA paper, www.skybrary.aero/bookshelf/books/1568.pdf.
8. United Auto Workers, www.uaw.org.
9. United Steel Workers, www.usw.org.
10. Fatigue, Sleepiness, and Medical Errors, Ashish K. Jha, Bradford W. Duncan, David W. Bates, Agency for Healthcare Research, www.ahrq.gov/research/findings/evidence-based-reports/chapter46a.html.
11. NUREG0700, www.nrc.gov/reading-rm/doc-collections/nuregs/staff/sr0700/nureg700.pdf.

9

Controlling Combustion Risks: Policies

Fuel and Combustion System Policies: The Backbone of Sustainability

There comes a time in the life of a fuels and combustion equipment safety and risk management program when thought must be provided to make things sustainable. The immediate fixes must become institutionalized. Knowledge-based practices need to become rule based. In this chapter we summarize and reinforce 10 important concepts sprinkled throughout the book and frame them in an approach for developing sustainable policies.

Real-Life Story 36: High-Fire Light-off Destroys Two Boilers

A boiler explosion occurred at a manufacturing plant that wiped out both of the plant's packaged 20,000-pound per hour capacity, 125-psig water tube boilers. Although a contractor's technician was there commissioning a new gas valve at the time of the incident, he was somehow not injured. The end result was nearly a $1 million loss, including the replacement of both boilers.

In this case the boilers were laid out such that the two fuel trains (one for each boiler) ran down in between the two identical, side-by-side boilers. The incident occurred as a contractor was commissioning a replacement fuel valve on one of them. This client had replaced the original jackshaft single-point linkage-based parallel-positioning system by separate actuators (one fuel and one air) on one of the boilers to try to gain some efficiency. Since then the boiler had been experiencing capacity problems. It appeared not to be putting out enough steam and that the fuel valve was not driving to high-fire properly.

Fuel and Combustion Systems Safety: What You Don't Know Can Kill You!, First Edition. John R. Puskar.

The manufacturer of the fuel valves and actuators was called because the control of these was via proprietary software. They could not provide a service technician for at least three days. This was just not acceptable given the plant's production demands. A local service contractor was brought in to assess the situation. Although the service contractor did not do the original retrofit and was not an authorized dealer of the valve and actuator system, he felt confident that he could perform the fix. The technician diagnosed the problem as a failed fuel valve actuator, and a new one was air-freighted in. He changed it, but of course he did not have the right software to commission it. This valve had a feedback position indication and had to be synchronized properly so that the software knew what position it was in at all times. This was all unknown to the technician. He took the old valve off, installed the new one, saw a little light come on, and assumed that the installation was OK. The particular style of valve installed, did not have external position indication, so it was not possible to tell whether the valve was open or closed by looking at it from the outside.

The technician thought that he was done, so he hit the start button. The boiler purged, the pilot lit, and then BOOM! The boiler being serviced opened up in all directions. The suddenly expanding boiler sidewalls sheared its fuel train and the sister boiler's fuel train. The second fuel train was within 3 feet because the two fuel trains were positioned between the two boilers. The fuel train on the boiler being serviced sheared after the safety shutoff valves, so no gas escaped. The safety shutoff valves closed immediately, but the sister boiler's fuel train sheared before the shutoff valves. This resulted in a fire ball emerging from the general area between the boilers. One of the plant's operators had the presence of mind to run in immediately and shutoff the main fuel isolation valve controlling gas to the boiler room. This minimized the fire damage.

When I arrived on site and learned that the incident occurred during startup, I suspected that it could have been a high-fire light-off. Remember that most equipment, especially boilers of this size, must be lit-off under low-fire conditions. After confirming that all utilities and piping systems were in a safe state, I asked that the piping immediately before the main firing rate fuel control valve (the one serviced) be cut open. Sure enough, the fuel valve was wide open. This valve should have been in a low-fire position.

The system was receiving low-fire air and high-fire fuel flow. This was too rich (too light), so it flowed and accumulated for up to 10 seconds before some part of the mixture was lean enough to be ignited by the pilot. Once this occurred, the entire volume accumulated ignited, creating the explosion.

Lessons Learned There are a lot of proprietary fuel/air ratio control systems out there that have very limited repair and replacement options. If a failure occurs, you could be down a long time waiting for the right technician or parts. A review of serviceability and available replacement components should be a serious reliability risk review item.

External position indication for fuel valves is a valuable feature and a good specification issue to remember. Having a chance to see the valve position would have at least given the contractor a chance to determine quickly if safe conditions were occurring during the light-off sequence. If the valve had external position indication and the contractor did a dry-fire light-off (no fuel) while watching this valve, the incident could have been avoided.

Just because someone works for a boiler service contractor or burner service company does not mean that the person knows what they he or she is doing. Make sure that if someone is invited in, they have the proper knowledge and experience and software tools, if necessary, to service the equipment.

9.1 POLICY COMMANDMENTS FOR FUEL AND COMBUSTION SYSTEM SAFETY

The items presented in this chapter are not only related to what one might consider to be traditional policies such as lockout/tagout and near-miss issues, but also specifications, procedures, and practices that all provide a base level of rules for long-term fuel and combustion system risk mitigation.

I have named the 10 policy issues discussed in this chapter my *policy commandments for safe fuel and combustion systems*. These commandments are shown in Table 9.1 and explained throughout the chapter.

9.2 COMMANDMENT I

Have meaningful unique fuel and combustion system specifications for related equipment, its installation, and its commissioning. Every corporation surely has something they call specifications for the purchasing and installation of equipment.

TABLE 9.1 Policy Commandments

Commandment Number	Commandment
I	Have meaningful fuel and combustion system specifications for equipment, its installation, and its commissioning.
II	Create and maintain equipment-specific lockout/tagout instructions and line breaking/gas piping permits systems.
III	Perform interlock and safety device testing including valve tightness testing at least annually.
IV	Review all fuel and combustion systems for code changes and upgrade continuously to stay current.
V	Maintain a best-practices and lessons-learned program for fuel and combustion system issues to include root-cause investigations that are widely communicated.
VI	Develop detailed startup/shutdown procedures, including prestart walk-around protocols.
VII	Implement regular training (knowledge and skills) with validation processes.
VIII	Create a system to screen and validate fuel and combustion equipment vendors.
IX	Conduct regular fuel/air ratio validation and optimization programs.
X	Implement a management-of-change process that includes fuel and combustion systems.

The problem is that these are likely to be somewhat general and related to production equipment that is not at all like what is really needed for fuel systems and combustion equipment. The trend today is to make manufacturing divisions and plants autonomous and separate from corporate headquarters organizations. This works very well for many things; however, in my opinion it does not work for vital safety issues such as fuel and combustion system safety. The consequences of not having detailed and relevant fuel and combustion system specifications that are consistent across the entire organization can be costly and even tragic. There are good reasons why broad-ranging corporate specifications for fuel systems and combustion equipment make sense. If these do not exist, people will need to learn the same lessons over and over. In many cases, repetitive learning is OK. For example, when it comes to picking a landscaping contractor for a plant site, the consequences of not doing it according to corporate specifications are not severe. However, do you really want people to keep installing fuel train components that have been problematic and cost millions of dollars in damage at several sites? Is it really OK that many people have to learn this continuously? How many lives might this cost?

The fact is that fuels and combustion systems are too complex and critical a matter not to have good guidelines spelled out across all facilities. Even seemingly simple matters such as which brand or vendor of fuel train components can have a big impact on reliability and serviceability. If, for example, there is consistency among components, fewer spares will be required for the entire site. I have seen situations where this was not dealt with in a planned fashion. It then takes tens of thousands of dollars' worth of components sitting in storerooms to provide the overall site with reliable systems that have spare parts. Make sure that specifications consider the brands and styles of combustion components that already exist at the facility. Such things as flame detectors, automatic fuel valves, regulators, pressure switches, and burner management systems can be common for many types of boiler and furnace systems.

Specifications for combustion systems and components should also consider plant operating environments. For example, wet areas, such as those found in food manufacturing where sanitation procedures are carried out regularly should specify at least NEMA 4–rated electrical components (review NEMA ratings and requirements when specifying equipment). NEMA 4 components can also provide protection for dusty areas such as those common foundries.

Another example of environmental influence that can drive specifications is the effect of ambient temperature on the style and type of automatic shutoff valves chosen. It has been my experience that hydraulic actuators for valves work best for commercial applications where conditions such as ambient temperatures and thermal cycling are not severe (less than 140°F peak). I have found a lot of leaking hydraulic actuators in circumstances where ambient temperatures are normally elevated. In these cases stepper motor–based actuators may provide better service, although these also have recommended maximum operating temperature ranges. These are just a few examples of dozens of important issues that can affect fuel and combustion equipment safety and reliability for many years to come that have to be considered in specifications.

Another consideration where specifications can be important is minimizing the chances that you end up with proprietary software that is not open protocol. Many vendors are now trending toward the installation of PLCs or proprietary fuel/air ratio control devices that are specific to their brand. In some cases, for example, there is no one who can set fuel/air ratios or replace components unless they come from that vendor and are one of their service people. This might work well in urban metropolitan areas. However, it does not work well in the middle of nowhere (which I have found out seems to exist in many states and parts of the world).

Finally, make sure that specifications are reviewed by your insurer and that purchasing protocols call for the plant's property insurance team to provide approval of what is purchased and installed. Insurers have standards or code compliance issues that they also want equipment to meet. This is usually presented as a desire for something to meet Factory Mutual or Underwriters' Laboratories requirements or NFPA 85 or 86 standards. Most equipment vendors provide these types of listed fuel trains or components as an option. Make sure that all sites have an awareness of insurer requirements and get them (insurers) involved early during purchases and upgrades of combustion equipment.

9.3 COMMANDMENT II

Create and maintain equipment specific lockout/tagout instructions and line breaking/gas piping permit systems. The second commandment recommends having an equipment-specific special process for combustion equipment lockout, tagout, and installing blinds. This should include placarding and tagging of components and valves for better identification. OSHA 1910.147 identifies zero-energy-state requirements widely referred to by most maintenance and safety professionals. The zero-energy state for machine tools used in manufacturing is a well-understood topic. However, it is often not clear how this needs to be applied to combustion equipment and fuel systems. In my opinion, any isolation for equipment where a human being is to enter a firebox should include a blind, not just a closed valve. This might be beyond zero-energy-state requirements, but blinds are the only isolation that is absolute.

Isolating energy sources on thermal processing equipment can be complex enough. For example, heat-treatment ovens may have many fuel trains and atmospheric gases that need to be isolated. *Placarding* can help to provide schematics to be used as a guide to facilitate correct isolation procedures. A placard or small diagram is attached to a piece of equipment to indicate where the key piping and electrical isolation points for energy control are located on the machine. This may be laminated and posted in a protected area.

When combustion equipment is to be isolated, make sure that the fuel valves are also locked out. After reviewing thousands of combustion systems, my experience is that most maintenance professionals are very attentive when it comes to locking out electrical equipment (with actual locks). However, I have seen many jobs where there is only a paper tag on the fuel valve which says that it is locked (tagged out but not locked out). Many people believe incorrectly that the automatic valves in the system, and possibly in the double block-and-bleed systems that sometimes exist on fuel trains, will protect them, so that manual valves do not need a lock. Fuel trains should be isolated with blinds if possible and, at the very least, locked manual isolation valves. Many special fixtures are available to lock out manual fuel train isolation valves. Each site should have a sufficient number of these to ensure that this task can be accomplished easily. You have already seen that troubleshooting or electrical work could result in errors that cause automatic fuel valves to open. It is also possible that the valves could leak through in the closed position.

IMPORTANT

Verify that actual locking out of piping systems and fuel valves is taking place when LOTO activities relating to combustion systems are occurring. Make sure that the staff has the hardware they need to do this. Lockout fixtures are available for all types of valves.

Some organizations have gone beyond having a lockout procedure and actually have a lockout permit system. This can be especially important for large boilers or complex ovens and furnaces with many zones. This provides an additional element of oversight and verification of conditions. The permit protocols I have seen include a review of the lockout by a person knowledgeable as to combustion equipment, who reviews and signs off to approve the way in which things have been done. Other possible key elements to successful isolation and lockout permit programs for fuel systems and combustion equipment may include the following:

- The permit starts and ends at specific times unless it is renewed. If it is renewed, this also takes two signatures.
- Once actual locks are in place and equipment is isolated, there is a final field validation walk-down step to verify that the plan and instructions were implemented properly.

9.3.1 Managing Lockout/Tagout in a Plantwide Shutdown or Turnaround

In the electrical utility and chemical process industries there could literally be hundreds of isolation and lockout points for servicing equipment on a particular outage. Most organizations that do frequent large-scale isolations maintain a LOTO office stocked with fixtures, locks, and keys for just about any type of situation. In these cases there is also often tracking software in place (e.g., TagLink[1]) for managing what is locked out, by whom, and for how long. The rules of engagement for these types of processes usually include paper LOTO permits that contractors or work crews need to carry when they are performing work, which they are able to show in the case of an audit. These permits can be considered legal documents in case an incident occurs and could be produced as evidence in litigation.

9.3.2 Line-Breaking Permits

Many best-practice organizations have a special process for line-breaking (piping system opening) permits and gas piping repairs. They come into play whenever a line or pipe is to be opened for any reason. Opening a line can be removing a flange, removing a plug, or even opening a valve. It's not just cutting or unscrewing a section of pipe. Key considerations for line-breaking permits include the following:

- It is necessary to verify what's in the line and that it is safe to open it.
- Whether liquid can be in the line must be determined. If so, consider splash shields or containers to drain it. Will residual coatings on the inside be a problem in later cutting or welding operations? What will be done to mitigate fire hazards?
- The method of opening the line may be important, whether it is with a saw, torches, drills, or even just unbolting flanges or removing a cap or plug. Your site and all users must define what actually constitutes a line-break event for your operations.
- Validation of zero energy (especially with fuels) may require a need to purge with nitrogen to remove flammable materials. There should be a comprehensive protocol that complies with NFPA 56 in place for purging and reenergizing gas lines.
- Start and end times must be specified (permit clearing process and timing may need to be addressed).
- The permit form should include a requirement to have sketches and a written description of the work, along with the name of someone who would validate the process and provide field verification that the plan was implemented properly.

9.3.3 Hot-Work Permits

Another special permit that is usually a part of line breaks is for hot-work areas. Hot-work permits exist for areas where flammable materials are present and where the hot-work is a potential ignition source. Hot-work permits are most common where welding or torch cutting or an ignition source may be intrinsically at the scene of the line break. There is a lot of good information about hot-work permit processes at www.osha.gov and at www.csb.gov. Important features of hot-work permits usually include the following:

- A trained fire watch person is assigned to stand with an appropriate fire extinguisher and watch for something that might smolder or catch fire.
- The fire watch person may have to attend the site during all breaks and lunches.
- The fire watch person may need to attend the site for some minimum amount of time after a job is over for the day.
- A discussion would be held prior to the work regarding the specific hazards and the plan for reporting a fire were one to occur.

- A discussion would be held about protecting certain equipment with fire-resistant blankets if needed. For example, work over or near a cable tray might involve measures to protect the cables from molten metal (e.g., weld spatter dripping onto conductors).

9.4 COMMANDMENT III

Perform interlock and safety testing of fuel trains and devices (including valve tightness testing) at least annually. The third commandment is that the organization must comply with nationally recognized codes and standards that call for combustion equipment safety devices to be tested and inspected at least annually. You may think that if equipment has been running successfully, someone must have been checking and testing it. Certainly, in the case of boilers, the state or insurance provider must be doing checks if they're required—right? Not quite.

Remember, many states and provinces have boiler laws that require annual jurisdictional inspections. These are usually only water-side pressure vessel inspections by persons authorized to do such inspections. They are usually not checks of safety devices on fuel trains or those related to combustion systems. In a limited number of states and municipalities that have adopted the ASME CSD-1 code, annual safety interlock testing may be verified by the inspector; however, this is only on boilers up to 12.5 million Btu per hour. Even in these circumstances, what passes as evidence that a complete and thorough testing job has been done is inconsistent. There is no uniform enforcement of this code; it's pretty much up to the individual inspector asking questions or asking to see testing records.

For ovens, furnaces, thermal oxidizers, and all combustion equipment other than boilers, state or provincial inspections are usually not required, although inspections may be required by local jurisdictions. In these cases, code-required inspections and testing of safety devices are also the responsibility of owners.

No one will come directly to your site and force you to carry out safety interlock and valve tightness testing. In some cases an insurer can threaten to drop your coverage as motivation. Only your desire to keep people safe, combined with the fear of having an incident and not having records to confirm that you were in compliance with nationally recognized standards, can motivate you to do so.

If you start an inspection and testing program, you must also be prepared to handle the reporting and managing of defects. This implies having a plan to follow for reporting defects that includes how you will report and track them, who gets notified, how they confirm the notification, and how you will handle getting them repaired safely. There's a lot of liability in knowing that you have a failed component, not making an immediate repair, and putting everyone at risk. You need to make sure that there are clear lines of communication and a plan for servicing failed components.

Don't let your corporate legal counsel tell you that it's better not to document findings. I have worked closely with one of the country's top plaintiff's attorneys, Robert Reardon,[2] considered one of America's "super lawyers." He was a lead attorney in the Rhode Island night club fire case in 2003, when 100 people were killed and on the 2010 Kleen Energy explosion case, where six people were killed at a Connecticut power plant under construction. He has indicated that it's always better to be aware of your problems and be working on them than to have no program or to claim ignorance. Ignorance of the law is no defense. In fact, having no program and no ongoing effort to create one makes you and your organization vulnerable to intentional torts and various forms of negligence—and even criminal prosecution.

9.4.1 Frequency of Testing

Most codes and standards have language regarding the frequency of recommended testing that is similar to that noted below for NFPA 86. This is what serves as the basis for requiring at least annual testing and inspection of this equipment.

NFPA 86, the Standard for Ovens and Furnaces, in Section 7.4, Inspection, Testing, and Maintenance, states:

> **7.4.3** It shall be the responsibility of the user to establish, schedule, and enforce the frequency and extent of the inspection, testing, and maintenance program, as well as the corrective action to be taken.
>
> **7.4.4** All safety interlocks shall be tested for function at least annually.
>
> **7.4.5** The set point of temperature, pressure, or flow devices used as safety interlocks shall be verified at least annually.
>
> **7.4.9** Valve seat leakage testing of safety shut off valves and valve proving systems shall be performed in accordance with the manufacturer's instructions.
>
> **7.4.9.1** Testing frequency shall be at least annually.

Remember that codes provide minimum standards. Almost all codes indicate that interlock and trip testing of safety devices and valve tightness testing should be done at regular intervals, and most indicate that this should be at least an annual process. The most successful organizations do this more frequently. Everyone wonders what the optimal timing is for these types of activities. There has been electrical component failure rate data from NASA and other organizations that have been used in studies to determine what frequencies work best. One such study evaluating this topic, published in 2001,[3] led the sponsoring company to adopt a testing period of three to six months instead of on an annual basis. What seems to work best is to start off with an annual program [unless more frequent testing has been identified as beneficial by a hazard and operability study (HAZOP), operating data, or standards and codes]. Analyze failure data and identify the items that have the highest failure rate. These may be candidates for more frequent testing.

For example, it has been my experience that airflow proving switches have the highest failure rate among safety interlocks, followed by high gas pressure switches. Air proving switches are susceptible to damage and premature failure because they are exposed to a regular flow of combustion air that can have any number of contaminants. High gas pressure switches, on the other hand, are normally closed—they come out of the box with the contacts made. If no testing is ever done, there's never a reason for the contacts to be exercised; hence, they tend to fail closed. Because these devices seem to fail more frequently, many organizations test them more often than annually.

Many organizations with somewhat complex equipment, which might include multiple-burner systems, multiple fuels, or atmosphere gases, have reviewed testing scope and frequency in the context of a HAZOP review. There are lots of ways to do this and many books on the subject. The best overall references I have found are the ANSI risk management standards package (Z690.1, Z690.2, and Z690.3-2011), available for purchase at www.ansi.org. I have found a simple starter process to be a tool published by Zurich Insurance called the Zurich Hazard Analysis (www.zurichna.com) process. All of these tools seem to get back to the same thing: You identify things that can go wrong and then you and your team think through the probability of an event and its severity. The probability and severity would be broadly categorized by the team in some agreed upon format. It's best if you have a cross-functional experienced team taking part in this so that you get a lot of different viewpoints. The end result of a successful HAZOP is to have identified an overall level of things that you need to implement for an acceptable level of risk. This could be through things that reduce probability or severity or both. The frequency of testing safety devices is obviously a layer of protection that can be examined for reducing the probability of a failure.

9.4.2 Defining Priorities and Managing Defects

Let's assume that you understand what is required to put together an equipment evaluation and testing program. You have the scope and the frequency well understood and your inspection and testing protocols consider the following elements:

1. Maintenance practices
2. Operational issues that can affect safety
3. New technologies that can enhance safety
4. Equipment installation issues
5. Code compliance issues (gap analysis items)

There are still a number of things that you will need to address and have in place to make the program successful: defining priorities, developing testing procedures, and having methods/protocols in place to document and manage the findings.

Defining Priorities Let's face it; all findings (defects) are not equal. Priorities can be ranked as A, B, and C levels or given names such as *urgent*, *managed*, and *good practices*. Some inspection and test findings may put someone's life in immediate danger. In the case of these, everyone needs to stop at once as soon as the finding is made and it needs to be addressed immediately. Examples of these could be water-level controls on boilers that have failed but still allow boilers to continue to operate, or airflow proving switches that are stuck closed with contacts made allowing equipment to run but no longer be able to detect a low-airflow condition. There is no situation where equipment with failed safety devices should continue to operate. This will have to be part of a testing program's standard operating procedure and needs to be communicated to all involved. This means that testing might have to take place during off-hours or when equipment availability is not critical so it can be down when repairs are made.

Other findings can be managed carefully and corrected over time. Still others may be nice to do someday but possibly over an extended period of time. These lower-priority issues move the needle toward safety but perhaps in an indirect manner.

Developing Testing Procedures Detailed testing procedures need to exist for all combustion equipment and components. These should include pictures, schematics, and the most basic of steps. Starting equipment conditions and safety protocols should also be included. Best-practice organizations have used single-point lesson format teaching tools for each separate component procedure. Without written testing procedures there can be no hope of consistency. These procedures should include criteria for what a good device is and what a bad device is, and why. Procedures should also include such things as tools required, and comments about documentation and managing defects found.

Managing Defects Found There are several key needs regarding documentation. These include knowing that the equipment testing has been completed according to the scheduled and planned frequency and identifying that all of the most severe (urgent) findings were made known immediately to someone on the leadership team when found at the site and repaired.

Remember, equipment generally has to be out of service while this type of testing and inspection is taking place. It's a great idea to have a meeting with all stakeholders planned on the day of the start of the work. It's then also important to repeat this meeting as a close-out process when the testing is completed so that all stakeholders are completely informed of the results and possible ongoing needs.

Part of the report-out system should include field data sheets filled out and signed by the technicians who did the work and then also signed by a supervising manager who reviewed the results. This is to create accountability for two issues: incomplete improper inspections and urgent issues not being repaired. The paper trail accountability approach can also help to identify who is responsible for safety defects found that are deemed to be manageable and not corrected that day. Once problems are found, they can create liability concerns if they are mismanaged.

It has already been stated that there will be some items that, if found failed, will require the equipment to be shut down immediately for repair. There might be temporary work-arounds for some items, but this is rare. In most cases the equipment will just need to be down and locked out, production be damned. This is, of course, some of what needs to be communicated before testing starts. It also means that its especially important to plan on having spare devices on hand.

All defects will need to be managed carefully, including the B and C priority items. It's very possible that items that are today a B-level priority can in short order become urgent A-level items. For example, a hot spot on the skin of a furnace could be burning off paint and initially be considered a B-level priority. This is important, of course, but perhaps not urgent. If this is ignored completely over time, it can continue to degrade and become an urgent matter. No one knows how long this might take, so follow-up actions might include regular monitoring of such conditions. Level B and C items need to be well documented, managed, and scheduled for repair and not ignored.

9.5 COMMANDMENT IV

Review all fuel and combustion systems for code changes and upgrade continuously to stay current. The fourth commandment is that the organization must work toward keeping its equipment's level of safety compliant with the most current codes. An organization must decide on an overall global basis, as a policy, that it will continue to maintain a level of safety of its equipment that continues to advance and evolve. This happens by default with cars, for example, because they age and get replaced on a frequent basis. New technology, such as antilock brakes, comes with the new vehicle. In the case of combustion equipment, the replacement cycle is not very frequent. The improvements are also not as easy to see and understand. Therefore, for some reason many organizations seem happy to say "We're grandfathered," meaning, in essence, "We met the code when the equipment was installed and we don't have to change a thing."

But just because you don't have to enhance the equipment's level of safety doesn't make it right. The best approach is one where a policy is made which says that upgrading and staying current with codes and standards is the right thing. Remember, NFPA codes and standards related to fuels and combustion change every three years. The changes reflect new technologies and safety issues that are usually an attempt to avoid defects, failures, and in some cases, catastrophes that have occurred. Deciding to stay current means that an evaluation is made of what changes occurred in relevant codes, and a comparison is made with the current equipment (gap analysis). Best-practice organizations that conduct "gap analysis" processes understand their overall level of code and standard compliance on an enterprise basis.

The gap is the difference between the state of the existing equipment and the latest versions of the relevant codes or standards. In doing this, a judgment as to the level of risk abatement that can occur given the specific equipment and its use needs to take place. This must be followed up by what the financial impact is for compliance, timing for repairs, and even training and management of change issues related to the upgrades proposed. The findings can be entered in a database, sorted, prioritized, and used to plan capital allocations. Maybe there's a flexible three- to five-year time horizon for compliance. The important thing is that a compliance target exists at all.

9.6 COMMANDMENT V

Maintain a best-practices and lessons-learned program for fuel and combustion systems issues to include root-cause analysis. The fifth commandment is that the organization is to develop best practices and lessons-learned communications and then disseminate them widely throughout the organization. Best practices and lessons learned are ways to develop corporate memory for both good and bad practices that are discovered over time. Best practices would be things discovered specific to your operations related to fuel and combustion systems that can be documented and passed on for safety, reliability, reduced emissions, quality improvement, or cost savings. This is typically something one particular plant or site does very well that others need to consider. The idea is to document this, provide internal company subject-matter experts for reference, and then challenge the others to get to the same level.

The most successful programs include corporate fuels and combustion-related websites that have best-practices and lessons-learned content. These websites usually also contain testing and inspection checklists, procedures, statistical information on findings, and even online Web-based training content.

It's also vital that the organization conduct root-cause analyses of all incidents that occur and also report and tracks near-misses. These well-documented incident studies are the heart and soul of "lessons learned." Some organizations simply make repairs after an incident and move on. They do not have a specific policy or practice for identifying root causes, and if they do, they do not communicate the results very effectively to a wide audience so that learning can take place. In fact, many don't understand the concept of what a root-cause analysis actually is. In the fuel and combustion system world there really are no minor incidents. A small poof or delayed ignition one day could be a major explosion the next day.

What does it mean to do a root-cause analysis? There are many definitions and ideas about what this is. There are formal processes, and good reference books are available. The idea is to go through a formal process to find out what happened to cause the near-miss. It usually starts with laying out facts and then moves on to identifying issues that lead to the problem, things that made or could have made the incident worse, and things that could have prevented the incident or reduced the probability of it occurring.

One good process is called TapRoot (see www.taproot.com). This is a very thorough process that requires training by TapRoot. There is also a good reference book on the subject called *The Root Cause Analysis Handbook*.[4] Another excellent resource is NFPA 921; the guidelines for fire and explosion investigations.[5] You should review this document before any investigation, as it provides guidelines and recommendations for the safe and systematic investigation of fire and explosion incidents. In any case, document a process you want all to follow, make sure that it gets published, and show examples. Have a company champion for this area who has had some training or experience to help.

9.7 COMMANDMENT VI

Develop detailed startup and shutdown procedures for all combustion equipment to include a prestart walk-around. The sixth commandment is to be sure that the site has a specific startup and shutdown procedure for all its combustion equipment. These procedures should include the very important prestart walk-around. On the surface, perhaps to those who do not regularly work around fuel systems and combustion equipment, it seems that this is adding complexity for no reason. One might think: "OK, so what do you do besides hitting a start button?"

Some of the reasons why detailed startup/shutdown procedures are important is that things don't always go right when combustion equipment is started, and the consequence could be catastrophic. Sometimes the equipment does not start at all. Sometimes the flame is not correct, or the temperature won't come up, or a strange noise occurred that is a precursor to an explosion.

You also have no idea who will be starting the equipment. What is their level of training and state of mind? Are they new? Are they tired or stressed? People that may not at that moment be at their best could benefit from having things spelled out directly in front of them.

Many types of equipment get started in manual mode under low-fire conditions. Once things are verified to be correct, the equipment is put into automatic control. The operator needs to understand specifically when it is acceptable to do this and how to do it.

The best startup/shutdown procedures I have seen come in spreadsheet format and are customizable. These have pictures of control panels, fuel trains, and other important parts of the equipment. They also have provisions for identifying key safety issues for every step of the process and even what PPE (personal protective equipment) may be required by those conducting the startup for each step. Once you have a format, it can be used over and over to establish consistency throughout the organization.

9.7.1 Prestart Walk-around

Startup and shutdown procedures are specific to each piece of combustion equipment. They differ among styles of boilers and types of furnaces and ovens. However, prestart walk-around protocols can be common to all types of combustion equipment. The items that follow have been developed from years of reviewing thousands of pieces of equipment. They do not include everything that might enhance safety in your specific situation, but they will certainly help. You can always add to the list of items to be considered.

It is good practice to perform a walk-around prior to startup with a clear focused protocol to guide you. This should be done especially after a unit has been down due to maintenance, repairs, or on extended downtime. This walk-around should attempt to identify any deviations from the normal condition of the equipment. In this section we provide a basis for this protocol. *Note*: The items below are not specific to any particular piece of fuel-fired equipment but should be used as a guideline only. Equipment-specific walk-around protocols/checklists should be generated at your site for each unique piece of equipment and even every fuel system.

Elements of a Generic Prestart Walk-Down Checklist

1. *Reviewing for bypassed safety devices (jumpers).* No one can understand the condition of equipment safety devices by opening a control panel and looking in. However, you might be able to see some unusual things in the control panel that encourage questions to be asked. The objective of a visual review of wiring would be to look for obvious evidence of electrical jumpers at

safety devices. Sometimes you don't have to be an electrician to identify suspicious wires.

There are a few simple things that you have to consider before you open an electrical panel (even a control panel). First, there are lockout and arc flash protection protocols to consider. I will not discuss these in detail, but you should make sure that your facility is in compliance with these (see NFPA 70, the National Electrical Code, for more information about arc flash protection and requirements). Special PPE, such as safety glasses or a face shield and flame-retardant clothing, may be required before your panel can be opened. An old trick that many electricians do when opening panels is a good thing to pass on here. Many old timers turn immediately as a panel door starts to open. This is in case the opening of the door triggers an event such as a wire coming loose and short-circuiting. This should at least help to protect your eyes from the arc flash explosion that could result.

Once the control panel door is open, look for things like wires that are a different color than the other wires, wires that appear to be attached temporarily using alligator clips, wires that are of a substantially different gage than the others, and wires that are not neatly within the wire trays and are obviously just lying in the box. Even though most of the wires are terminated with crimped connections so that they fit neatly under a screwhead, what if there are also wires where the ends of the insulation were just stripped and the bare wire shoved under a screwhead? If you find any of these conditions, before proceeding you may want to stop and have an experienced electrician review conditions with an accurate wiring diagram to make sure that nothing is bypassed and all is well.

Besides the control panel, make sure to look over all the field devices carefully as well. I have seen both leads of a switch terminated together at the switch, cardboard jammed into relays to keep them open, and even a small stick of wood jammed into an air switch to keep it made regardless of the airflow.

2. *Checking for correct set points.* Temperature and pressure set points exist for many fuel train safety devices. These provide protection only if they are set correctly. Set points should be derived during commissioning and be well documented. These should then also be marked on switches with permanent markers or spots of paint. If you cannot find any of this documentation, it will have to be re-created through recommissioning by a qualified service contractor.

   If you find devices that are obviously set wrong, have someone investigate and make the settings correct before moving on. One such easy-to-understand identification of incorrect set points is if they are at one extreme end of the component scale (either at zero or at the maximum setting). It's never a good design practice to have a component that has to be set this way to operate.

   I once saw a 30,000,000-Btu per hour steam boiler that required 23 inches w. c. of natural gas pressure at high-fire. The high gas pressure switch was found to be set at 7 pounds or 194 inches w. c. Obviously, this high gas pressure switch was useless in this condition since by the time the gas pressure approached something more than eight times the high-fire requirement, the flame would probably have blown out. Sure, the flame detector should pick this up and shut things down safely, but one of our layers of protection would be gone without the high gas pressure switch being able to function as intended.

3. *Reviewing valves.* Another important startup walk-down issue includes reviewing each of the automatic gas shutoff valves to make sure that they are indeed closed prior to light-off. These are usually required to have external position indication. Make sure that you understand what the open and closed positions look like. It may be as simple as a small window through which you look in at the actuator that says OPEN/CLOSED. Or it might be a small LED that is lit. Make sure that these valves are closed prior to light-off.
4. *Reviewing main manual shutoff valves.* Remember that all manual shutoff valves for fuel trains should have handles and should be accessible. Try to close the main manual shutoff valve to make sure that it is operable. You should also remember that lubricated plug valves and many others have the means on the valve stem to indicate whether or not they are open or closed. Never just trust a valve handle position. If a manual valve cannot be closed, you could be at risk of not being able to isolate the fuel train should you have some kind of leak or problem once the system is started.
5. *Reviewing valve actuators.* Remember that automatic valve actuators can sometimes give you clues about their condition. This is especially true with hydraulic actuators. Remember that the top half of some hydraulic actuators is a fluid reservoir. Once this fluid begins to leak, a potentially dangerous scenario can take place. These leaks never get better; they can only get worse. Enough fluid can leave the system so that the actuator can only push the valve body to a partially opened position. This can restrict flow in a way that can't be detected because of the gas valve's position relative to the low gas pressure switch. Once enough fluid has left the reservoir the actuator can no longer push the gas valve open. The equipment will not light or function at this point. This then becomes more of a reliability problem than a safety problem. You should report and replace failing hydraulic actuators immediately.

   I also like to grab these actuators and see if I can wiggle them to verify that they are attached securely. I sometimes find the actuator holding screws not tightened firmly. If you can wiggle the actuator with respect to the body of the valve, you have a problem and need to tighten the setscrews holding the actuator to the valve body.
6. *Reviewing instrument tubing.* Instrument tubing, including that used for sensing lines and transmitter lines, should be reviewed carefully. See if there are any obvious kinks, and make sure that they are connected firmly on both ends if possible. Grab the connection fittings and make sure that they are at least hand-tight on the switches or devices.

Also, remember that valves in sensing lines for safety devices need careful attention. In general, the best practice is to avoid shutoff valves in safety device sensing lines. Many codes and standards do not allow valves in sensing lines—and for good reason. Process industry equipment sometimes has valves in transmitter lines for calibration and testing. If valves exist, make sure that they are open. Consider implementing a program to lock, seal, or tag the shut off valves in the open position and formally manage access to and use of these shutoff valves.

7. *Checking linkage connections.* Loose or pulled-off fuel, air, or valve linkages can also be a dangerous issue. Linkages can also become corroded and lead to the need for excessive friction and force to get things moved. This will surely break linkages or actuators over time. It may not seem that important, but believe me when I say that a regular shot of oil to linkage swivels and other moving linkage and exposed actuator parts can make a big difference. Look closely at linkages during startup to see if they are binding or really loose. Grab them; give them a wiggle to verify that they are firmly connected. Are any of the holding clamps or brackets loose? Do match marks still line up?
8. *Reviewing explosion reliefs and/or latches.* Process equipment such as ovens and furnaces are supposed to be designed to either contain an explosion or to allow it to dissipate through explosion relief panels or doors. Many ovens accomplish this with special calibrated friction latches on these doors. Prior to startup, make sure that the doors can open freely and that the latches are not bound. Manufacturers of such doors call for regular lubrication of the latches and periodic checks of the opening force required.

   There are also supposed to be labels on explosion relief panels, along with tethering to restrict movement. Make sure that no one has restricted the movement of these panels with conduits or things stacked against them.
9. *Reviewing purge air availability.* It's important that the intakes of combustion air or purge fans not be restricted. This means that filters or screens, if they exist, are in place, unobstructed, and clean. If airflow dampers are in place, care should be taken that they are set correctly (mark placement ahead of time). Remember, many codes do not allow dampers in air systems that can be shut off to the point where they compromise the purge airflow in the closed position (see NFPA 86).

   A thing to remember in cold climates is to verify that ice or snow has not accumulated to restrict air intakes. I have seen this be a problem with combustion air intakes to boiler rooms that were low along the ground as well as during icing weather at process plants where air intakes were high in the air and subject to direct precipitation.
10. *Reviewing air system ductwork.* In some cases, ovens, furnaces, and boilers have combustion air and flue piping duct systems Walk these systems and pay close attention to flex connections, damper positions, and inserted data or control elements (i.e., thermocouples, tubing, etc.). Are there any holes or tears in the fabric connections? Are fasteners or bands still in place securely? Is the

ductwork crushed, restricted, or compromised in any way? Are access panels securely in place and doors closed?

11. *Reviewing fans.* Are fans free of excessive vibration? Are belts on? Are belt guards on? Vibration is always an indicator of trouble with rotating equipment. You should gain an understanding of the level of vibration that's considered normal.
12. *Checking vent terminations once operational.* If the fuel train vent terminations are accessible, verify that gas is not flowing from them prior to and immediately after startup or during operations. Flow could be an indication of a regulator or switch diaphragm rupture. If the bleed vent line between the safety shutoff valve and blocking valve is flowing, it could be that the bleed valve (if it exists) is leaking through in the closed position. You should err on the side of safety and shut the unit down if such conditions are found, verify what the issue is, and make the needed repairs. Remember to use extreme caution in approaching vents that may be relieving gas.

Real-Life Story 37: Failed Instrument Tubing Could Have Led to Catastrophic Results

A chemical plant was starting up a process heater that had natural-draft firing. It is commonplace in such a situation for the fuel and air controls to be put under manual control while the unit is heating up. Many fired heaters are lighted manually, and then an operator watches everything carefully, including the draft, airflow, and fuel flow. Once combustion conditions have been stabilized and the equipment is warmed up, it is common for systems to be put into automatic control. In this case, the operator did not notice that the oxygen was not reading correctly. He saw the temperatures coming up, thought all was well, and put the unit into automatic mode. This meant that instead of the fuel and air controls being set where he wanted them, a microprocessor based controller (PLC) would take over and reposition things according to the information collected from field data instruments.

Generally, two styles of flue gas instruments are used for to measure oxygen and control fuel/air ratios: an in situ unit and a sampling unit. In situ units have the oxygen sensor itself sitting in the flue gas stream. Sampling units have a pump that draws out a sample and passes it by an oxygen sensor that is sitting somewhere in a box outside the flue gas stream. In this case, a sampling unit was being used. Unfortunately, its sample line suction tubing was compromised. Hence, it sucked in pure air that had about the normal 20.8% oxygen (normal atmospheric conditions). Once the controller was in automatic and read 20.8% oxygen, it started dumping fuel into the burner far out of proportion to the air because it thought it needed to get to about 6%. This older-style unit did not have cross-tie limiting controls that would have noticed that the fuel and the air were way outside safe ranges. The operator and many other personnel were startled by black smoke pouring out of the unit. The operator immediately shut the equipment down and avoided an incident.

Lessons Learned Prestart walks-downs of equipment are very important. If this had been done and there was a protocol for reviewing instrument tubing, this problem might have been caught prior to it becoming an incident. This story should remind you of some of the benefits of upgrading equipment and having HAZOP processes that explore vulnerabilities and create countermeasures to avoid issues. If this had been done, the need for cross-tie limiting might have been identified. Even if cross-tie limiting exists, there must be safe ways to test instruments and verify that the commissioning was correct and that these systems continue to function correctly over time.

9.8 COMMANDMENT VII

Implement regular training (knowledge and skills) with validation processes. The seventh commandment suggests that your organization have processes to train staff and document the training. The need to fix people through training is the most important part of fuels and combustion system risk mitigation. Well-trained people can safely operate equipment that is less than desirable, but the opposite is not true. So with that in mind, you need specific policies in place to make sure that training is achieved clearly and consistently for all personnel who interface with fuels and combustion systems in any way. If you think about it, this should mean at least some type of short video or presentation for all employees. Remember, codes such as those of NFPA 86 and ASME identify what is considered training, and these also provide some direction for curriculum and training objectives.

9.8.1 Creating the Training Program

In putting together a training program, basic things that need to be decided include curricula and hours per employee. New employees should also be screened for their level of competence when it comes to operating or maintaining fuel systems or combustion equipment. I have done this by creating written tests so that baseline levels of knowledge could be established. This helps you provide only the training needed instead of making everyone sit through hours or days of instruction they don't need. Today's world also contains great online tools and platforms (e.g., www.exambuilder.com). I have used this extensively as a tool to facilitate consistent testing at multiple locations throughout an enterprise.

Creating curriculums can be done by interviewing current and former operators, engineers, and maintenance staffs to get an understanding of all the knowledge and tasks required for the particular fuel or combustion system or job function. You can then take all of these topics and tasks and create a matrix to weigh their importance among themselves and define the number of hours that each might require. Course material may be commercially available for some of these. In other cases, vendors of components or equipment might have materials. In most cases there will also be a need to create some custom content.

You should also consider becoming familiar with the learning taxonomies that fit best. Bloom's taxonomy[6] is a classification of learning objectives published in 1956. It provides insights into what learning systems work best for different objectives. For example, there may be information that people need to memorize. That is one type of learning, but using some of this information to solve problems is another. You must consider these types of things when preparing training course materials for fuels and combustion systems so that the end result, a safe working environment, is really the outcome of the training effort.

There are already tools for boiler operators to meet certain minimum levels of competence. There are certifications offered by organizations such as the National Institute for the Uniform Licensing of Power Engineers[7] and the American Society of Power Engineers.[8] Consideration should also be given to what supervisors and managers need to know to direct these operators. Some of these programs provide guidance on this as well. I am not aware of parallel already established programs for ovens and furnaces.

Many organizations do some element of their training online with Web-based approaches. Web-based training is great for general awareness-level training, but it cannot be relied on as a complete training program for many somewhat complex concepts or for the transfer of skills. Hands-on training specific to your equipment will also be needed. A comprehensive training program should include combustion and fuel system basics, equipment-specific issues, and mock hazard drills. Startup/shutdown procedures should also be taught to staff and drilled regularly.

Training must be tracked and managed. Many organizations manage training with learning management systems. These are online Web-based databases where online training is logged automatically. If there's ever an incident, you can bet that OSHA and plaintiff's counsel will request copies of training logs immediately.

Real-Life Story 38: The Inexactness of a Detailed Procedure

Misinterpreting what seemed like simple detailed procedures for changing strainers led to an explosion at a Midwestern hospital. No one was injured. There was $60,000 in repairs for re-skinning the boiler and making repairs. It could have been much worse. The hospital's boiler system included four boilers each capable of about 20,000 pounds per hour of steam. Each boiler's primary fuel was natural gas, with No. 2 fuel oil as backup.

As is often the case, there was a warning incident, in this case a "poof incident," followed the next day by the explosion. Poofs are considered by some not to be a real explosion, but just a minor excursion of air/fuel ratios followed by an accelerated burn. Some people who don't like to use the term *explosion* just say that there was an "event." The same operator was on duty during both events. When asked what he was doing when the incident occurred, he mentioned that he was changing over the strainers. The plant had a very detailed meticulously documented procedure in place for online changeover of fuel oil strainers (two strainers in parallel, switching the dirty one to the clean one). Every fuel oil valve was tagged with a brass tag. A chart on the wall of the boiler room next to the strainers explained step by step how to switch over and what the valve lineups should be. He even took me over and showed me. He said: "See, its simple. You first close both supplies to the strainer you want to change and then you open the valves to the strainer you want to use."

It was clear when he showed me this that he had interrupted the fuel flow to the burner and then reestablished it. If they are sighted correctly, most optical flame detectors must see no flame for 3 continuous seconds before the BMS indicates a flame failure and closes fuel valves. In this case fuel was probably interrupted for just about these 3 seconds. This was long enough for the flame to extinguish. It is likely that unburned fuel flashed off inside the boiler and that vapors then accumulated and "poofed" or exploded. This was probably also the cause of the poof the day before. This man had been an operator for only a few weeks. It's likely that other operators understood what they could and could not do despite the exact wording in the procedure.

Lessons Learned One of the lessons learned is that training needs to include the principles behind the words. It also speaks to validating procedures, done as a walk-through, together with every person around at least once. If you interrupt fuel flow, even for a few seconds, you disrupt burner operations. The low fuel pressure switch or the flame detector may stop firing if they are set right and working correctly. If not, there could be an explosion similar to this one.

Training must include the transfer of both knowledge and skills. The skill parts are things that people must do with their hands—in this case, cleaning strainers. Make sure that the transfer of hands-on skills is part of your training curriculum and that there are ways to validate these, such as shadowing, evidence books, the use of simulators, and/or mock drills.

Whenever fuel oil is burned, strainer maintenance is very important. The strainers require regular cleaning, and most facilities use a duplex system, so that when one side is clogged, a parallel, clean unit can be switched into service. Make sure that there are proper procedures for this process that do not involve interrupting fuel flow.

Fuel oil needs to be sampled and evaluated if it is held in long-term storage to verify that it is not contaminated and is ready to burn. Also, most sites conduct practice burns of alternative fuels on a regular basis to validate operator skills and proper equipment functioning. Remember also that safety devices on backup fuels, such as fuel oil systems, need to have the same regular code-required testing as well as valve seat leakage testing even if the devices are not in regular use.

9.9 COMMANDMENT VIII

Create a system to screen and validate fuel and combustion equipment suppliers and service vendors. The eighth commandment is about having methods in place to screen and validate the competence of vendors that provide you with fuel or combustion equipment services or products. Validating vendors for critical combustion system work is vital to maintaining ongoing safe facilities. In my career I have seen just as many catastrophes caused by vendors and contractors as by in-house personnel. This is somewhat stunning given the common belief that if they are vendors in this field, they should know more.

Thinking that you can outsource fuel and combustion equipment services completely, with no in-house knowledge or skill, is a mistake. If heat processing of any kind is important to your world, someone in the organization needs to be smart enough to look ahead, ask the right questions, and interpret what is going on. This is especially the case when it comes to working with purchasing personnel who want to accept the lowest bid. This is like buying surgery. Do you really want the lowest-priced surgeon?

A good practice is to start with a process for complete and detailed validation of combustion system–related vendors. These vendors could be for projects that might include routine tuning of burners, purging and then reenergizing fuel lines, verifying the function of the safety interlocks, troubleshooting combustion problems, and redesigning gas trains. For combustion equipment this needs to go far beyond what is considered normal vendor validation. If the low-price bidder has no experience with the combustion equipment in your plant, you may be paying for them to learn your equipment, so errors are more likely. Ask to see at least the following: (1) training records of personnel who would come to your site, (2) experience with systems exactly like yours, (3) years of experience of the personnel who would come to your site, (4) schools and workshops attended by the personnel coming to your site, and (5) references from satisfied customers. Be keenly aware of proprietary systems that you might have and the possible need for someone to be certified in a particular service. This can be a problem for some brands of burners and combustion control systems.

Real-Life Story 39: Near-Misses Screaming for Attention

I arrived at this site to find a mangled D type of water tube boiler. I was allowed to step under the caution tape to get closer. There was still debris all over the floor and windows blown out. Luckily, no one was hurt, but it was clear that this was many hundreds of thousands of dollars in damage. It all started because of a desire to operate more efficiently. Someone promised the site an increase in fuel efficiency if oxygen in the stack could be more tightly controlled. Oxygen trim systems have been around a long time, and many of them work very well. This one was, however, rotten from the start. This boiler had been experiencing frequent flameouts since the system was installed. In most of these cases the operators reported that gray (and sometimes black) smoke was coming out of the stack immediately before the boiler would shut down. The trim system was designed to measure oxygen in the stack continuously through a new sensor and tweak a damper and the fuel and air controls to optimize the fuel/air ratio at all firing-rate conditions. Everyone in management and associated with the site knew that there was a chronic problem here.

The system had not worked properly since it was installed. Service people had been out at least three times in the past month, all for the same reasons. Instead of focusing on the root cause, it was decided that an alarm and a shutdown of the boiler would be installed as an interim solution. If the carbon monoxide level in the stack hit a certain level, an audible horn would sound, if an even higher level were achieved an automatic shutdown would occur. Sure enough, soon after this, the boiler was setting off the alarm and the new safety system was shutting it down regularly. After one such shutdown, the boiler operator approached the boiler's control panel, hit the start button, turned his back, and started walking away. This man was about 6 feet away from the front of the boiler when a terrific explosion occurred.

You can understand what happened here by recalling the fire triangle. The fact that smoking had occurred out of the flue stack and that the alarm and shutdown had occurred were all evidence that fuel, in the form of carbon monoxide, unburned fuels, and other hydrocarbon derivatives, had surely existed in the firebox at the time of shutdown. We know that the ignition temperature of natural gas is about 1100 to 1200°F. We also know that the flame temperature of natural gas is near 3500°F. It's very likely that there were surfaces close to the burner that were well over 1200°F at the time of the shutdown and explosion. When the operator hit the start button, the boiler's combustion air/purge fan started. This brought in fresh air that diluted the firebox mixture to its UFL (upper flammable limit), completing the fire triangle and allowing ignition of the contents.

Lessons Learned If the earlier near-misses had been more fully investigated, the root cause could have been identified and the incident avoided. Obviously, after several issues with the equipment, someone should have been there to remove the faulty oxygen trim system from service and put the boiler back into its original safe condition. A significant number of the incidents I have investigated have been after equipment service or the installation of new components. Improper commissioning is a huge factor in combustion system incidents.

This is another story that reinforces the need to have a program to validate service vendors and their people. One of the important lessons to learn here is how to treat equipment when it has shut down after running fuel-rich. You can see that attempting a light-off right away can be the wrong thing to do. The act of introducing purge air could be disastrous if there is a fuel-rich condition upon shut down. In this case, the right things to do include getting people in the area evacuated and then avoiding any kind of startup until conditions have cooled in the firebox. It's not clear how long this could take, but it is most certainly hours, not minutes. This particular incident has made for many great mock hazard discussions: "OK, everyone, black smoke is pouring out of the stack and the equipment has shut down. What do we do next?"

9.10 COMMANDMENT IX

Conduct regular fuel/air ratio optimization and validation programs. The ninth commandment suggests that the organization have a policy for the regular periodic optimization and validation of fuel/air ratios on combustion equipment. These programs should include detailed procedures for when this is to happen (frequency) and how it is to happen. Any issue that interfaces with changing fuel/air ratios can put your facility at risk. Besides conducting this process on some regular schedule, it may also need to occur when repairs mean that someone has had to interface with these systems (e.g., replacing firing rate control systems or gas valve or actuator components).

Burner tuning (fuel/air ratio adjustment) is somewhat of an art. However, there is still room for a number of important well-defined steps that should be taken during this process to reduce risks and ensure the safety of all involved. The following are some simple points to consider when writing a procedure or specifications for this work.

1. *Define the purpose of the tuning (fuel/air ratio adjustments).* You don't just bring in a vendor and say: "There it is; tune it." There has to be a reason. It can be for efficiency enhancement, for meeting environmental requirements such as NO_x, or even for meeting equipment capacities such as processing rates or temperatures.

   If your site is under an EPA consent decree for maintaining a particular emission limit or if the site has continuous emissions monitoring, the contractor needs to know this. If you are found to be out of compliance, you may have to report yourself. Make sure that you review all environmental permit conditions before conducting any fuel/air ratio adjustments, and contact your corporate environmental staff as well to ask them if there are or could be any impact from what you are doing.
2. *Define the documentation and processes that will be required of the service provider.* Service providers should be documenting "as found" and "as left" conditions. The client should include information about what firing-rate steps are to be used as check points. It is typical for this to be 10 steps (incremental firing rates of 10%) from low fire to high fire. Make sure these levels are held for at least 5 minutes once reached before a reading is taken. It's also important to ask that readings be taken and verified for the firing rate increasing and decreasing to assure repeatability.

   Define the beginning state of the equipment (i.e., must have been running for at least 8 hours).

   One also needs to define what criteria are to be provided. The usual information is firing-rate position or load, carbon monoxide, oxygen, and oxides of nitrogen, NO_x levels, and flue gas temperatures. Contractors should also be documenting the type of instrument used and evidence of the last calibration.

   Contractors should provide evidence of what changes were made. This needs to include anything that was done to change PID loop settings. For programming changes to a PLC, site management needs to make sure that they have a copy of the program prior to any changes. The new setup should also be printed and copied and then sent off-site for storage. This is important because if problems were to arise, such as a compliance issue or an explosion, all of the changes can be documented.
3. *Identify the conditions that will be required of the equipment before testing can begin.* The Environmental Protection Agency has guidelines for stack testing for combustion equipment.[9] These guidelines were written such that one could be assured of accurate and repeatable results. The guidelines cover things such as the condition of the equipment prior to testing. In many cases, for example, it requires for having the equipment at normal operating conditions for a number

of hours before the test. Even though the fuel/air ratio setting that is likely to occur for your purposes is not for compliance, it makes sense to spell out conditions like this before testing can begin.

Real-Life Story 40: Troubling Tuning Job

I arrived at the site to find a boiler opened up completely, with tubes exposed in the front and rear. It was one of the largest Scotch marine fire tube boilers with which I had experience, with an input of 1200 BHP (boiler horsepower). At the time of the accident, I was told that the boiler exploded as a result of tuning problems. The boiler had access doors in the front and back to provide access to the tubes. In this case the right-side door (as you face the burner) was blown open. Fire tube boilers with these types of doors often have specially designed cast iron fixtures that act as bolted clamps to hold doors closed. In some cases, doors are also bolted with brass nuts. When force is applied to these fixtures or to the brass nuts, they are meant to shear. If they don't shear, they can break the door hinges and the door could fly off instead of opening, relieving pressure and then being restrained by hinges. Flying metal doors are never a good thing.

Interviews with the staff indicated that a boiler technician from an outside contractor was on site making fuel/air ratio adjustments on each of the plant's three similar boilers. He had done boiler No. 1 the day before, and he was now on boiler No. 2. He had left the boiler house to go to another part of the facility when it was reported that black smoke was billowing out of boiler No. 1. He returned and set the fuel/air ratio again. The fuel/air ratio was controlled by a small PLC that was proprietary to the fuel valve manufacturer. It came as a packaged fuel/air ratio and firing-rate control system. The technician proceeded to work on boiler No. 2 and went to lunch at about noon. It was reported about 12:20 P.M. that black smoke was again billowing out of boiler No. 1. The technician was notified and was walking out of the onsite cafeteria when the explosion happened.

Upon investigating, it was found that the boiler technician had failed to recognize that the root cause of the smoke was not some mystery computer glitch but a mechanical failure. It turned out that the air damper actuator system had broken and was no longer supplying air in proportion to the fuel being delivered. Previously, the air damper had been binding. This binding and the extra torque required caused the damper actuator to fail. In fact, when the damper system was moved by hand, there was obvious severe binding coming off low fire. A disassembly of the actuator revealed that a roll pin in the coupling from the actuator to the damper blades had sheared. A close-up review of the failure indicated that it started as a small crack that had been cycled for some time until it finally fatigued and broke. The roll pin failure meant that when the actuator turned a certain amount, the air dampers did not follow.

The reason for the binding was bent damper blades that did not move freely with respect to one another. I was told at the site that a can of lubricating spray was left in one of the damper air boxes when the boiler was taken out of service previously and then brought back into service over a year before. The can had exploded when the boiler heated up for the first time and the force had bent the air damper blades.

These blades were about 36 inches long and 4 inches wide. These were never really straightened out to move as freely as they did when they were new.

Lessons Learned It's important to check fuel/air ratios as firing rates are both increased and decreased. The flue gas composition should be consistent on the way up and the way down. Sometimes this technique reveals problems with dampers and actuators. There are so many things that can be done incorrectly when setting fuel/air ratios that it is really important to have detailed specifications for the way you want it done. Every site should have a comprehensive protocol for setting fuel/air ratios.

This incident led to a recommendation to ensure that fuel valves and air dampers be disconnected from their operators and checked to move freely at least once a year. It's also very important to make sure that all linkages and damper bearings are lubricated adequately. In this case, several of the damper bearings were heavily rusted. This did not help anything and created more stress on the roll pin.

9.11 COMMANDMENT X

Implement a management-of-change system that considers fuel and combustion systems. The tenth commandment asks that the organization have a management-of-change process in place for fuel and combustion systems. Management of change means a lot of things to a lot of people. In general, the concept is one where an organization consciously documents and validates incremental changes in the context of how an accumulation of these little things can affect the overall organization or major systems in ways that were not intended. An example in the combustion world could be adding exhaust fans to a boiler room. If the impact on the entire ventilation system is not considered and reviewed, the additional negative pressure created could cause naturally drafted boilers to back-draft, bringing flue products into the boiler room. This could be deadly under certain conditions.

Changes to fuel and combustion systems that are significant and need to be managed include at least the following:

- New boiler, oven, furnace, or thermal oxidizers are installed.
- Operating conditions of equipment are changed (i.e., operating temperatures or rates of production).
- Equipment is upgraded (e.g., new burner, temperature controls, or fuel/air ratio controls).
- Gas piping is added or changed.
- Gas pressure is changed in the facility.
- Building ventilation systems are changed.
- Electrical systems are upgraded in ways that might change voltages or grounding.
- New paint formulations are now being used on drying ovens.

Management of change (MOC) happens best when there is a process. Your organization should consider a careful review of NFPA 654, the Standard for the Prevention of Fire and Dust Explosions from the Manufacturing, Processing, and Handling of Combustible Particulate Solids. It requires that facilities implement MOC systems to prevent dust explosions by addressing the items listed below. These MOC items are also of concern to fuel and combustion systems in general.

1. *The technical basis for the proposed change.* The technical bases for proposed changes to combustion equipment need to be spelled out clearly. For example, it would be important to document and communicate the reasons that burners were being retrofit. Is it efficiency, firing-rate capacity, reliability, or emissions? An example of such a project would be the retrofitting of burners to get lower emissions. Many geographical areas are classified as "nonattainment" for NO_x and therefore have severe restrictions on how much NO_x can be emitted. Hence, the overall goal of reduced NO_x should be communicated clearly as well.

   In this particular case there are a number of choices for that technologies that could be used. Some of them could be technologies that are not compatible with the site's maintenance capabilities or local service contractors. This can affect the long-term reliability the equipment on which these burners are installed. This type of information allows all of those involved to contribute ideas from their own experiences that could make the project more likely to succeed.
2. *The safety and health implications.* Changes in fuels and combustion systems almost always have safety and health implications. For example, running a new gas line through an existing facility begs the question: What if there's a leak in the future? Is the new piping in any traffic areas? Where is the shutoff for this line?

   In the case of installing new combustion equipment, even a hot water heater, one then has to ask: What if flue gases enter the room? Is there an air handler that could spread these flue gases to other parts of the building?

   If there are unintended consequences to safety and health, the new process or equipment changes will quickly be abandoned. Staff members who look after worker safety and health will need to be part of the planning process. Also, the maintenance and production personnel who will look after the revised equipment should be part of the review process to assure their approval.
3. *Whether the change is temporary or permanent.* If equipment modifications are not permanent, the timetable for the temporary changes needs to be obvious to all. Short-term modifications have a way of becoming permanent, so all the management of the change process steps still need to be followed.
4. *Modifications of operating and maintenance procedures.* Have all modifications of operating and maintenance procedures been considered? Have those involved had a say? Has documentation been created and put into place well in advance of the new system startup so that from the first day, safety and reliability can be assured?

The staff coming back to equipment after an outage where changes have been made will need to have a new set of operating procedures right from the start. If there are, for example, new buttons to push or a different sequence of operation, or new displays, all concerned must have the updated reference materials, including startup/shutdown procedures. Maintenance personnel will need things such as wiring schematics that are accurate, copies of PLC code, and even burner commissioning data.

5. *Employee training requirements.* All of the updated information described in point 4 does no good if people do not know about them. Although all good engineers love to have a paper trail of information about new equipment, the front-line operators and maintenance people will also need to be trained. Best-practice organizations have operators from all shifts trained by the startup crew, including designers and engineers who can answer questions, to avoid later issues. Training should include mock hazard drills for upset conditions and plans to validate knowledge transfer.
6. *Authorization requirements for the change proposed.* This MOC policy should include requirements for special authorizations to make modifications to critical fuel and combustion systems. This would create a framework to make in-house experts aware of high-risk modifications that were proposed or might take place. It should be known, for example, that special approvals are required to change PID loop settings on a digital controller that changes fuel/air ratios.

   You might cringe at the thought of adding an additional layer of management to some aspects of the operations, but because of the significant safety issues that fuel and combustion systems involve, spelling these out will clearly be of value for mitigating risks. Best-practice organizations have processes where a dedicated team of in-house experts screens all significant combustion system retrofits or new systems through a checklist review process to include approvals by insurers and even third-party independent combustion experts.

NOTES AND REFERENCES

1. TagLink, www.lockout-tagout.com.
2. Robert Reardon, www.reardonlaw.com.
3. G.J. Kauffman, T.L. Oakey, T.R. Price, and R.W. Johnson, "Combustion Safeguards Test Intervals: Risk Study and Industry Survey," *Process Safety Progress*, Vol. 20, No. 4, December 2001, 257–267.
4. *The Root Cause Analysis Handbook*, ABS Consulting, Rothstein Associates, Inc., Brookfield, CT, 2005.
5. NFPA 921, Guidelines and Recommendations for the Safe and Systematic Investigation or Analysis of Fire and Explosion Incidents, National Fire Protection Association, Quincy, MA, 2011. www.nfpa.org/categoryList.asp?categoryID=667&URL=Publications/NFPA%20Journal%C2%AE/July%20/%20August%202004/Features&cookie%5Ftest=1.

6. Richard C. Overbaugh and Lynn Schultz, Bloom's taxonomy, Old Dominion University, ww2.odu.edu/educ/roverbau/Bloom/blooms_taxonomy.htm.
7. National Institute for the Uniform Licensing of Power Engineers, www.niulpe.org.
8. American Society of Power Engineers, www.asope.org.
9. U.S. EPA, Promulgated Test Methods, www.epa.gov/ttnemc01/promgate.html.

10

Controlling Combustion Risks: Equipment

Controlling Equipment Risks from the Start

It's intuitive that controlling equipment risks involves regular safety testing and maintenance of equipment. However, much of the safety and risk management of fuel-fired equipment needs to occur in the design and specification of equipment, along with its installation and commissioning. In this chapter we address these issues as well as ongoing safety device testing requirements.

Real-Life Story 41: Missing Death by Moments

The security video first showed just a flicker of light, then smoke, then a fireball, and finally, metal flying. What happened is that a security guard, in his twenteeth year working at a food related manufacturing plant, had just made rounds, checking the access doors of one of the thermal oxidizer control rooms to make sure that the room was locked. This was a large unit that sat outside. Ducts leading into it came from the processing plant that delivered fumes to be destroyed. The control room was a small unattended room where all the combustion controls, gas valves, regulators and so on, were housed to keep them out of the weather. This room exploded violently into a fireball just seconds after the guard checked the door and walked away. The flying metal and huge fireball would probably have severely injured or killed him just seconds earlier. Instead, no one was injured. The overall cost to the site in equipment damage and lost production was over $1 million.

The thermal oxidizer system was recently installed at this food related facility to get rid of fumes that were low in volatility but high in odors. I thought the smell around the

Fuel and Combustion Systems Safety: What You Don't Know Can Kill You!, First Edition. John R. Puskar.

plant was pleasant at first, but I guess if you lived around the corner and smelled it every day, you would not be happy. In this case, the company decided that the use of a thermal oxidizer with a natural gas injection system that included regenerative beds for heat recovery was the best fit. You might hear these types of systems referred to as regenerative thermal oxidizers (RTOs). The technology involves first burning the fumes and then swapping heat from one set of heat recovery beds to another so that even though you are destructing fumes at 1500°F, the exiting flue gases are in the 350°F range. The heat recovery beds looked like the back end of over-the-road semitrailers. They were box like structures filled with a special ceramic material stacked so that air can flow through. One bed heats while the other bed is in service transferring heat to combustion air.

These types of systems have been in widespread use for years. They work great when designed and maintained properly. However, there are lots of moving parts and lots of sophistication in some of them that requires special expertise and regular attention. In this case a number of large air valves had to swing to different positions hundreds of times a day for the heat recovery system to work properly. This system was also unique in that there was natural gas injection to the low-fuel-value fume streams. This meant precision timing for opening and closing of fuel valves and lots of finesse for safety to be achieved. This natural gas injection occurred intermittently timed with the large air valve opening and closings. The mixed stream of fumes and gas was routed to a small burner and then, finally, to the heat recovery regeneration beds. Many opportunities existed for things to go wrong, and they eventually did.

I arrived to find a peculiar site. The damage was limited to the control room, control panels, and fuel trains. Although the control room was heavily damaged, some aluminum-bodied fuel train valves were melted, and others were intact. Similarly, some control panels were destroyed, with others left intact. The burn patterns and evidence at the scene suggested that an explosion worked its way back from the natural gas injection point into one of the automatic gas train valves that cycled for injection. This burst one of the valve bodies and no doubt could also have moved the gas piping and caused some of it to shear. In my opinion, some combination of these events is what released considerable gas into this room, creating the larger explosion and subsequent fire.

Most investigations of this type hope to find something that appears to be the smoking gun and principal cause. Then there are usually things that might not have been a direct cause but made the situation worse and contributed to the overall level of damage. This was no different. The primary cause was a failure of a valve proving device, allowing the fuel valve to stay open, injecting gas when it should not have been. This older style unit did not have other layers of protection (found in later newer units) to stop this incident. There were also components installed in ways that did not meet NFPA 86 requirements. Upgrading these would have made the systems more robust but might not have prevented the incident. It was also clear that the site lacked knowledge and the maintenance expertise required for proper maintenance of some of the airflow transmitters and control components. There was no comprehensive understanding of what the critical safety devices were, let alone a program to regularly test them.

Lessons Learned It was clear that the maintenance team at the site did not completely understand all of the safety devices that existed on this unit, nor what needed to be tested and when. In this case it was clear that the large air valve position switches and airflow transmitters were also critical for safe operation and needed to be tested regularly.

It's your responsibility to understand the entire scope of safety device testing for the unique equipment that you have installed. The equipment manufacturer should be the first stop for this information if the company is still around. If not, a HAZOP evaluation by experienced and knowledgeable people can be priceless. Remember, the requirements can go far beyond basic fuel trains and combustion systems.

Management-of-change processes could have helped the site to understand that the new RTO installed had more sophistication than others installed at the site. RTOs in general are often much more complex than most boilers and process furnaces. For example, setting proper fuel/air ratios can be very different. There has to be enough air to handle the burner fuel and organic compounds that act like supplementary fuel so that these can be adequately destroyed. This is different from a boiler or process furnace, where all that must be provided is air for the burner fuel.

The arrival of any new equipment should include an analysis of what it will take for it to be safely maintained from the day it arrives. If your maintenance and engineering departments do not have the expertise to understand and maintain them, consider using a specialist engineering firm with combustion expertise to ensure reliability and safe operation or invest in more training for in house staff.

10.1 CONTROLLING EQUIPMENT RISKS

Now that we've looked at people and policies in earlier chapters, the final critical element of understanding and managing fuel systems and combustion equipment safety is making sure that equipment risks are controlled. The first step in controlling equipment risks is making sure that the equipment is specified and installed properly. Then a comprehensive preventive maintenance and safety device testing program needs to be in place. In this chapter we help you understand the importance of key installation issues, assessing immediate hazards that may exist, and providing insight into test methods for safety interlocks and fuel valve seat tightness.

Much of what is said here comes from firsthand knowledge of mistakes being made. This is an expensive but effective way to learn things that is not recommend. I and others have paid a lot for this education. Installation issues can be broken down into two major themes: scoping and commissioning.

10.1.1 Up-front Scoping Issues

Scoping can be related to buying new equipment or retrofitting old. Most of what follows applies to both new and old equipment. However, let's cover a few issues that deal specifically with scoping a retrofit project. Retrofits are a different matter. You are asking someone to try to fix what might be years of neglect and a history of changes to a piece of equipment. For example, in the case of a boiler that is getting a new low-NO_x burner, several issues can bring disappointment. Sometimes conditions inside the boiler itself may have changed over the years, such as the spacing of tubes between water tube passes, which can allow for leaked carbon monoxide–rich combustion products to the flue gas stream. This is not a fault of the burner, but in some cases it may not be discovered until the burner commissioning process. Whose fault is this? As a customer, how do you feel when you are told that the boiler still does not meet emission expectations after $250,000 has been spent, but it is not the vendor's fault?

What about a furnace or oven where the contractor starts to tear into that "rats' nest" of a control panel or field junction box you have, just to find that things are not as they seemed? What happens now? Is it fair for the contractor to be responsible for 25 years of someone else's sins and problems? It has been my experience that where customers pay for an up-front study and design work, the outcomes are much more predictable and everyone is happier. The results of the study could be that the decision to retrofit is abandoned. If so, consider it money well spent. Also, please remember that even with an up-front scoping or design engineering project, there will no doubt still be defects that are hidden and waiting for the equipment to be disassembled and out of service or, worse yet, started up to be discovered. It's just that there will be fewer of them.

When buying new equipment it's all about deciding the big picture issues, such as what combustion equipment you should buy, what size or capacity it should be, and then deciding how and where it will it be installed. These are just a few questions to ask before a firm order is placed for equipment or installation contracts are let. Project scoping issues related to the up-front installation of combustion systems include the following.

1. *What Btu input capacity is really required*? Who has determined the input capacity? What were the assumptions made? How much can the needs grow without a major retrofit to the system? Is the capacity in alignment with sparing philosophies now in place?
2. *What will the fuel be? If it is natural gas, what is the incoming gas service pressure and flow capability? Should you consider alternative fuels*? Someone will have to decide what the primary fuel will be and if there will be a backup fuel. Then someone will have to report to the utility or pipeline what the pressure and flow capability at the meter will need to be. This is where their responsibility will stop. This decision drives fuel pipe sizing throughout the facility. Then comes the matter of where shutoff valves, purge points, and blind installation facilities should be located. How will the system first be energized (i.e., gas put into the piping)? Will this require complete facility shut down? Will the utility need to be called in? If they are and require a complete pressure test of the entire facility, are you ready for this? Are our purging and gas introduction procedures in accordance with NFPA 56? Should we include dual gas feeds, a dual regulator system, or network the piping distribution systems for reliability?
3. *Have insurers been consulted*? This has been mentioned several times and cannot be overstressed. Insurers have a right to understand the risks that you and they are taking. They often have staff that can make great recommendations

based on loss histories from many years, many clients, and many industries. You don't want to give them reasons to deny a claim.

4. *What is the impact on the rest of the facility from the ambient heat given off by the equipment*? Regardless of what you are putting in, ambient heat will be generated. In some cases, ventilation equipment must be installed to remove this heat. If so, make sure that the ventilation systems are compatible with combustion air requirements and do not affect the ability of flue products to get out of the space safely. I have seen heat-generating equipment in sensitive machining areas where tolerances of ten thousands of an inch were expected. The impact on these areas from the ambient environment was not acceptable, and expensive equipment moves then had to take place.
5. *Who will repair the equipment if it goes down*? Do you have adequately trained staff to handle the new technologies that may be coming with the new equipment? Have you considered your reliability needs adequately and planned effectively for spares? Have you gone through an extensive management-of-change assessment?
6. *Is there a need to upgrade fire walls, floors, roofs, and ceilings*? Is the area where the equipment is to be installed sufficiently fire resistive? Does the area satisfy local building codes for fire safety, exit egress, emergency lighting, and so on? Do you need to add sprinkler heads or change their temperature ratings? Where flues pass through ceilings and roofs, has building code–related clearance to combustible requirements been considered? Have the manufacturer's instructions been evaluated for the clearances required for maintenance? NFPA 54 has restrictions on where gas lines can be run. Have these requirements been investigated? For fire tube boilers, where one needs to plan for the possibility of a future tube replacement, the tubes are as long as the boiler shell and need to be installed as one long pipe.
7. *Where will the flue products, excess heat from hoods, relief valve discharges, and or gas vents go*? You have to plan for where flues and other terminations will be routed and have the potential to discharge. You must be careful with clearances to building air intakes and even building openings such as windows so that contaminants cannot easily be reentrained back into a building space.
8. *Do sufficient utilities exist to support the device to be installed (i.e., electrical power, compressed air, etc.)*? There may be impacts to site utilities that you had not planned on. For example, NFPA 86 requires consideration for nitrogen to be able to service removing atmospheres from heat-treating furnaces. If you are in the heat-treating business and are adding furnaces, have you considered this impact? There may also be effects on such things as compressed air for opening and closing doors, and electrical power for motors and quench pumps.
9. *Do smoke detectors, alarms, or automatic shutdowns need to be installed anywhere for code compliance or changes made to be compatible with existing systems*? If existing smoke detectors or alarms exist, they may need to be modified. If they don't exist, should they? In the case of boilers that come under the scope of ASME CSD-1 criteria, the facility might have to be

modified so that the required boiler emergency shutoff panic buttons exist at boiler room doorways.

10. *Are all performance expectations of the equipment clearly spelled out, and is a validation process planned and understood by all*? The contractor will want to be paid for everything before the paint dries. When will you be willing to release what part of the funds? What will the finish line be? If, for example, the boiler is supposed to make a certain amount of steam, do you have accurate metering in place to measure this? If the boiler is installed in the summer, is there a place to vent or send the steam so that the upper end of the performance requirement can be checked? Are there guarantees from the furnace manufacturer for temperature uniformity in certain zones? How will you check this other than your parts coming out wrong when all the staff is in place and trying to operate for production, or do you have custom-made oven or furnace survey instruments on hand?

10.1.2 Commissioning Issues

OK, so a new big piece of equipment has just been installed at your site. Commissioning and startup are the things that really turn this device and everyone's expectations into reality. If it's done correctly, it will serve you safely and reliably for years to come. If it's done incorrectly, things will be damaged right from the start and you will have a big headache for the rest of your career.

One good way to help ensure commissioning success for a retrofit project such as new burners being installed is to take extensive operating information about the equipment before allowing a project to be started. This could mean documenting temperature profiles and fuel/air ratio verification, including fuel and air pressures, at a number of firing rates on an oven or furnace before anything is even touched. After all, if you don't know what you started with, its difficult to understand where you ended up.

There are two primary parts to commissioning—safety and performance—and there needs to be a plan to validate both of them. Let's start with the safety plan. Many explosions have occurred during commissioning. Think about it: It's the first time everyone's assumptions really have a chance to be validated. Did the right equipment get bought, installed correctly, and wired properly? Does the PLC program really work? And so on.

The objective of the safety plan should be to get all safety devices validated and burner fuel/air ratios set properly so that performance issues can be ironed out without an incident. This will involve the verification of the correct set points and calibration of all temperature control and safety devices, whether or not they were part of the project.

Once the electrical power system is energized, cold air curves for the burner should be set and equipment should be dry-fired with no fuel. Plans should include steps to light only pilots first, then individual burners on low fire, then cycling individual burners. There should be many carefully thought out steps and daily safety briefings to make sure that nothing goes wrong.

The performance plan depends, of course, on expectations that were agreed upon at the time the scoping was done. This is largely going to be about such things as temperature control, firing capacity, temperature profiles and uniformity, and perhaps even emissions. If you are commissioning ovens or furnaces, you might need to have on hand survey instruments that are sent through to identify temperature profiles. This finish line for commissioning should be well defined up front before work begins. If there are emission targets, make sure that the test methods and conditions are also agreed up on up front.

In my opinion all commissioning steps should be done with a customer's best and brightest personnel shadowing the contractor's commissioning team very closely and taking lots of notes. There should be nothing going on that you as an owner don't know about and are not fully involved in.

Real-Life Story 42: Poor Commissioning Creates Problems Later

When most people think of devastating explosions of the industrial type, they usually think of large petrochemical complexes or massive steelmaking plants. But this case involves a simple commercially available baking oven in service at an iconic American company that you would recognize immediately. This type of oven was similar to thousands currently in use all over the world. The one that exploded was one of dozens crammed into a relatively small room. Upon arriving at the site it was clear that one oven was obviously missing its door. There were also a lot of very upset people very nervous because workers had just been injured and the entire facility was shut down. The explosion had blown off the oven door, shooting it across the room and smashing one of the workers into the wall. He had a broken arm and was bleeding from his right ear. Fortunately, he would recover, but at the time it was a scary scene for those workers close by.

A closer examination of the oven indicated that it had an indirect heat exchanger that was supposed to prevent flue products and any unburned gas from entering the actual baking chamber. Holes were found in the heat exchanger on this unit. It had been operating under conditions in which the combustion products were not being removed properly. The design of the oven included a special exhaust fan tied into the stack, called a *draft inducer*. It's sometimes there to pull flue gases through heat exchangers that might have a lot of piping and small passages (i.e., considerable pressure drop). In some cases you can't put a lot of pressure into the forced-draft burner fan, so you also have to suck the flue products out and send them on their way. This draft inducer was found to be operating backward. This happens because if the wiring leads on a motor are not terminated properly, the motor runs the wrong way. It happens often on pumps and fans. In the commissioning process it's always something to check. The unit continued to operate, but not well. This lack of effective flue gas removal had caused the heat exchanger to overheat, and eventually it failed and leaked. This allowed products of combustion (fuel rich) to be recirculated into the baking chamber. Then all it took for the explosion to take place was completion of the fire triangle with air and temperature (ignition source) for the explosion to take place. The fuel rich combustion products could have occurred easily because the room had many ovens and an obvious lack of combustion air. All the burners probably ran somewhat rich.

A review of other ovens at the site unearthed numerous problems, including safety interlock devices that were failed. The facility had no testing program in place. It was clear that management did not have a healthy respect for this equipment or staff that knew anything about servicing it. Many facilities that have combustion equipment do not have enough of it to really need a full-time knowledgeable staff person for preventive maintenance and regular safety testing. The equipment seems to continue to work just fine for years. The problem is that when this lack of maintenance and testing catches up—and it will—the end result is usually very ugly. This client lost hundreds of thousands of dollars in production over the days and weeks of repairs and interruptions to make the facility's equipment safe. It's very fortunate that no one had to pay the ultimate price to learn these lessons.

Lessons Learned Careful attention needs to be paid to commissioning activities, including measuring drafts along with understanding combustion air requirements. The initial commissioning issue (induced-draft fan running backward), in this case eventually led to the heat exchanger failure and explosion. This facility now has developed a process by which draft readings are taken in flues as a regular part of preventive maintenance procedures to make sure that all is well. There is also a renewed emphasis on the testing of combustion-related interlock safety devices such as flame detectors and high-temperature limits as required by NFPA 86 and OSHA bakery standard rules [29 CFR 1910.263(l)(9)(ii)].

10.2 TESTING OF FUEL TRAIN SAFETY INTERLOCKS

10.2.1 Training and Certification for Technicians

Now that you understand the installation and commissioning paths to minimizing fuel and combustion system risks, it's time to consider the testing of safety interlocks and fuel system—related valves. Before we explore the technical parts of this, let's look at who should be performing this work and what their qualifications should be.

There is no nationally recognized certification program in the United States for those who would attempt to do safety interlock or valve tightness testing on a regular basis. Several provinces in Canada have extensive qualification programs, as do some European countries, and these can serve as a good model. Some states and jurisdictions require mechanical contractor licensing. Some of these states also require natural gas piping and boiler installer certifications. These qualifications certainly help for commercial buildings and residences. However, large industrial boilers and combustion equipment other than boilers contain many interlocks and devices that require very specialized skills and knowledge. This is part of the reason why some of these required certifications do not apply to industrial plant work. You will need to check state and local laws to determine this.

NFPA 54 requires that all installation, testing, and replacement of gas piping, appliances, or accessories, and repair and servicing of equipment, be performed only by a qualified agency. A qualified agency is a person or company experienced in the work to be done and the precautions needed, who meets all other requirements. This could certainly be and often is in-house staff if they have the knowledge and skill sets required.

There are differing philosophies as to the necessary skill levels and the type of personnel who ideally performing should be this kind of testing at a facility. I have seen numerous types of programs within large commercial organizations in industrial facilities. The best chance of success is with a small group of people who receive intensive training and get the chance to do this type of work frequently. You can't give someone a few days or even a week's worth of training and then only allow them to do the actual work once or twice a year and expect good, safe and consistent outcomes. There is just too much to know. People need constant practice in using these skills.

The most successful organizations provide extensive training programs and a lot of hands-on skill work before turning technicians loose to do this kind of work. This is also where contractor qualification programs can be very important so that you can be confident in the skill levels that you are getting if you hire this work out.

IMPORTANT

There's more risk in the wrong person conducting safety interlock device or valve seat leakage testing than in living with the hazard that is being addressed. Mistakes made testing these devices have caused explosions and left equipment unsafe to operate. MAKE SURE THAT ONLY TRAINED PERSONNEL WHOSE EXPERTISE HAS BEEN VALIDATED DO THIS WORK.

10.2.2 Testing Methods

There are two tests that must be conducted for each safety interlock device. The first, the *interlock trip test*, verifies that a trip occurs and that the device responded as expected. This trip that occurred usually involves the tested device closing the safety shutoff and blocking valves to stop fuel flow to the burner. The second, the set-point test, verifies that the device trips at the point at which it is set. This is something that needs to be watched for and recorded as the technician is performing the test. A failed device can be one in which the trip does not occur or the trip occurs at a set point other than that indicated on the device.

A lot of preparation needs to take place before testing of this type can proceed safely. This preparation includes having proper set-point documentation for each of the components and having the right tools available. Most burner manufacturers don't provide specific set points for devices because they don't know the specific application for their equipment. Instead, they provide guidelines. You'll have to refer to the burner manufacturer's literature for this. In some cases you may have set points from the original commissioning data at the time of installation. If not, existing set points will need to be validated in the field by a skilled technician with a flue gas analyzer during a recommissioning process.

You must also make sure that you have spare parts available in case you find a defective device. Remember, any failed device requires that the equipment be taken out of service until repairs are made. This must be a hard-and-fast rule clearly understood by everyone, regardless of production demands. Sometimes this style and type of testing must be scheduled specifically for times when repairs can be made.

Next we discuss techniques and procedures that focus on the type of switch (normally open and normally closed) instead of its specific application (with the exception of air flow switches because these have been found to have failed and been set improperly on many occasions). The following is to provide you information from an awareness perspective and does not replace the need for a knowledgeable person preparing specific procedures for specific equipment applications. Conducting these tests in the field requires a working knowledge of how these switches are applied as well as the consequences of failed devices or devices set improperly.

There are many ways to test pressure switches. In some cases a precision pneumatic calibrator is used that provides a test pressure, and in other cases, system pressure is used. Similarly, some protocols call for switches to be tested in place and others involve removing switches for bench testing.

The most common pressure switch test method is to leave the device installed and use the set-point adjustment method using system pressures.[1] This test is described in the literature that comes with many pressure switches in the maintenance sections of their documentation. However, you will need to determine what is most appropriate for your site and your system's specific needs. The set-point adjustment test involves temporarily changing the set point to a condition where it would normally be known to trip and then moving it back after the trip. Make sure that proper set points are marked on the switch to begin with so that there is no confusion getting back there. For example, if a high gas pressure switch is installed in a system where the normal operating pressure is 18 inches w.c., and the set point is 22 inches w.c., the set point would be rolled down to 18 inches while the unit is firing. If the unit did not trip at this pressure, and the pressure were being determined accurately with a manometer, for example, the device would be considered failed. Remember, this trip needs to occur on the first attempt, not after exercising the switch or banging on it until it works. If the switch is failed, it should be replaced immediately.

Now that you've planned for failed items and have the appropriate information, personal protective equipment, the right tools, and spare parts, and have communicated with all stake holders, you're ready to begin. We'll be examining evaluation and test methods for purge timing, pressure switches, position switches, flame detectors, temperature switches, and other devices. This will be a generic description to give you an overview of this type of testing. There are many specific considerations that can only be made in the field, depending on the type of equipment being examined and the type of device being tested. One of the first things you should think about is making sure that there are job safety analyses completed for each procedure.

10.2.3 Verifying Purge Timing

Let's first review purge timing issues. The objective of purge timing is to make sure that adequate airflow moves through the firebox to dilute the contents to below its LEL prior to introducing an ignition source. NFPA 86 specifies that at least four fresh air volume changes are required to pass through a firebox before an ignition source can be introduced for ovens and furnaces. It's different for boilers. Boiler purge times can be identified in documents such as ASME CSD-1 and NFPA 85. It's also different for single-burner boilers than for multiple-burner boilers. There is quite a lengthy explanation in NFPA 85 that you will have to consider carefully before properly understanding the proper purge time for a boiler. Purge requirements can also be different where codes other than NFPA are enforced (international sites).

Most codes and standards require that purging airflow be proven to the burner management system by at least two methods during a specific purge process. In some cases, this could be an air pressure switch and a motor starter contact. To assess this properly, you need to understand what the requirements are for your specific situation.

If the airflow goes below the minimum flow during the purge time sequence, the equipment usually is designed to lock out, requiring the purge to be restarted. Most burner management systems provide a countdown for purge timing once the purge timer has started. In most cases the purge time doesn't start simultaneously with hitting the start button. Air dampers have to open to their maximum position and switches have to be made indicating that proper airflow is occurring before timing is begun.

Purge timing systems are manufactured as resettable mechanical timers and as solid-state devices. There have been many cases in which mechanical timers were turned back for the convenience of service personnel and never properly reestablished. Obviously, solid-state devices are preferred, as they are harder to tamper with. These usually have a purge timer card that can be removed and replaced to change purge times. In some cases, BMS systems have DIP switch settings to change purge times (DIP switch, small manual switches set in a line with others that allow for configuring).

10.2.4 Pressure Switches

At least three pressure switches are associated with most fuel train and combustion equipment safety systems, including the low gas pressure switch, the high gas pressure switch, and the airflow proving switch, which may be a differential pressure switch. In addition to this, pressure switches are also commonly used for boiler firebox pressure, draft, and exhaust fan proving. Pressure switches usually have one sensing line, although there are combination high and low gas pressure switches which have a common body but sensing lines to each of the two different sensing diaphragms.

Differential pressure switches have two sensing lines. They are the only type of switch that can actually be used to sense flow if installed across an orifice plate or fixed pressure drop of some type. You must be cautious when using only a static pressure switch to identify airflow. These have been known to be incorrectly installed and applied, for example, between a fan and a damper. In this case, a damper can be closed and considerable pressure will build, but no flow occurs. The burner management system may continue to think there is airflow because the device sees pressure. NFPA 86 has provisions for applying pressure switches to prevent this.

Low gas pressure switches, airflow switches, and high gas pressure switches operate somewhat differently. Low gas pressure switches and airflow switches are normally open switches, which means that when they are in an unpowered condition, the contacts inside are not touching each other and therefore are not making a continuous electrical circuit. In the electrical world, these contacts are known to be open. When a low gas pressure or air pressure switch set point is made, the contacts are forced together and remain there, making a continuous electrical circuit for as long as the pressure on the diaphragm is above the switch's set point.

In the case of a high gas pressure switch, the contacts are together, or closed, making a continuous electrical circuit in an unpowered, uninstalled state. This is called a normally closed switch. If the set point is exceeded when the switch is in use, the contacts separate, which opens the electrical circuit. Pressure switches should never have their set points set to the extremes of the switch scale. If this is where the switches need to be set for the unit to operate, the ranges for the switches have not been selected properly.

Some switches are indicating types, allowing you to see the effect of pressure change. Such is the case for some mercury switches, which allow you to see the set points and the contact conditions. These have a small glass tube (Exhibit 10.1) with liquid mercury that can bridge the gap between two small exposed wires on one end of

EXHIBIT 10.1 Pressure switch that contains a glass tube with mercury. Honeywell has new non-mercury replacements.

the glass tube, making a circuit. When the tube tilts and the mercury rolls to the other end, the circuit is broken or made to “open.”

You must be very careful to install mercury switches level. If the switch is not level, it won’t be accurate. Many people who install and service gas-fired equipment don’t realize that there is a small metal arrow hanging inside many styles of these switches that points to a target. This target device is installed so that you can determine whether or not the switch is level in the field. It’s a good idea to have this verification as a regular checklist inspection item for older systems that might have these installed.

Many food manufacturing facilities, hospitals, and schools have a significant number of mercury switches. Due to the toxic nature of mercury, these facilities should replace these switches with non-mercury equivalents.[2] There is considerable risk of having an accident and damaging the delicate glass tube that holds the mercury when these are removed or replaced. This can allow a mercury release, considerable cleanup expense, and the possibility of some very bad publicity. If you have to dispose of mercury switches, do so in accordance with environmental regulations.

Switches should be selected with consideration of the environment in which they will be operating. When switches are misapplied to their environment, it can lead to premature failure. For example, humid environments such as might be encountered in a food operation, or very dusty areas (cement plants), should have NEMA 4–rated switches (Exhibit 10.2). These provide an enclosure that is resistant to some of these types of environments. You should review NEMA enclosure rating specifications to understand what can work best for your facility’s environment.

EXHIBIT 10.2 Dungs’ NEMA 4–rated pressure switch.

You should also be aware of the need for venting of switches. NFPA 54 requires that if a device has a vent connection, a vent pipe needs to be installed. Many of the mercury switches discussed previously incorporate diaphragms or housings which typically require vent piping. Diaphragm pressure switches also usually have vent connections.

One should also consider using ventless pressure switches if possible to simplify installations and reduce costs. Ventless pressure switches have limiting orifices installed that are designed to control the maximum amount of gas discharged in the case of an internal diaphragm failure. The manufacturers of these switches indicate that they should only be used in well ventilated areas.

10.2.5 Airflow Switches

Airflow switches are by far the most commonly failed switches. At least half of all airflow switches I have tested or seen tested (thousands of pieces of equipment) were found to be failed or set at a level lower than design requirements. This means that inadequate airflow could establish them as being okayed for ignition to occur or continue. Inadequate airflow switch set points are dangerous.

> **IMPORTANT**
>
> In my 30 years of reviewing combustion systems I have found that by far the greatest number of failed safety devices are airflow switches. They are usually not set right, and most often, stuck in the made position from accumulated dirt in the combustion airstream.

Airflow switches can also be tested using the set-point adjustment method, temporarily moving the set point of the switch back toward its minimum to see if you get a trip and then moving it back to the proper setting.

A method of establishing proper airflow switch pressure setting requirements for most boiler systems and industrial process ovens is to measure the combustion air pressure at the switch when high-fire-rate airflow is occurring during the purge process. If, for example, this air pressure is measured at 20 inches w.c., the switch should be set at slightly below this, retaining a factor of safety. You will have to verify that the setting does not make for nuisance trips while still providing adequate protection. An experienced burner technician should assist in this process. Success in this process assumes that the airflow rate for the purge is correct. Purge airflow rates should be measured with instruments and verified to be adequate for the timing that is provided as part of this process.

10.2.6 Flame Detectors

At least three different technologies are in use to establish whether or not a flame exists: thermocouples, optical detectors, and flame rods.

Thermocouples A thermocouple is created from two dissimilar metals that are connected together. When that junction is heated, an electrical current develops. Thermocouples rely on this simple principle. Thermocouples are the least expensive and simplest form of flame detection, but they have very limited use. One of the biggest problems with thermocouples is their response time, due to the very small electrical signal that is generated. In most thermocouple-based flame safety systems, this small signal holds a gas valve open. Thermocouple-based flame detection systems used on residential hot water heaters can take up to 3 minutes to sense the presence or absence of a flame and close the gas valve. This long response time is one of the reasons why you would only usually find thermocouple-based systems on very small appliances such as a small capacity domestic hot water heater.

Flame Rods Flame rods work on the principle that flames conduct electricity. They also rectify an ac signal to a partial dc signal. Flame rods consist of an electrode and a ground. The ground in some designs is at least four times the surface area of the electrode. The electrode and the ground are positioned close together. The gap between the electrode and the ground is then immersed in a flame. The flame actually conducts electricity like a wire between the two. The flame rod, the flame, and the ground then make up an electrical circuit.

One of the failure modes for flame rods is overheating, which causes a metallurgical change in the crystalline structure of the material that renders them incapable of then conducting electricity. Most flame rods are made of a special stainless steel and aluminum alloy called Kanthal. When Kanthal is overheated, the aluminum comes out of solution and moves to the outer surface, forming aluminum oxide, a non conductor. One can't see this occurrence; however, when it does happen, the circuit is discontinued, even when it's immersed in a flame. Another mode of flame rod failure is physical sagging of the rod until it is no longer in the flame. This occurs especially if the length of the rod deviates from the burner manufacturer's recommendation.

IMPORTANT

Remember that flame rod systems have an electrical current when operational. This could be a shock hazard if they are handled or work occurs near them.

Flame rods need to be inspected visually, as do igniters. In some cases, igniters and flame rod systems are combined. Flame rods can sometimes be corroded, carbonized, or eroded. The ceramic insulator portion also needs to be reviewed for cracking or shorting. Similarly, cabling needs to be inspected for shorting or evidence of breakage. Flame rods can be tested during the pilot light-off sequence. You simply turn off fuel to the pilot system once its lit to validate that no flame is indicated on the burner management system display.

Optical Detectors Many different types of energy emanate from a flame. Visible light is actually a very narrow band, even though it is what our eyes can readily detect. The infrared and ultraviolet wavelengths have much larger signals than does visible light. There are optical flame detectors that use each of the three signals. However, the most optical popular flame detectors rely on either ultraviolet or infrared energy.

An important thing to remember is that if used on an oil flame, optical detectors must be of the infrared design. Infrared signals have a longer wavelength. These can penetrate what are typically more smoky fireboxes that occur when fuel oil is used. If you use an ultraviolet detector on a fuel oil system, you're liable to have more nuisance trips, due to the loss of signal. For most natural gas burners, an ultraviolet detector is preferred.

The most notorious and hazardous failure problem for optical detectors is that they could stay indicated as being on when a flame was not actually present. Manufacturers designed a self-checking feature into optical detectors to overcome this potential failure mode. This is a special feature or option that must be requested for it to be part of a device. Self-checking detectors usually come in a different style of housing. However, you can really determine if they're in use only by researching the full model number and serial number of the device. One version of a self-checking detector involves an internal mechanism that periodically blocks the incoming firebox flame signal. The electronics are synchronized to sense when this is occurring. If there continues to be a flame signal when one of the blocking cycles has occurred, the system interprets the detector as being stuck in flame mode. In this case the system would be shut down.

Some models of flame detectors of the self-checking type also have an arrow on their back side, which indicates a preferred installation orientation. If you do not install them this way, or install them in ambient environment above the operating temperature recommended, you're likely to experience a premature failure rate.

Something to keep in mind for all styles of combustion control components is the presence of ambient environmental condition limitations. In the case of flame detectors, the maximum ambient temperature for most non-self-checking detectors is 215°F. Self-checking detectors can be much more sensitive. For many of them it is not recommended that they be installed in an environment above 145°F. This is why you will often see that an insulating bushing or small tube has been installed to provide cooling air connected to the site tube in these detectors.

Optical detectors are installed as part of a flame detection system. The system consists of the detector itself and a small tube or pipe nipple that's installed at the outlet. The longer this sight tube is, the more narrow the field of view of the detector and the more difficult it is to see the flame. Some optical detectors are also installed on a swivel joint. In the case of swivel joints and any new installation, it's important that the detectors be sited properly. One of the important tests for establishing the proper installation and siting of an optical detector is the pilot-spark pickup test. This is usually done at commissioning and startup, but it's a simple test and can easily be completed annually to verify that nothing has changed since commissioning.

A pilot-spark pickup test establishes that the detector will not pick up a spark signal from the igniters without there actually being a real flame from a pilot burner. A pilot-spark pickup test is done by shutting off the gas as close to the burner as possible. This includes the main fuel and the pilot gas. A normal startup sequence is then initiated. After the unit purges and when the ignition system starts the spark igniter, the flame signal needs to be monitored. This can usually be done at the BMS display. If any flame signal is detected, the detector is not installed properly. I have also seen this test fail because galvanized or stainless nipples were used as sight tubes. These provided reflected light to the sight tube. Any signal recognized might satisfy the burner management system and allow the main burner gas valves to open even though the main flame is not lit. This can release a considerable amount of gas that would not light immediately. Instead, it could accumulate and ignite all at once in an explosion.

Infrared flame detectors need to be sited and tested with a hot refractory saturation test as part of their setup and commissioning. If these are not installed properly, they can saturate with signal and not recognize a flame loss. This test procedure is commonly found in manufacturer's installation instructions for infrared detectors.

Optical flame detectors can be tested by removing them while a unit is firing. Just remove the detector and stand away from the sight tube. The burner management system should recognize the loss of flame signal and usually close the main gas valves within 5 seconds. Optical detectors are supposed to sense the absence of a flame within 4 seconds. Gas valves are then supposed to be closed within 1 second of the command being issued.

Real-Life Story 43: Proper Flame Detector Commissioning Is Vital

I was called in to investigate a boiler explosion at a hospital. The boiler in question had just been retrofitted with a new burner and was slated to have new controls installed shortly. Fortunately, the explosion didn't hurt anyone and the boiler suffered only minor damage, warping the boiler skin and putting it out of commission for about 2 weeks.

Operators reported that the boiler experienced a flame-out condition immediately before the explosion. The failure theory was that the flame detector failed to recognize the loss of flame within the required 4-second period and that fuel oil continued to flow into the boiler with no flame. It vaporized on the hot surfaces and accumulated to a flammable mixture inside the boiler firebox and flue passages. This accumulated mixture was then ignited by the hot refractory in the boiler. If this theory of the incident were true, I wanted to know a couple of things. First, why the flame-out? Second, why did the flame detector not respond properly?

The first question was answered after I learned that the boiler was knowingly being operated without the installed draft control being functional. This boiler was connected to a tall old masonry stack designed originally for solid fuel firing. This stack was shared with another boiler. Tall stacks of this type tend to provide much more draft than is needed for a boiler firing on gas or oil. This and the fact that another boiler shared the stack made properly functioning draft controls essential. Maintenance logs had noted that the boiler's draft control damper was not in operation and had not been for some time. No one had taken steps to remedy the problem. Logs also indicated occasional flame instability issues and trouble lighting. The incident occurred on a cold day. In cold weather tall stacks draft more than on warm days because the difference between the specific gravity of the vented flue gases and the ambient air is higher. It was very likely that this draft control issue is what caused the flame-out. Too much draft can blow out a burner flame like the candles on a birthday cake or, at the very least, cause flame instability.

I asked to see the flame detector. It was unscrewed from its housing and the sight tube moved around as this occurred. It was clear that the detector site tube was not precision aligned and locked into place. I looked at the lense and it was fouled and discolored. I was shocked to see that it was the old original flame detector with the new burner. When the lens was examined closely, the discoloration was found to be from fuel oil inside the lens. The lens was removed and oil seeped out. Combined with the flame detector being misaligned, this detector was in no shape to do its job properly.

Lessons Learned There's no substitute for proper commissioning. This must include basic sighting of flame detectors and conducting pilot spark pickup and saturation tests, for example. Equipment should never be operated knowingly with important systems not functioning. Draft controls can be very important for flame stability and proper flame patterns within boilers and other equipment. Draft controls are especially important where multiple boilers have flues combined in a common stack and where stacks are tall and have considerable draft. Flame-outs and flame instability issues should be considered near-misses. Root-cause analyses should be conducted for these types of events. This might have led to the discovery that the draft control was impaired.

10.2.7 Position Switches

Position switches (sometimes called *limit switches*) are used in a number of ways within fuel trains. These switches usually have a small mechanical arm that extends from the device. The arm may be in direct contact with something that moves, such as a valve stem or linkage.

Sometimes these little arms have small wheels attached to minimize friction and to help the device and the arm move more smoothly. They are used within safety shutoff and blocking valves and control valves, on linkages for position verification, on dampers, and even on some furnace ovens or doors. Position switches are often used on firing-rate control valves to ensure that the flame is at a minimum stable light-off position (low fire start). Sometimes this switch is buried in the housing of an actuator as an integral piece of its construction. In other cases there is a switch mounted directly on a fuel valve and a small tab in contact with the lever or linkage to perform this function.

Position switches are also often an integral means of valve stem and actuator position indication for safety shutoff and blocking valves. In this service these switches are called proof-of-closure switches. These validate to the burner management system that the shutoff valve has indeed done what it's been asked to do and that the stem has come into contact with the seat.

Testing position switches is often done by moving the switch or the switch contact so that the device "makes" or not when the BMS expects it to provide a signal. Testing these switches is very specific to the service they are in and requires knowledge of the device and the overall control system.

10.2.8 Temperature Switches

All thermal processing equipment has some type of operating temperature controller and should also have some type of a high-temperature limit device. These are usually a thermocouple tied to analog or digital controller or temperature switch. Operating temperature controllers do the normal control work to meet desired set points. High-temperature limit devices are there to ensure that maximum temperature limitations have not been exceeded. This would come into play in the case of an operating controller that fails or whose calibration has changed. Codes require that thermocouples used for operating and high-temperature limits must be distinct and separate. Codes also require that high-temperature-limit devices have manual reset capabilities.

High-temperature-limit testing can be done by temporarily rolling down set-point temperatures on the device. This can also be done by removing the device and using some type of known calibrated heat source. Another test for the thermocouple-based high-temperature-limit controllers is to remove one of the wires of the thermocouple, simulating an in-service failure of the thermocouple wire. The open circuit should cause the controller to think that the temperature has risen to a maximum temperature and cause the system to lock out. This is called *upscale burnout protection* and is another requirement of the controllers used in this service by codes.

10.3 REBUILT SAFETY CONTROLS

Several organizations currently offer rebuilt or reconditioned or remanufactured versions of obsolete burner management systems, flame detectors, and other components. In many cases, rebuilders have removed the FM/UL label, since only the manufacturer can provide this labeling and accreditation on a new component or on a factory-rebuilt component. Component manufacturers often don't rebuild this equipment nor authorize anyone else to do so. Most codes require components to be listed and labeled for the service intended. If the code listing and label are gone, your equipment is not code compliant.

One of the problems with rebuilding these devices is that they are typically obsolete, meaning that they are no longer supported by manufacturers. This means that original equipment components for the replacement of defective parts are not available directly from the manufacturer, so substitute repair components (e.g., resistors or other electrical or mechanical devices) need to be installed. In some cases only defective parts are replaced with something similar, and not everything on a circuit board or within a device even after the device has "seen" thousands of cycles.

ASME CSD-1 (a boiler code) prohibits the use of rebuilt safety components. Many insurers do not consider rebuilt safety devices to be insurable objects. In my opinion these are things you should stay away from both for boilers and for other categories of combustion equipment. Prohibiting these should be a conscious corporate-wide strategy and the subject of an information piece to all concerned to educate personnel about how these can be spotted and why to avoid them. Imagine your plight after an incident when the forensic investigation indicates that your equipment may have such a component installed. Why give the insurer a reason to fight your claim just to enjoy a small amount of savings? In my opinion it's simply not worth it.

10.4 VALVE SEAT LEAKAGE TIGHTNESS TESTING

It's important that all valves, manual and automatic, be tested for their function and seat leakage. In this section we address seat leakage issues related to automatic fuel train valves. There are usually five or six automatic fuel valves that are most commonly used on fuel trains for combustion equipment of over 400,000 Btuh per hour input. These are the safety shutoff valve, the blocking valve, the vent valve, and the pilot shutoff valves and sometimes a vent on the pilot shutoff system as well with larger boilers. The vent and pilot valves are usually of the solenoid type on smaller equipment. Where extra reliability is desired, these can be upgraded to other styles of valves. Solenoid valves are usually the lowest-cost automatic valves available for this application. Although they generally do a great job, it's important to recognize that they are not designed to last forever. Codes such as NFPA 85 and NFPA 86 have recognized this. These codes both call for the use of more than one automatic valve in pilot applications, and NFPA 85 (for boilers over 12.5 million Btu's per hour) requires that a double block-and-bleed system also be used on pilot systems.

Leaking manual and automatic fuel shutoff valves can have serious consequences. Remember, two of the key functions of a fuel train and its automatic valves are to keep fuel out of the firing chamber when combustion is not occurring and to be able to shut firing down immediately in the case of a problem. Automatic fuel shutoff valves must be ready to do these two things reliably and completely all the time, including being able to close completely within 1 second of being asked to do so.

If the safety shutoff valves or pilot valves leak when in the closed position, they can allow fuel to accumulate in the firing chamber, making for a flammable mixture. Consider the case of a valve leaking at 1 cubic foot per hour. If it feeds a relatively small firebox, say a 200-housepower fire tube boiler, the mixture in the entire firebox could be at a flammable range in a matter of minutes. This creates the potential for a number of things to go wrong, such as a less than perfect purge, a spark, and/or the mixture finding its way out of an exhaust opening or flue stack to an ignition source.

In the case of leaking vent valves (from double block and bleed systems), consider that the burner pressure will be disrupted. Gas that leaks through the vent valve after the main regulator will make for a lower pressure and a leaner flame. Lean flames are often unstable flames and are subject to going out. This means that the low gas pressure switch and/or the flame detection system will need to be adjusted to work correctly. Every time this equipment operates in this condition, the site operates at a greater risk of catastrophe.

Real-Life Story 44: Paint Oven Explosion Shrink-Wraps Trucks

An automotive manufacturer had an explosion and fire at an assembly plant and ended up with a destroyed paint drying oven and a number of new trucks wrapped in sheet metal. No one was injured in the incident. The incident occurred shortly after the unit was shut down.

Most vehicle paint ovens are designed to apply indirect heat to vehicles after they have been painted. In the case of this design, burners fired into a chamber from which hot air is recirculated into an interstitial space that forms a long tunnel. Vehicles moved down a long conveyor through various zones. Initial zones were designed to put a thin skin on the paint to make it more immune to dirt and finish imperfections that might occur as the paint fully cured. Later zones were soak zones for fully curing the paint, followed by cooling zones.

In this incident, the explosion occurred from a buildup of natural gas in the interstitial space between the inner and outer walls of the oven where hot air only (and not combustion products) was supposed to be circulated. An investigation of the fuel trains found that even though the unit was off, gas still leaked past the main automatic shutoff valves and that in fact some of these valves were not even fully closed. This meant that after a flame-out or shutdown condition, gas could still enter the firing chamber with there being no flame.

After the accident, more of the valves were recovered and tested. Many were found to take 5 seconds or longer to close completely, and some of the valves did not close completely at all. When contacted, the valve manufacturer indicated that during the assembly of valves with certain date codes, a particular lubricant had been used on the stems. The lubricant had been found to harden over time and did not allow the stems to move freely, thus leading to slow closing and sometimes sticking valve stems. The manufacturer indicated that a recall was initiated some time ago, but this was only to their distribution channel. They acknowledged that in some cases, users of these valves may not have been known, so could not be contacted.

Lessons Learned Make sure that all valve performance parameters are evaluated, including closing time and seat leakage. Valve closing time (1 second) is more a qualitative check than a quantitative measured check, but nonetheless is one that should be carried out. You can do this by observing the valves in action (ideally, they should have an external position indication that you can see). You can verify that closing is occurring almost instantaneously from the time the closing action is initiated. Fuel valves that fail to close within the time required or don't close completely create a potential for explosions since fuel can continue to enter the combustion chamber with no flame present.

There have been a number of fuel train components and devices where maintenance warnings, recalls, and/or safety alerts have been put out by manufacturers. Some of this information is available at the U.S. government's Consumer Product Safety Commission website. In other cases, manufacturer's websites list this type of information. It's a good idea to do some research with the component manufacturers representing equipment in use at your site to see if you have any of these devices.

10.4.1 Valve Seat Leakage Tightness Standards

Before discussing leakage testing further, let's take a look at how valve seat leakage or tightness is customarily measured and rated and what codes and standards require. There are several standards for valve seat leakage. Different valve manufacturers apply different standards for their products. NFPA 86, the Standard for Ovens and Furnaces, in Section 7.4.9 requires that valve seat leakage testing of safety shutoff valves and valve proving systems must be performed in accordance with the manufacturer's instructions. There is also a table in the NFPA 86 (appendix) that has acceptable leakage rates from published standards. If you then refer to some of the most popular valve manufacturers maintenance documents (Honeywell,[3] Maxon,[4] and Dungs),[5] they provide instructions on how to do valve seat leakage tests of their valves, along with published leakage pass/fail criteria.

Leakage criteria for valves are also published by a number of organizations (ASME, ANSI, FCI, and API are a few). One of the key selection criteria for valves is their leakage class rating. If you explore valve manufacturers' failure criteria, you will see that the differences usually coincide with the ANSI leakage class design ratings. Generally speaking, the less leakage, the more expensive the valve. The class rating and design leakage that is suitable for your application may depend on things like the size of the firebox or combustion chamber, the amount of time that the equipment is down and vulnerable to a leak, the extent of the use of blinds within your fuel train design, and a number of other factors that are reliability- and risk-related, such as downtime costs for your process and equipment sparing philosophies.

10.4.2 Automatic Valve Seat Leakage Testing Procedures

The primary test for automatic valve seat leakage is called a *bubble test*. Seat leakage tests are usually done with valves installed, in place, using gas line pressure as it exists. Remember, the valves to be tested would be those on both the main and pilot fuel trains. Make sure to follow valve manufacturers' protocols for this test. The following is simply a summary of the basic principles behind a bubble test and is not expected to replace a valve manufacturer's instructions. *Always remember to have the manual burner firing isolation valve closed whenever conducting valve seat leakage testing. This can prevent accidental opening of safety shutoff or blocking valves to allow gas to enter the combustion chamber. Also, valve seat leakage testing should be carried out only by qualified personnel after adequate training.*

Bubble test protocols usually start with gas pressure being applied upstream of the valve to be tested from the fuel train. Then a test port somewhere downstream in a closed piping section is opened. This test port is usually piped with a $\frac{1}{4}$ inch hose, 0.032-inch wall thickness, to a small cup of water, with the tip of the hose immersed in this cup of water. If the valve leaks through in the closed position, it flows into the small tube and creates bubbles when it comes out the tip into the water. The bubbles are counted over some time period. Success is for the valve to be leaking at less than some number of bubbles per minute. Remember that this is a live gas test where gas can come out of the tube you are holding, so the area should be ventilated and there should be no ignition sources around.

If you ever want an idea of whether or not valve seat leakage testing is occurring at your facility, take a close look at the valves test ports. Virtually all automatic valves have test ports for seat leakage testing. One port is for the space immediately before the stem-to-seat connection and the other port accesses the space immediately after.

If the test port plugs appear to be rusted into place, seized, or do not appear ever to have been removed, it's an obvious sign that the annual tightness testing required is not taking place.

In double block and bleed systems it is also important that a pressure hold test be conducted on the vent valve to assure that it is not leaking. Vent valves cannot be bubble-tested like most other valves, since most codes and standards prohibit manual shutoff valves downstream of these valves. Instead, they must usually be tested with a pressure hold test. This kind of test is done with the equipment in an off and safe state, with the gas to the burner isolated at the firing valve. In this test the vent valve needs to be closed by a qualified and properly trained technician. Then gas is trapped between the safety shutoff and blocking valve and the vent valve and the pressure is noted. Success is indicated by no degradation of this pressure over several minutes.

IMPORTANT

Always make sure that that the fuel train is isolated so that no gas can enter the burner and combustion chamber through the main or pilot sections when valve tightness testing is taking place. This testing process and the manipulation of fuel valves should only be done by qualified personnel.

Some test techniques call for vent valves to be closed temporarily and/or in some cases safety shutoff valves need to be opened temporarily. I want to repeat again that there should always be a downstream burner firing valve or pilot valve closed to prevent an accidental release of gas to the burner while testing. In fact, if this valve does not exist you cannot properly do bubble testing. Again, refer to the valve manufacturer's instructions for test procedures.[6]

Many sites have purchased new equipment configured with testing mode switches that allow valves to be manipulated with the equipment off, without the need for electricians to enter panels to install temporary test wiring to change valve positions just for seat leakage testing. In many cases, customers have also retrofitted existing equipment to include these switches. The use of these test mode switches is a well-known practice in many industries. In fact, typical schematics for accomplishing required switching are included in the appendix of NFPA 86, the Standard for Ovens and Furnaces. When retrofitting a panel to accomplish safer and more efficient testing capabilities, make sure to have your modifications reviewed and approved by the authority having jurisdiction (usually, your insurance company). Most major insurers are familiar with this option and have readily provided approvals in the past.

At the end of the day your facility might have several different styles, sizes, and types of valves to develop procedures for and test. I have found that valves usually either leak or they don't. Those that leak seem to leak excessively and are easy to find. Remember, regardless of what is deemed to be an acceptable leakage rate, you still need to apply an element of risk analysis for your particular situation. This means that if, for example, the combustion chamber is somewhat small and if the equipment is down for considerable amounts of time without a zero leakage manual gas valve closed, even a small leak may accumulate to a flammable mixture. Codes and manufacturers' standards are the minimum; you can always apply leakage criteria that are more stringent than those that manufacturers consider acceptable.

Real-Life Story 45: Small Pilot Gas Valve Leaks Cause Big Problems

A packaging company's largest plant was put out of commission when an explosion and subsequent fire destroyed one of its main ovens. Luckily no one was injured, but economic losses were in the many hundreds of thousands of dollars. Packaging plant issues are troublesome because it means that whomever they serve could also go out of business. The oven had just completed a 72-hours run cycle. The explosion occurred unexpectedly 45 minutes after the shutdown. A worker near the oven was knocked off his feet, two small fires had to be extinguished, and an adjacent oven was damaged. Thankfully, that was the worst of it.

The explosion was found to have occurred because of an accumulation of gas in one of the oven zones from a leaking automatic pilot solenoid style of gas valve. The leaked unburned fuel accumulated into a flammable mixture and found an ignition source.

Lessons Learned Fuel valve seat leakage testing is a necessity and this must include pilot systems. Even though the pilot system might only be piped with $\frac{1}{2}$ inch pipe at a low pressure, a small leak that accumulates over time can still be deadly. It's also important that you consider code changes such as the need for two pilot shutoff valves in series even if your equipment came with only one. The addition of a second pilot shutoff valve reduces the risks of valve leaks at the price of a minor investment.

10.4.3 Reducing the Risk of Automatic Valve Seat Leakage

You can improve the reliability of valves and reduce the chances of seat leakage by using extra care when installing them. The greatest risk of valve leakage is when there has been some disturbance to the gas piping systems. This is when debris could have been loosened and is on its way to contaminate a valve. This could be a repair somewhere else in the system or a replacement of a component, say a regulator. Some valve manufacturers recommend that if valves are replaced, they should be seat leakage–tested within a short time since the action of the replacement disturbs the gas piping and creates a risk of contamination. Also, be aware of overtightening aluminum-bodied valves and the overuse of pipe dopes. These could also put particles and debris into a position to end up in the valve seat-to-stem interface area.

Pay special attention to the age and condition of all solenoid valves. One review of client solenoid valve failures indicated an unusual rate of failure for those that were installed for 15 years or more. This client chose to proactively replace solenoid valves of this age rather than risk more random failures.

Also, make sure that you look at and really use the recommendations that are in many solenoid valve manufacturer's installation guides. There are great things in there: for example, one popular valve manufacturers recommendation is that for maximum useful life solenoid valves should be installed in the horizontal position with the valve stems facing up. Also, consider adding sediment traps in cases where long risers from vents might cause condensation and debris falling back into the valves. You might also want to head the ambient operating temperature recommendations. Many solenoid valve manufacturers[7] have maximum ambient temperature ratings of only about 140°F.

10.4.4 Manual Isolation Valve Leakage Testing

OSHA 1910.147 (lockout protocols) requires equipment to be at a zero-energy state. In regard to the fuel piping systems, this means that manual isolation valves must hold and be leak-free or you must install blinds or appropriate double block-and-bleed systems to work downstream from them safely. Single closed valves should never be considered a positive means of permanent isolation. There have been cases of valves appearing to be closed properly because they are leak-free but later found to be actually seated against debris. In these cases there is a tremendous risk of the debris moving and then the valve leaking. Single closed valves should be considered only as a temporary means of isolation for getting blinds installed after careful planning and depressurization of systems.

Leak-checking of manual isolation valves in the field, for a quick qualitative measure of the valve's condition, can sometimes be done with a pressure hold test. If a more quantitative and positive understanding of a valve's condition is required, standards such as API-598[8] exist, which provide bubble test criteria for different valve classes. A tag on the valve somewhere should indicate its class rating.

10.4.5 Lubricated Plug Valve Maintenance Issues

One of the key findings in 30 years of safety inspections for fuel systems is that more than 60% of lubricated plug valves that I have checked have leaked through when in the closed position. In some of these cases, those plug valves were either new or in relatively new installations. I have found the primary cause of this to be inadequate sealant installed in these valves. This is a serious problem when maintenance personnel are locking out and isolating gas piping systems because they believe they are getting to a zero-energy state when there is actually still gas leaking past the closed valve.

IMPORTANT

Conducting a plug valve maintenance program requires installing sealant properly in a valve using a special sealant gun. It is very important to know what type of sealant material to use on plug valves. Do not use just any sort of "grease," as you might for automotive applications. The wrong lubricant will not seal and may make matters worse. Contact the valve manufacturer or a maker of specialty sealants to better understand what the proper material is and how to inject it.

It is vital that you have the proper type of sealants for all valves in your system and be knowledgeable on how to install them. Valve sealants are very specific to applications and valve manufacturers (remember, it is not automotive grease). Choose the sealant gun gage so that you can watch and identify when the valves are filled. You will need training to understand what the gauge is telling you while sealant is being injected. There are a number of conditions that can occur, such as sealant bypassing the plug and filling the pipe, or sealant not going into the valve at all (failed button head fitting). Make sure that you operate the valve frequently while installing the sealant so that the material gets spread across the entire plug face. If a considerable number of valves are to be sealed, compressed air–powered sealant injection tools are available.

Before you start servicing valves make sure that you review where upstream valves exist and verify that they function. If you should break a valve casting or have the piping rupture or leak while you're servicing a valve, there should be a plan and understanding of how gas to your work area will be isolated.

Real-Life Story 46: Where Do We Start Digging Up the Gas Line?

I was involved with a client that had a utility natural gas service entrance that ran 3000 feet underground before an 18-inch main stubbed up to serve the 4,000,000-square foot manufacturing complex. There was a lubricated plug valve on the line where it came up out of the ground. The client had tried to operate this valve and found it to be seized and not capable of being operated. The fact that this valve was seized open presented a fire hazard because if there were a problem, the gas could not be shut off.

Replacing this valve would be a nightmare since it would mean that the utility would need to shut off their service at the street. They indicated that if this were to occur, they wanted a pressure hold test of the entire 3000-foot-long underground line. If there was any sort of minor leak, the entire line might need to be excavated to find it. This could take the plant down for weeks.

The solution to this problem was injecting the valve with a special solvent provided by the valve sealant manufacture. This solvent was made to dissolve the waxes that can remain over time and harden as a residue inside lubricated plug valves. After several days of injecting the solvent and coaxing the valve carefully, it was eventually freed, cycled, and then injected with fresh lubricant. Everyone at this site now understands the importance of regular maintenance for these valves.

Lessons Learned Regular lubrication and exercising of key lubricated plug valves, including main shutoffs, is vital to their remaining functional. Codes and valve manufacturers call for annual lubrication injection to such valves. Function testing of main incoming service valves needs to be done carefully so as not to jeopardize functioning gas systems by removing flow. It is usually possible to find a time of minimal flow or equipment downtime and then try to move the valve a little to determine if it can move or if it is seized.

Special solvent materials and techniques are available to help some seized valves become functional again. A traditional replacement of this plants main gas valve would probably have meant hundreds of thousands of dollars in lost production because of the time required to find and repair leaks in the underground section of the system.

10.5 REFRACTORY AND EQUIPMENT OUTER SKIN PROBLEMS

Refractory is an important consideration for all combustion equipment. It takes many forms: from fiber insulation to bricks to pourable and castable materials. Refractory is an entire specialty world unto itself. It's what protects the vulnerable carbon steel of which most equipment is made from the 3000 °F + flame temperatures. If refractory was not present, these flames would immediately destroy the oven, furnace or boiler structure and walls. This section is not so much a primer on refractories as it is a place to raise your awareness of this material, how it is used, and more important, how to recognize some obvious defects.

When first approaching a boiler or some other types of combustion equipment, it's always a good idea to do a walk-around. This is as simple as it sounds: Walk around the equipment and look for "red flags" such as paint that had been burned off or warpage of what should normally be flat metal surfaces. Paint that has been burned off usually indicates a refractory problem on the inside of the unit. Repeated overheating from flame impingement, as well as over-accelerated startups, can destroy refractory. When equipment is started up too rapidly, the steel expands faster than the refractory, and the refractory can crack and break.

Once a metal plate has been warped, you could be in for dangerous conditions, such as buckling and movement of tubes inside the boiler or accelerated destruction of refractory. Once refractory is compromised, the damage only gets worse. You will eventually have a burn-through and flames or combustion products will come out or air will get into the combustion chamber. Neither is a good situation and both can be very dangerous. I have seen burn-throughs happen suddenly and literally cook conduits or fuel train components to the point where they fail.

Also remember that there's still plenty of asbestos containing refractory around. If you are servicing older equipment and have any doubts, make sure that the material is first tested by a qualified asbestos abatement contractor. When it comes to refractory, asbestos is not the only hazard. Crystalline silica[9] is another hazard to which workers handling refractory can be exposed. Before work starts, especially demolition, make sure that the materials and site are reviewed for asbestos or crystalline silica hazards. A qualified industrial hygienist should be involved to evaluate the working conditions and possible hazards before proceeding.

NOTES AND REFERENCES

1. Honeywell, C6097 pressure switch installation and testing, www.rapidengineering.com/pdf/third_party_literature/Honeywell/Honeywell_C6097_GPSwitches.pdf.

2. Honeywell mercury replacement literature, https://customer.honeywell.com/resources/techlit/TechLitDocuments/63-0000s/63-9727.pdf.
3. Honeywell V5055, valve installation and tightness testing, https://customer.honeywell.com/resources/Techlit/TechLitDocuments/95-0000s/95-6981.pdf.
4. Maxon valve, maintenance, www.maxoncorp.com/Files/pdf/English/Valve_Technical_Data/E-valve_maintenance.pdf.
5. Dungs, valve installation and tightness testing, http://mobile.dungs.com/fileadmin/media/Downloads/DBs_BMAs/261402.pdf.
6. Maxon Valve, maintenance, www.maxoncorp.com/Files/pdf/English/Valve_Technical_Data/E-leak_testing.pdf.
7. ASCO Solenoid valve engineering information, www.pacontrol.com/download/Solenoid%20Valves%20Engineering%20Information.pdf.
8. API-598, www.api.org.
9. OSHA crystalline silica, www.osha.gov/dsg/topics/silicacrystalline/index.html.

11

Global Perspective on Fuel and Combustion System Risks

Managing Fuel and Combustion System Risks Outside the United States

It's a big world out there and combustion equipment is everywhere. I have learned a lot by seeing what the state of the art is and is not in both developed and developing countries. In this chapter I provide insights from these experiences. I am bringing you the good, the bad, and the ugly so that you can take advantage of them all without the pain that others have experienced to gain this knowledge. This chapter is especially important if you operate equipment in developing countries. This can be an entirely different experience and one that requires considerable thought about fuel choices, installation issues, and training of staff. In my opinion, to be successful your focus has to be on simplicity. Real Life Stories in this chapter communicate this clearly. Don't be fooled by the title of the chapter. There's information here that applies for equipment operated anywhere.

Real-Life Story 47: Sometimes the Obvious Is Not So Obvious

I was out of the United States training a group of operators about fuel trains and combustion systems on their ovens. These had fuel trains at ground level and a small blower mounted higher, about waist level. The fuel trains had somewhat delicate plastic-covered, high and low gas pressure switches connected to the 2-inch main gas lines with little $\frac{1}{4}$-inch sensing line nipples. It would seem obvious that you would never want to climb around on the gas piping and risk damaging one of the plastic

Fuel and Combustion Systems Safety: What You Don't Know Can Kill You!, First Edition. John R. Puskar.

switch housings or break one off the gas train, especially the low gas pressure switch. There's always gas to the low gas pressure switch as long as the main isolation valve is open. Breaking this off means an immediate release until someone thinks to shut off the main isolation valve.

I explained during one training session that the switches are delicate and jokingly said through a translator: "You would never want to climb on this fuel train and put one of your feet on one of these as a step." I thought it was kind of a stupid thing to say, as it would have been exceedingly obvious to everyone already. One of the operators came to me afterward and thanked me because no one had ever told them that. They indeed did climb on the fuel train all the time to get to things on the roof of the oven, and they could very easily have stepped on one of the switches. They did not know that if they broke off the low gas pressure switch, fuel would flow into the plant unless someone shut the main isolation valve off. I looked closely at the fuel trains and burners and was shocked at the damaged airflow sensing lines and gas lines, which were no longer level. I was sure that some of the threaded piping was now probably leaking from being stepped on so much. What I thought was so obvious was really not obvious to people who don't live in my world. They saw the fuel trains simply as pipes with things occasionally sticking out of them for some reason.

Lessons Learned Don't assume that people you are training know anything about fuel trains and combustion equipment safety. Start training at a very simplistic level. Review equipment carefully and have fact finding and culture discovery sessions with plant personnel before designing content to make sure that what you will be discussing and how you will be presenting it are relevant and useful. I have found that equal parts of classroom and field trips where equipment can be visited and touched are a good recipe for success. Many operators are not oriented academically and learn better with their equipment in the field rather than in classroom settings.

11.1 GLOBAL PERSPECTIVES ON FUEL AND COMBUSTION SYSTEM RISKS

When you consider fuel and combustion system risks from a global perspective, you must consider differences in codes and standards, equipment, and even cultures. The differences between fuel and combustion equipment safety issues and risks between developed countries and developing countries are vast. So far, our discussions have been confined mostly to things the way they are in the United States. But as the business world becomes more global, companies are relying increasingly on production facilities in developing countries in Asia, South America, and Eastern Europe. In this chapter we provide information for risk managers, equipment suppliers, and others who work in the combustion field on the world of fuel and combustion system risks outside the United States.

The fuel safety and combustion equipment problems that developing countries face will not be solved overnight. However, companies with facilities in these geographical areas can help save lives and improve their chances for success by changing their combustion culture and developing maintenance and safety procedures that can be embraced worldwide. Remember, problems with fuel-fired equipment can jeopardize many people's lives. This equipment is unforgiving and you get only one chance for success.

11.2 HIGHLIGHTS OF THE EUROPEAN COMBUSTION WORLD

Outside North America, the only well-developed and well-known set of guidelines or rules for combustion equipment comes from Europe. The European Committee for Standardization or Comité Européan de Normalisation (CEN) developed the EN Standards (www.ENStandards.org) to serve the common interests of countries within the continent.

CEN is a nonprofit organization whose mission is to foster the European economy in global trading and to ensure the welfare of European citizens and the environment. CEN seeks to do this by providing an infrastructure for the development, maintenance, and distribution of standards and specifications.

CEN was founded in 1961. Its 30 member nations work together to develop EN standards in various sectors. Their goal is to build a European internal market for goods and services and to help better position Europe in the global economy. Some of these standards are voluntary, whereas others have been made effectively mandatory under European Union (EU) law. More than 60,000 technical experts, as well as business federations, and consumer and other societal interest organizations, are involved in the CEN network, which reaches over 460 million people.

11.2.1 Popular EN Industrial Standards

The following safety requirements parallel the NFPA codes and standards (mainly NFPA 54, the National Fuel Gas Code, and NFPA 86, the Standard for Ovens and Furnaces):

Ovens and Furnaces

EN 746-1	Safety requirements for industrial thermoprocessing equipment
EN 746-2	Safety requirements for combustion and fuel-handling systems
EN 746-3	Safety requirements for the generation and use of atmospheric gases
EN 746-4	Particular safety requirements for hot-dip galvanizing thermoprocessing equipment
EN 746-5	Particular safety requirements for salt bath thermoprocessing equipment
EN 746-6	Particular safety requirements for material melting, remelting, and liquid-phase maintenance of thermoprocessing equipment
EN 746-7	Particular safety requirements for vacuum thermoprocessing equipment
EN 746-8	Particular safety requirements for quenching equipment
EN 1539	Dryers and ovens in which flammable substances are released

Boilers These standards are somewhat parallel to ASME CSD-1 and NFPA 85:

EN 12952 Water tube boilers and auxiliary installations
EN 12953 Shell boilers (firetube)

Piping This is somewhat parallel to NFPA 54, the National Fuel Gas Code:

EN 15001 Gas infrastructure/gas pipe work in buildings

11.2.2 Other Important Fuel and Combustion system Standards Organizations

Canada: Canadian Standards Association (CSA; www.csa.ca) The Canadian Standards Association is a not-for-profit, membership-based association serving business, industry, government, and consumers in Canada and the global marketplace. CSA develops standards for products and their installation and is also a testing laboratory for certifications.

Great Britain: British Standards Institute (BSI; www.bsi-global.com) BSI Group is a leading business services provider to organizations worldwide, offering a range of services for management system certification, product testing and certification, and standardization.

Germany: (TÜV www.tuv.com) TÜV is a global provider of technical, safety, and certification services. It is headquartered in Cologne, Germany. The TÜV Rhineland Group employs more than 12,000 people in 62 countries and generates annual revenues of $ 1.1 billion (40% outside Germany). TÜV provides a listing service for many industrial products, including boilers and furnaces.

France: Bureau Veritas (www.bureauveritas.com) This organization is accredited in a vast number of areas to verify and give official acknowledgment that a system, product, person, or asset complies with a specific requirement for which certification is required. Certification usually includes on-site audits, standardized testing and inspections, and then surveillance audits during the certification period of validity. A design review phase may be included where applicable.

Russia/Ukraine: (GOST www.gost.sgs.com) Practically all goods exported to Russia are subject to some form of mandatory certification requirement, such as GOST-R. Since 1993, GOST has been the Russian system of mandatory quality certification for all products produced in and imported to Russia. The purpose of the system is to protect the health and safety of Russia's population by excluding potentially hazardous or unsafe products from entering Russia's commercial space. The registered GOST-R sign (i.e., mark of conformity) demonstrates a product's compliance with Russian quality standards. The rules and regulations for product testing and compliance with the GOST requirements are established by the Russian state authorities. The requirements include technical product testing and expert evaluation by an independent third-party testing facility accredited by the Federal Agency for Technical Regulation and Metrology. A violation of the certification terms is a violation of Russian laws.

11.3 FUEL SYSTEM RISKS IN OTHER COUNTRIES

If you are a supplier or organization currently working in or contemplating working in a developing country, you should pay close attention to this section. In the beginning of your engagement or experience with a different country, ask the following key important questions: What are this country's fuel and combustion equipment standards and codes, if any? Are standards of other countries acceptable? What will be our guidelines for designing, purchasing, installing, operating, and maintaining equipment?

In developing countries you will usually find a mixture of equipment from the United States and Europe. Which standards are followed usually depends largely on the equipment vendors. If a European vendor has penetrated the market area, equipment is usually configured and installed according to EN standards. If the equipment is provided by a North American supplier, the ASME/NFPA codes are usually followed. As long as you choose a standard to follow (e.g., NFPA, ASME, or EN), you will do well. The best advice is to work with whatever standard is most prevalent or best known in the area and make sure that you are in communication with insurers and have their approval. Table 11.1 lists requirements in other countries that are known to the author. The list is not complete, and the information should be verified before starting a project.

11.3.1 Keeping Things Simple

Many companies doing business outside North America and Europe do not think to take special precautions to address the risks that exist with fuel and combustion system equipment. They may have been lulled into a false sense of security because they've never had problems in the United States, related to fuel systems or combustion equipment. In developing countries, however, they may be dealing with different fuels, minimal operations and maintenance skills, and a very different safety culture.

The state of development of infrastructures and support systems in some developing countries is close to what existed in the early twentieth century in the United States. Fuel and combustion system problems in developing countries have caused explosions and fires that have completely destroyed new ventures. For example, on May 1, 2006, in Kashipur, northern India,[1] seven people lost their lives and many others were trapped when a boiler exploded inside a paper mill. In northern China[2] on February 29, 2013, thirteen people were killed and 43 injured in Hebei province when a chemical plant explosion occurred. I am not saying this was the case in these two incidents; however, I have witnessed poor safety attitudes and inadequate respect for combustion equipment in several developing countries. In many of these countries

TABLE 11.1 Boiler Standards and Practices in Some Latin American Countries

Country	Boiler and Furnace Standards	Inspection and Certification
Bolivia	NFPA 85 and NFPA 86 used; IEC 6151 is sometimes also used. Mexican Standards also used: Norma Oficial Mexicana 020-STPS[a]-2002, Recipientes Sujetos a Presion y Calderas req. de Seguridad, Norma Oficial Mexicana NOM-122-STPS-1996 Condiciones de Seguridad e Higiene Para El Funcionamiento de Los Recipientes Sujetos a Presion y Generadores de Vapor o Calderas. Chilean standard also used: Norma Chilena 933 Req. Seguridad para Hornos.	No government inspection. ASME stamp required for pressure vessels and boilers. Many companies require third-party certification.
Mexico	NOM[b]-020-STPS-2002, Recipientes Sujetos a Presion y Calderas Req. de Seguridad NOM-043-SEMARNAT[c]-2003.	Operating license required; five-year hydrotest. Environmental checks by a laboratory registered with EMA.[d]
Peru	Petroleum refineries: furnaces: API RP 530, 530M, 532 and STD 630; ASME ANSI A 58.1, B16.5, B31.1 and B31.3; ASME Boiler and Pressure Vessels Code, Sections I, II, IV, VIII, and IX; AWS D1.1; ICBO UBC-79 or equivalents. Boilers: ASME Boiler and Pressure Vessel Code Other Industrial Installations Furnaces: No Peruvian Standard Boilers: ASME Boiler and Pressure Vessel Code. NFPA 85 and 86 are not adopted in Peru.	OSINERGMIN, for refineries and oil storage. Ministerio de Trabajo (Ministry of Labor): other industrial and commercial installations.

[a]STPS, Secretaria del Trabajo y Prevision Social, (the Labor Department).
[b]NOM, Norma Oficial Mexicana.
[c]SEMARNAT, Secretaria de Medio Ambiente y Recursos Naturales (the equivalent of the U.S. EPA).
[d]EMA, Entidad Mexicana de Acreditacion.

some incidents don't even get reported. My experience is that at times you get the feeling that there is an expectation of "things happen"—clean the place up and call in the next bunch of people waiting for a job. This is reminiscent of what historians have said it was like during the early nineteenth century in the United States. In another example, in Bangladesh it is reported that there are more than 10,000 unregistered boilers in use. According to officials at the Bangladesh Boiler Directorate, the department is suffering from a severe shortage of boiler inspectors, which leaves these facilities vulnerable to accidents.[3] Accidents can create circumstances where new employees do not want to work at a facility and governments can deem the entire enterprise undesirable. Laws regarding criminal liability related to managers are also much different than they are in the United States. This can make the risk of jail a reality for things that could be beyond a manager's control.

IMPORTANT

Remember that systems in developing countries may serve you better if they are simple and robust. Think of support, repair, and operational environments. There is such a thing as too much technology for some applications and cultures.

Although it is human nature to want to use the most efficient technology available, this desire needs to be tempered with the realities of the level of knowledge and skill of the site's operations and maintenance staff. As the risk or safety manager of an American company doing business in the developing world, the two keys to success are training and keeping things simple.

Real-Life Story 48: Quality Craftsmanship?

A maintenance manager at a plant in a developing part of the world was so proud of the recent repairs that had been made to a water tube boiler's outer back wall skin that he called me over to bask in his glory. "You must admit our craftsmanship is superb," he said. "Yes, it's a very nice job. How old is the boiler?" I asked. "Oh, it's about 10 years old. We replace the skin in the back in that very spot about every two to three years like we're supposed to." "Wow! What do you mean by 'like we're supposed to'"?, I replied. He responded: "Well, the back wall corner area burns out and the sheet metal skin of the boiler starts to get very hot. The paint burns away and the metal starts to warp, so we shut the boiler down and change it." I reviewed the flame, the oil, and the atomizing air settings. It was obvious that the flame was pushed back into the corner of this boiler on mid to high fire. The regular burning out of the back wall and the subsequent routine replacement of the skin was caused by a simple flame adjustment problem. Clearly, the burner had not been set up correctly when it was commissioned.

Lessons Learned Remember, flame impingement is always a bad thing. Hot spots should never be considered routine or something that just happens with equipment. In some cases the wrong burners are installed or they are installed or commissioned incorrectly. When reviewing any site, ask about recent repairs and look for fresh paint or different-colored panels on a furnace or oven or boiler. Throughout my career I have seen numerous incidents of paint being applied over and over where the underlying issue was really refractory failures that were getting worse over time.

There are often a very limited number of competent service contractors in developing countries. It is always a good idea to provide extensive training and skill development to the site's own maintenance staff. This may be beyond what might normally be done at more traditional U.S. facilities. This incident is one more example of why root-cause investigations of failures and widespread communication of the results is important.

11.3.2 Design Issues to Minimize Developing Country Risks

A number of risk factors must be considered in the design of a new facility in a developing country: the type of fuel, the type of safety and burner management systems, training requirements for personnel, and the availability of spare parts. My experience has shown that many designers who deal with this type of equipment in the United States are ill-prepared to ensure safety and reduce risks in developing countries.

An American company may not have any operating experience with propane or fuel oil or with fuels that have a higher level of impurities and contaminants. Yet these are often the only fuel choices for facilities in developing countries. These fuels may require that designers consider special measures such as extra strainer capacity, special automatic fuel valves and burners that are of an older but simpler technology, and component vendors that are unfamiliar. Fuel/air ratio controls that are not microprocessor based usually make the most sense in some of these locations. In these situations, robustness and simplicity sometimes need to drive design, not energy efficiency.

11.3.3 Training: The Missing Link for Controlling Developing Country Risks

Human error is the leading cause of accidents both in the United States and abroad. Forward-thinking design and equipment selection cannot solve this problem completely. Very often, little thought is given to training in developing countries. Training is even more important in these countries than in the United States because in many cases there is little or no industrial infrastructure or knowledge of fuel-fired equipment or combustion systems. To complicate matters, often little or no information is available in native languages. I have carried out a number of training assignments in Mexico and South America and was surprised at how little has even been translated into Spanish, let alone the myriad of other South American, Asian, and Indian languages.

Another thing that makes training very important in developing countries is that more operator turnover occurs in the developing world than in the United States. Some facilities experience more than a 100% annual turnover rate. If new employees cannot be trained quickly and easily, the chances of a catastrophe rise dramatically.

Language, literacy, and the quality of translation may also be an issue. Even dialects within a language can be challenging. For example, I always thought Spanish was Spanish until I tried to have training programs in English for use in Mexico translated by a Spanish-speaking man from Puerto Rico. It was a mess. Even the Spanish and technical terms spoken in Argentina are different from some of the Spanish in Mexico. Different dialects may appear to be similar but have very different meanings for the same words.

Finally, translation of instructions and documentation should be done by people who fully understand the language in which the documents are written and the language and dialect spoken at the plant site. If the translation is done by a professional translator, it must be reviewed by someone (preferably an engineer) who is familiar with the equipment. It is not unusual for nontechnical translators to translate technical phrases word for word rather than by the meaning. We have only to look at the differences in the vocabulary in the English spoken in the United States, England, and India to understand this. An example of one embarrassing case was when in one class the term washer was misinterpreted and translated as a machine one cleans clothes with rather than as a fastener device placed under a nut.

When preparing combustion equipment training outside North America and Europe, it is also important to develop basic training materials that are also at the appropriate grade level. Remember, many operators may have had no formal schooling. The information provided must be very visual and at first cover only the most egregious of basic hazards. An example of how to illustrate very basic hazard recognition training material would be to show black smoke coming out of a flue and explain that this means that mixtures are too rich and/or that the combustion air is restricted.

Also be careful when selecting illustrations. An American presenter was giving a seminar in the Dubai and used a picture of a skunk to clarify the smell of the odorant used in fuel gas. He was not aware that skunks are found only in North American and parts of Europe. The students did not understand what was being presented.

Culture also needs to be addressed. In many cases, workers are from rural areas and might be somewhat new to industrial equipment. A facility in India experienced a serious fire and explosion related to leaking propane from the fuel train of a process oven. Workers at the plant did not understand that the smell they were frequently experiencing was actually propane leaking from an incorrectly installed vent pipe in the system. None of the workers was trained to know about the hazards of propane and how to detect it. They burned wood or coal at home and had not experienced propane before and had no way to know what the smell was or that leaking propane was dangerous.

11.4 DIFFERENCES IN TYPES OF SYSTEMS AND EQUIPMENT FROM U.S. TO FOREIGN OPERATIONS

During my fuel system and combustion equipment testing and inspection work outside the United States, I have found many important differences in equipment and approaches to safety. The following random compilation includes some great technologies that are moving their way into the U.S. marketplace.

1. *Valve proving systems.* Valve proving systems, which leak-test automatic valves, are in widespread use throughout Europe and are just arriving in the United States. Valve proving systems come in several different designs (Exhibit 11.1). Some use an independent pump to pressurize gas between the automatic valves for a pressure hold test validated by a pressure switch. Others open and close valves and use special porting to route a fuel train's gas for a pressure hold test. Valve proving systems[4] effectiveness relies on the

EXHIBIT 11.1 Dungs' valve with a valve-proving system module.

sensitivity and function of the pressure switch used for the pressure hold test validation. These systems can replace the traditional double block-and-bleed concept common in the United States where they are approved for use.

2. *Modular fuel train components.* Modular fuel trains are cast-machined devices that allow for entire traditional fuel train components, regulators, high and low gas pressure switches, safety shutoff and blocking valves, and even sometimes gas flow or capacity control, to be installed in about 24 inches of pipe space in one contiguous device. These are generally used for small commercial or light-duty systems. In most cases, modular fuel trains are not capable of very high gas pressures, and the pressure drop through the device is greater than in full port valve systems.
3. *Vent valve discharge bubblers (flow indicators).* In many European installations vent valves are installed with discharge bubblers to indicate undesired flow from a leaking vent valve by gronely.
4. *Tubing.* In many cases gas piping is done with tubing instead of with traditional pipe and fittings. This yields fewer joints that may leak.
5. *Electrical connection plugs.* Electrical connections are often done with plug and socket fittings on components instead of more permanently wired terminations in protected junction boxes.

6. *More licensure of boiler operators and those servicing equipment.* Some eastern European countries have very strict licensing requirements for boiler operators and exercise considerable control over such things as the number of operators on a shift, training requirements for supervisors, and maintenance and safety testing requirements. There are also many countries with strict requirements for those that service combustion equipment.
7. *Gas filters.* In many cases, gas filters are used rather than strainers. These gas filters have fiber-based filter media and not the metal mesh or screening found in a strainer.
8. *Earthquake fuel shutoffs.* Earthquake fuel shutoff valves[5] are used much more frequently and are more widely accepted than in North America for main incoming fuel shutoffs to facilities.
9. *Emergency fuel shutoffs.* The use of an automatic main fuel shutoff valve for an entire site is not unusual. These are usually associated with several emergency stop buttons located in key parts of the facility for shutting down fuel.
10. *Gas leak/fire detectors.* It is not unusual to find gas leak and/or rate-of-rise heat detectors above or near fuel trains on boilers and other systems. These are usually connected to alarm reporting systems or automatic main fuel shutdowns.
11. *Measurements (SI/English).* Be certain that your instrumentation matches the system in use and the predominant language and culture at that site. It is a good idea to use gages that have both metric and English scales. Since there is likely to be a mixture of equipment, there is also likely to be a mixture of measuring systems. This is more probably where equipment is obtained from different countries, such as the United States and a European country.
12. *Metric threads.* Metric pipe threads are straight and not tapered like the National Pipe Threads (NPTs) used in the United States. You cannot mix metric threaded equipment with NPT equipment. In many cases, European components use bolted-on flanges that contain O-rings, and not ASME-type bolted flanges. You must be aware of this difference when designing or servicing components and piping systems. Also, while the flanges may look the same, there are differences in the bolting pattern, and one of the flanges may have to be drilled to make the bolt holes match up.

11.5 FUEL ISSUES

Fuel types vary widely in the developing world. In the United States, we use primarily natural gas. In the developing world, propane and fuel oils of many types are used. Not only is there diversity in the types of fuels, but there could also be diversity within the fuel that you are buying.

11.5.1 Fuel Oil Contamination

In some cases fuel oil in developing countries is a buyer-beware commodity. If the oil has been trucked or sent by rail or pipeline to a terminal and a local distributor using a series of tanks, contaminants are common. The major contaminants with fuel oil are BS&W[6]: (basic sediment and water). Solids include ash, carbon residue, asphaltenes, sediment, trace metals, aluminum, and silicon from catalyst beads. You should demand specification sheets for each load. You will also want to perform sampling and testing of loads. You must establish some sort of quality control for what you purchase. The testing is not complicated, but it must be done. One bad batch of fuel oil can put you out of business for weeks while systems are flushed out and cleaned.

11.5.2 Propane (LPG) Combustion System Concerns

In some developing countries, propane can differ from month to month with variations in its composition, and this can have a dramatic impact on preventive maintenance needs. Variations in composition may require that you frequently observe flames, measure flue gas compositions, and adjust air/fuel ratios. In the developing world, propane systems are usually a commercial grade of propane. Tanks are usually above ground, but I have found widespread use of manifolded gas cylinders rather than liquid propane tanks. Instead of having a truck refill the tanks, the used cylinders are replaced with new cylinders.

Cylinder safety is another important topic. This begins with proper unloading and handling of cylinders and then securing them carefully. Cylinders must also be carefully evaluated for their condition, especially on the bottoms where contact with soils can cause corrosion. All cylinders bear date codes. Most cylinders have pressure vessel recertification requirements that go by these date codes. Verify that yours are not overdue. Cylinder manifolding also means that you will have many threaded fittings where leaks can occur. Many propane cylinder installations are done in cabinets or enclosed areas. In these cases, even small leaks can accumulate and be a problem.

The following U.S. fuel codes and standards from NFPA will give you some basis for reviewing and understanding the systems you have: (1) NFPA 58, Liquefied Petroleum Gas; (2) NFPA 59, the Utility LP-Gas Plant Code; (3) NFPA 55, the Standard for the Storage, Use, and Handling of Compressed Gases and Cryogenic Fluids in Portable and Stationary Containers, Cylinders, and Tanks. Note that there is some confusion outside the United States on the relationship of NFPA 58 and NFPA 59. The vast majority of propane systems covered in this book comes under NFPA 58. NFPA 59 covers utility gas plants, plants that mix propane and air to reduce the heating value to that close to natural gas and supply gas pipeline distribution systems of 10 or more users. In some countries, NFPA 59 has been adopted to cover industrial systems, and must be followed.

NOTES AND REFERENCES

1. Hindustan Times, New Delhi, India, May 1, 2006, www.highbeam.com/doc/1P3-1030404581.html.
2. Ningbo News, www.whatsonningbo.com/news-7140-13-dead-43-injured-after-fatal-chemical-factory-blast-in-hebei.html.

3. People Daily, http://english.peopledaily.com.cn/200504/12/eng20050412_180632.html.
4. Dungs, www.dungs.com.
5. FM Global, www.fmglobal.com/shamrock/P0290.pdf.
6. Fuel oil specifications, Wartsila quality requirements and recommendations for heavy fuel oil, www.pomorci.com/Edukacija/80-100/Fuel%20oil%20Quality.pdf.

12

Business Contingency Planning

Even the Best Still Have an Occasional Issue

Everything presented in this book is focused on helping you to reduce the probability and severity of a fuel or combustion system accident. However, nothing can bring all of this to zero risk. For example, there will always be things beyond your control, such as weather events. It's my intent to help you to respond in an effective and timely manner and to know something about what to expect should there be an incident at your facility.

Real-Life Story 49: When Is a Flood a Flood?

In its most basic sense, flood damage is really, generically, water damage of any type. A $450 million disaster that occurred at a major midwestern electric generating station in early 1999 began when a toilet overflowed. Of course, as Murphy's law goes, the toilet was over a control room and the perfect storm of water flows occurred, sending water down conduits and into control cabinets where the most sensitive and important PLC-based BMS safety devices existed for the plant's boilers. The following is from an expert in a rate case regarding this incident. It summarizes what he believed had happened and is based on a transcript of his testimony:[1]

> In early 1999, the utility shut down its generating unit for a forced outage due to a ruptured steam line. When it did so, it shut off the gas flow to the boiler and placed a "hold" on the gas valve. Four days later it initiated a heat-up of the generating station. Initiating a heat-up includes sealing the boiler, establishing a vacuum, and opening gas

Fuel and Combustion Systems Safety: What You Don't Know Can Kill You!, First Edition. John R. Puskar.

> valves to introduce gas to the igniters. Flames from the burners then begin to heat the boiler. The unit's operators control this process using a computerized (BMS) Burner Management System. The Burner Management System was a PLC based fail-safe system that alerted employees to any unsafe conditions and automatically closed all gas valves if the flame went out or if any potentially explosive or unsafe conditions developed.
>
> While the utility employees were starting the heat-up process, two independently-contracted workers were attempting to weld a feed water heater. In attempting to establish a vacuum as part of the heat-up process, employees noticed that the weld repair was not complete and, in fact, the employees were told that the repair would take at least another twelve hours. So, the shift supervisor decided to stop the heat-up process, and he instructed the control operator to "take all the fuel out of the boiler.
>
> Later that afternoon, around 3:00 p.m., the toilets in the control room began to overflow. The cause of the overflow was a wastewater sump pump operating while the main sewer line was plugged. Two outside maintenance contractors were on site attempting to clear the sewer line, and had removed the toilet from the control room's restroom.
>
> The water from the control room's restroom traveled down drains, electrical conduits, and other openings in the control room floor to the computer room located several floors below. The water in the computer room caused electrical shorts to occur in the Burner Management System. A technician was called to the plant to assist in repairing the Burner Management System.
>
> The Burner Management System was under repair for more than eight hours. Despite the fact that the plant was supposed to be shut down during this time and, thus, no fuel was supposed to be in the boiler, fuel was flowing into the boiler. The flow of gas into the boiler continued to increase until 12:30 a.m. on the day of the incident, when the explosion occurred. After the explosion, a utility employee saw a fireball in the rubble of the boiler, which indicated that gas valves were open and gas was continuing to flow into the boiler. Upon seeing this, the employee shut the main gas valve to stop the gas flow.

The local newspaper reported the following[2]: The force of the blast woke up sleeping neighbors up to 20 miles away and sent flames shooting 200 feet into the air. By the time the fires died down, an 11-story boiler had been reduced to a five-story pile of rubble. Employees working that night said that it was a miracle that they had survived.

Amazingly, all that sheer devastation was caused by a boiler that wasn't even turned on. Only 12 workers were on site at the time of the blast. Things might have been different had the incident occurred during the day shift, when 135 employees would have been on the job.

This case resulted in lawsuits that went on for years between the utility, insurers, and the BMS manufacturer. There were different opinions as to the direct cause of the incident. However, it cannot be disputed that water in the wrong place at the wrong time had something to do with this disaster.

Lessons Learned Flooding is not just torrential rains and rivers running over their banks. You can see by this incident that you've got to open your mind to the concept of flooding as just general water damage. In this context a flood can be from a pipe break, it can be toilets overflowing, and even someone spraying down equipment with a garden hose.

At a pharmaceutical plant, where cleanliness is, of course, paramount, I was told of a situation months earlier when cleaning crews were doing a routine spraydown of everything, using with hoses. One of the cleaners hit a Scotch marine fire tube boiler front control panel and its combustion air motor. The motor single-phased dramatically, reducing the speed and airflow. Black smoke started pouring out of the stack. Luckily, it was shut down immediately with no further incident. I have been in food plants where workers with hoses were hosing down areas near high-voltage gear. I was so scared of what was being hit with water that I wanted to run.

Make sure that manual isolation fuel valves are locked out throughout periods of control panel troubleshooting. Anything can happen when people are troubleshooting panels and PLC programs. Repair practices have to include complete safety interlock trip testing when coming back from suspected damage to control panels.

Make sure that experienced staffs completing startups after troubleshooting first do a dry-fire startup (i.e. a startup with no fuel). Startups should then proceed with pilot fuel only. Once pilots are deemed to be correct, low-fire start should be begun and the system held at low-fire for additional trip testing before firing rates are increased.

12.1 BUSINESS INTERRUPTIONS FROM FUEL AND COMBUSTION SYSTEMS

Throughout this book there have been a lot of examples of how things can go wrong with fuel and combustion systems. But still, in the back of your mind you may be thinking, "That won't ever happen to me." Even with the best possible people, policies, and equipment, accidents can and do happen, and some events are simply out of your control. Depending on your location, your facility could be at risk from floods, earthquakes, tornados, and hurricanes. Regardless of whether we're talking about human-made or natural disasters, your facility needs a plan in place to deal with emergencies that is specific to fuel systems and combustion equipment.

12.1.1 Outages and Incident Action Plan Items

There are a number of factors related to fuel systems and combustion equipment that should be considered when it comes to business contingency planning. In a simplistic sense they would include loss of the main fuel supply, loss of boilers or a main steam header, and loss of furnace or oven capacity. But there are other, more subtle related events that can be expensive and dangerous lessons when it comes to fuel and combustion system contingency planning. Following is a list of such considerations.

1. *Losing the natural gas system supply.* In the case of fuel systems, natural gas can become unavailable through an economic or physical anomaly. There was a period in the 1970s and 1980s when people frequently spoke of firm and

interruptible gas. The utility could have called you and said that suddenly the gas you were buying was going to cost a lot more or they could have decided that there was not enough supply available so they were shutting you off or curtailing you. It's rare to hear of these types of supplier-related disruptions these days, but in some parts of the United States, depending on the supply piping infrastructure, curtailments do still occur. In most cases you can buy your way out of a curtailment or at least get some time to plan around it. Also, it is often not a 100% shutoff of gas but a request for a reduced consumption. You should make sure that you understand your site's gas contract, if you have one.

The greater risk is from an unexpected physical disruption, such as not having the gas available because of a piping, control or even a sabotage issue. That's right—sabotage, tampering, and even terrorism are things that we now need to have somewhere in the backs of our minds when it comes to fuel supplies, and me provide some thoughts on that as well. There have been numerous cases of main utility valves being mysteriously closed and not corrected for hours, causing lots of damage and disruption. This is why there has been discussion about protection at your main service entrance with fencing and locked gates.

Losing the entire gas supply main because of underground piping leaks can be a terrible thing in cold climates if the facility needs to be down for repairs and is vulnerable to freeze damage. I run across two or three sites a year that experience leaks in underground mains. Once this type of leak is identified, a utility will want to shut down the entire system immediately, to protect them from liability. In previous chapters we have discussed corrosion protection (i.e., having a system and maintaining it). There are other things that can help, such as having dual feeds and networking the gas piping so that it's a loop and not a single line. It's also possible to consider alternative fuels, such as backup propane air supply or even fuel oil. Propane–air mix systems have come a long way. They can make a mixture of propane and air available for injection into a fuel header that simulates natural gas in Btu value and function.

Losing gas service can also be caused by main incoming regulator failures. Some sites have installed parallel regulator stations to mitigate this risk. These systems remain closed off but ready to be moved into service simply by changing a couple of valve positions.

2. *Losing boilers.* This is where "sparing" philosophy comes into play. Sparing can mean complete 100% redundancy or something less like 66%, such as two pieces of equipment each capable of two-thirds of the total expected peak load. It's all a risk-based capital cost exercise to determine how many boilers of what size you should have. You should at least have a conscious awareness of the spare capacity that exists. Spare capacity and sparing philosophy should be evaluated and policy driven and not be something that just happened by accident. Now is a good time to see what the original design was at some sites versus what you now have. What's the right sparing philosophy for business conditions today? Have loads grown through the years so that now you actually are not spared according to the philosophy that once existed? What

about critical auxiliaries such as feedwater pumps and tanks? Are they spared? Do they need to be?

An obscure part of boiler systems that in my opinion is not given adequate attention is feedwater tanks. It is rare to find that these have been internally inspected and have had NDT testing done on a regular basis. There have been catastrophic failures of these that have made for tremendous damage and loss of life since in some cases they hold large quantities of saturated water. An opening of the feedwater tank pressure vessel, even if it is at 10 psig pressure, can make for a tremendous explosion. Besides the explosion risk, the lead time to get a feedwater tank can be 12 to 18 weeks. This can be very disruptive. The NBBI (www.nationalboard) can provide guidance on appropriate feedwater tank testing and NDT procedures to minimize risks.

Many sites install rental boiler connections ahead of time to prevent downtime in case they have to bring a unit onto the site. There is a huge boiler rental industry with at least six major nationwide players. Do yourself a favor and call one or two in to speak about this subject before you need them. They can probably do a walk-through review of your facilities at no charge and point out some things that can be done to make a rental boiler installation happen faster if this ever becomes a need. If you do ever engage them, be very careful about charges for things such as damage to their equipment. Make sure that there is a thorough investigation and documentation of the inside of the boilers ahead of time so that you don't get blamed for things like poor water treatment during the rental period.

Rental vendors will tell you they can have a boiler to you quickly, perhaps within hours. They do not tell you that you will have to provide connections for the steam, electrical, water, and so on. This can take days under the best of circumstances. Map out a place where boilers might sit, locate the utility tie-ins, and install additional valves if necessary to accommodate tie-ins. This is another matter that you need to consider in advance if the nature of your product or processes is critical. Hospitals especially need to consider being prepared for this contingency. Hospital accreditation requirements often call for alternative fuels, but what if fuel is not the issue—it's the boiler—what do you do then?

3. *Losing a furnace or oven.* In the case of furnaces or ovens there are often no good alternatives if a catastrophic outage occurs. In some cases, spare fuel trains and burners can be lined up in advance. However, in most cases, if a furnace or oven is damaged so that operation cannot occur, there is no simple backup plan other than finding another facility with capacity. Again, this needs to be a sparing philosophy determination. How many ovens or furnaces should you have?

Real-Life Story 50: Failing Furnace Gave Plenty of Warning Signs

Some incidents provide a very clear warning if you know what to listen and look for. In this case a paint oven operator at a manufacturing plant was annoyed because every time the oven lit-off he heard a poof and the doors at one end blew open. He thought he took care of the issue by jamming a chair into one of the doors so that it would not blow open. The chair kept moving around. In subsequent days he installed a piece of 1-inch angle iron he found lying around. This piece ended up bent and fell out after a few days. He called a millwright friend in maintenance and got a piece of 4-inch round bar stock and jammed it between a building column and the door. It worked. The door never blew open again. Instead, the inside of the oven blew out, brought down the line, and damaged a considerable amount of product that was in the oven being processed.

Lessons Learned There are a few obvious lessons here. First, explosion relief is a real issue. NFPA 86, the Standard for Ovens and Furnaces, requires 1 square foot of relief opening, preferably (in my opinion) somewhat near the combustion chamber for every 15 cubic feet of oven volume. NFPA 68, the Standard on Explosion Protection by Deflagration Venting, also provides valuable explosion venting guidance. Explosion relief is a designed place that is weak or hinged that allows the explosion to vent the explosion pressure pulse. Explosion relief hatches and doors are supposed to be tethered so that if something happens, metal will not be flying around.

Second, if a boiler or oven is poofing or hard starting, it's doing so for a reason. Remember, although most burner management safety control systems allow 10 seconds for a light-off on natural gas, it should never really take this long. If it takes more than 2 or 3 seconds to light a pilot or a main flame, you are on your way to problems. Light-offs should never be audible.

Third, it's clear that the operator and some of the other staff lacked basic training as to how the combustion and light-off systems worked and what trouble signs to watch for. There's no substitute for training, and that's why it's required by all of the relevant standards and codes. Personnel also have to be trained to recognize abnormal and hazardous operating conditions such as extended and audible light-offs.

4. *People smelling gas or unusual odors.* It seems trite, but I have seen entire facilities with thousands of people evacuated, costing hundreds of thousands of dollar in lost production, because someone thought that he or she smelled something. However you get there, whether because of a catastrophic explosion or just because someone smelled something, it's still an outage and you should be prepared for it.

IMPORTANT

Facility operations (i.e., industrial plants, schools, office buildings), are interrupted by gas smells somewhere every day. If not yet your facility, your day is coming. Be prepared; have a plan. Include where to evacuate, who should respond and how, and when will the fire department be called? Will it be a 911 call? A call to the utility? Do the gas shutoff valves function? Are they marked?

Incident action plans should include understanding what a site will do about gas smells in advance of when or if they actually occur. The site should have an action plan that includes understanding where main shutoffs and other important distribution isolation valves are located. The plan needs to ensure that these valves are marked, maintained, have handles, and be accessible if needed.

The plan should include steps such as who will be evacuated from what area. Where they will assemble. Who will check the source of the leak further, and how. Whether there is imminent danger to whoever would be called upon to go in and check. How this will be handled. If with an instrument, what instrument? Is the instrument intrinsically safe? Will the fire department and gas utility be called? Will a 911 call be made as soon as there is a confirmed smell of gas? What will the criteria be for bringing people back into the facility?

5. *Weather emergencies.* Another issue that needs to be considered is weather emergencies. In some cases one must plan carefully for what state fuel systems and combustion equipment will be left in (on or off) for situations such as floods, tornadoes, or severe wind.

   In many cases, production staff is ushered off a site when an evacuation occurs, but sometimes little thought is given to those who run central utility plants or critical production equipment such as major furnaces or ovens. The site managers will need to think about what the criteria are for when the staff should leave, When they leave, what will the policy be for the state in which the equipment is left? Will all boiler systems be shut down? Will fuels be secured? This all also applies to process ovens, furnaces, thermal oxidizers, and any other form of combustion equipment and its fuel systems.

6. *Heat treating and special process operations.* There are some very special fuel and combustion system processes and environments that need a real deep dive when it comes to the topic of contingency planning. Heat-treat areas are an example. When heat-treat atmosphere ovens are laid up, nitrogen often needs to be inerted into the atmosphere. This can be a considerable amount of nitrogen compared to what is sometimes available on site for normal operations. This also takes some time and must be done very carefully. Both having enough nitrogen on hand for complete system inerting and having emergency inerting procedures in place need to be part of the fuel and combustion system overall emergency plan. It's also important to review the nitrogen system capacity periodically versus the equipment installed. Do you still have enough to inert everything if you need to, or have furnaces been added without adding to the nitrogen system infrastructure? If there's a plantwide evacuation, who will do what with this equipment, and when? In some cases immediate shutdowns without proper cool-downs and orderly shutting down of systems can cause damage and be dangerous.

7. *Abandoned equipment and facilities.* In the case of buildings or to facilities that are to be abandoned or sit idle for periods of time, one also needs to

consider the state of shutoff valves to equipment and even whether or not a charge of natural gas is left in gas lines. A best practice is to remove fuel gases from abandoned lines and install a charge of nitrogen. Also, use blinds to close off unwanted sections, not just closed valves.

Policies might need to be established about securing equipment seasonally versus permanently. For example, it should be a policy that all space-heating equipment have fuel valves secured off during the summer. I have seen a number of cases in which fuel leaks have created problems during the summer months in places where they should not have occurred.

12.2 SPARING PHILOSOPHIES FOR COMPONENTS

The subject of sparing philosophy is vital to business interruption planning. It can be applied to fuel systems by adding robustness and options to fuel types, fuel suppliers, fuel piping systems, regulating stations within piping systems, and even the networking of piping systems within or outside a facility. It can be applied to combustion equipment by adding options to control systems, numbers of burners, and complete boiler or furnace system redundancy. Somewhere in this mix, a plan for having critical spare fuel and combustion system components will need to be considered.

If you ask a manufacturer or supplier or service person for spare parts recommendations related to fuel trains and combustion equipment, you will receive a list such as the one that follows. Consider it a starter list for your consideration in planning spare part needs for combustion equipment at your facility.

1. High and low gas pressure switches
2. Airflow proving switches
3. Flame detectors
4. Safety shutoff and blocking valves
5. Solenoid valves for pilot systems and vents (normally open and closed versions)
6. Water-level controls
7. Burner management system
8. Pump controller (for feedwater)
9. Boiler gage glasses
10. Combustion air fans
11. Relays and motor starters
12. Fuses
13. Control transformers
14. Temperature controllers (operating and high limit)
15. Thermocouples for ovens and furnaces
16. Gas pressure regulators (pilots and mains)

17. Modulating motors for gas and/or air valves
18. Linkage sets
19. Fan wheels
20. Combustion air blowers and motors
21. High-pressure cutout switches
22. Burner modulation controls
23. Boiler pump controllers
24. Safety relief valves

It would be difficult to have all of these for every piece of equipment. Now you can start to see why commonality among different pieces of equipment and vendors of safety devices and components is so important. If your facility delivers 10 psig natural gas to 50 pieces of equipment, it's likely that many of them have gas pressure regulators, high and low gas pressure switches, and safety shutoff valves that are nearly the same in function but possibly from different vendors. If one can do just a little bit of planning very few spare parts can serve the entire facility. The other thing to be aware of is that, in most cases, a 3- to 21-inch w.c. low gas pressure switch operates in the same manner regardless of its brand. Don't think that you need brand-for-brand redundancy. If you provide a list of all your devices with complete model and serial numbers to a competent supply house, they should be able to cross-reference like-for-like devices and help you to have an effective stock of spare parts at minimum cost.

IMPORTANT

When new equipment is specified or retrofits occur, make sure that fuel train and combustion system components are compatible with what exists at the site and may be in stockrooms and familiar to the staff. Most of this equipment is somewhat interchangeable. New styles and brands introduce complexity and confusion. These can be contrary to simplicity and the minimizing of risks.

12.3 FLOOD AND WATER DAMAGE

In the past decade, I have seen numerous facilities damaged or put into jeopardy by hurricanes and floods. Whether it is an industrial plant or an institutional facility such as a hospital, it can often be a struggle to recover and get back into operation. When putting together a business contingency plan, you need not only consider the equipment's vulnerability, but also have an idea of what it will take to recover equipment that has been damaged. The following considerations are provided to make your fuel and combustion systems more robust toward flooding incidents and your recovery more rapid and secure. As noted in the real-life story at the beginning of the chapter, "flooding" means simply that things got wet somehow, not necessarily by a storm of biblical proportions.

12.3.1 Preevent Design and Installation Issues

1. *Keep fuel train valves, piping, and components as high off the ground as possible.* Many fuel trains are installed a foot or two above floors. There is usually not a typical reason for this. In most cases, fuel train piping and components can be removed from packaged equipment and additional fittings installed to raise the equipment 4 or 5 feet off the floor. This can help to avoid damage from moderate flooding. Make sure that all operating devices and instrument are still accessible.
2. *Make sure that burner management systems are not obsolete.* Burner management systems are the main safety brain for many types of combustion systems. In For older equipment, the model may be obsolete. "Obsolete" does not, however, mean unsafe. Major manufacturers such as Honeywell[3] and Fireye[4] can give you information on model and serial numbers and pictures of their units that have been upgraded. The upgrades and new products they have delivered to the market place can make life better for your facility. Remember, BMS's are the heart of safety for combustion systems. It's a great place to make a safety investment. It's certainly a place where you want to understand what you have.

   Here's an example of why this is so important. The connecting pins and wiring on the back of an old Honeywell R-4138 BMS are completely different from those on the newer Honeywell R-7800. Hence, if an obsolete R-4138 is damaged or for some reason needs to be retrofit, it might take days to rewire the control panel. This could be the cause of considerable downtime. Users should be encouraged to replace obsolete equipment proactively instead of when an unplanned outage threatens a crisis.

   Other component obsolescence and recall concerns have come about from things such as valve actuators, valves, and pressure switches. A number of years ago, the Honeywell NaK actuators[5] were recalled because of the potential for sodium–potassium in the actuators to come into contact with water and ignite. Honeywell issued recalls for the units as far back as 1979. For a number of years they offered to dispose of them. If you ever find one, do not throw it into a dumpster. I was once speaking to a client about these when I noticed two of them sitting under his desk. He said they had been sitting there for years and did not even know why. Imagine what a spilled cup of coffee could have done!
3. *Consider NEMA weatherproof control cabinets and components along with TEFC motors, even inside.* A number of different control panel and component options are available. Check the NEMA ratings for what you are buying and get the most robust weathertight panels and component options available. The same applies to electric motor options for combustion air fans and feedwater pumps. One can get weather-resistant control panels designed for outdoor applications and install them inside if the operation is critical

and for some reason there is an expectation of water damage. One can also get totally enclosed fan cooled electric motors instead of drip-proof motors. These options make systems more flood resistant and robust.

4. *Keep spare components on site but in protected areas.* Some safety shutoff valve styles have a 4- to 6-week delivery. Many are also not serviceable. Similarly, some gas pressure regulators can be difficult to get quickly. You would do well either to have spares on site in a safe area or at least switch to more commonly available components. Also, make sure that spare components are stored somewhere away from the equipment so that the same event does not damage them as well.
5. *Prepare a list of vendors and suppliers for mission-critical components.* Many combustion equipment specialties are not readily available even when there's not a flood or emergency issue. People who do refractory or brick work, PLC programming and instrumentation related to burners, and who service burners can be difficult to find in a crisis. Think through where you might locate such people, even out of state, if you needed them.
6. *Run vent lines outside the building from all regulators and switches.* I have found many vent lines from regulators and pressure switches not terminated outside as is required by codes. In some cases, devices such as pressure switches and pilot regulators do not have any attached vent piping. This leaves them susceptible to being filled with water and damaged internally. If vent lines are installed, as is required by codes and standards, there's at least more of a chance to keep water out of diaphragms and sensitive component areas.
7. *Install dual fuel systems and/or at least connections for backup.* When installing new equipment or retrofitting, consideration should be given to making critical systems dual-fuel capable. In some cases the second intended fuel is fuel oil. This does not have to mean tanks and pumps. It could mean that an oil gun and oil train are present so that a tanker truckload of fuel can be delivered and stay there to run fuel out of the tanker unit on a temporary basis. In other cases you might want to consider fuel connections for propane or propane–air mixture systems to simulate natural gas.
8. *Have all documentation in order, including control and electrical drawings, lists of components, and set points for controls.* If an incident occurs, people may be there trying to get you back into service. They do not want to be burdened by trying to figure out wiring. Set points of control components will be needed to commission and readjust new components. You should have previous air fuel ratio setting information with gas and flows/air as well.

In many cases, equipment does not meet current codes. In some cases, state laws or the authority having jurisdiction may require that when substantial repairs are made, the entire piece of equipment must become compliant with current codes. This could add considerably to the complexity and work scope of trying to get your service restored. If this occurs, discuss the option of repairing the existing equipment to get back into production and then making upgrades on a mutually acceptable timetable.

12.3.2 Postevent Dryout and Recommissioning Issues

1. *Vent lines and regulators/switches need to be checked for water incursion.* Even if vent lines are installed, they need to be opened up to verify that nothing got in and that the components have not gotten wet somehow anyway. It's also important to make sure that relief valve vents from gas piping systems and even process or steam systems have not collected and held water. This can occur even from heavy rains. I have seen sites where these have collected water, then froze and compromised the relief valve's function. Many relief valves or piping systems have small drain ports in the body or in the discharge piping to help drain away moisture.
2. *Refractory dryout is critical.* Most fired equipment has some type of refractory. Proper careful dryout of castable refractory, bricks, and mineral wool in ovens is crucial. If you don't release the water slowly and carefully, your refractory can literally explode into pieces. This will keep the facility down a long time for repairs. Proper dryout may not be possible with the unit's existing burner or burners. You might need to bring in rental burners or heaters and let them run for days. This is a process that you want to start as soon as possible after water damage.

Real-Life Story 51: Getting Hot and Flipping Your Lid

I was called to the site of a refractory-curing explosion. It was at a primary metals facility where large, 16-foot-diameter 12-foot-deep vessels were recently relined with a poured refractory product. The poured material was about 18 inches thick on the bottom of this vessel. This refractory product had water in it, but a very small amount. There are very special dryout protocols that require specific rampup and soak times for each type of refractory. The soak times especially needed to be done carefully and in temperature stages according to a recipe.

There were also weep holes in the vessel to watch for steam escaping as water was baked out of the refractory during the heat-up process. No steam release means that the water in the refractory is not getting out. It's a much more sophisticated process than what appears on the surface. This work was being done by an outside contractor with temporary heat-up equipment designed specifically for these types of situations. There were three people around when the event happened. The heating equipment was manned by the contractor's operator and two plant maintenance personnel were there.

Suddenly the fiber-lined lid of the pot, weighing about 20,000 pounds, was launched into the air some 60 feet. It crushed some of the roof trusses, which stopped the vertical climb. It then dropped back down almost exactly back on top of the pot. While in the air, huge chunks of refractory spewed out everywhere. Imagine big chunks of sidewalk-sized debris flying around—that's what I was walking over everywhere. Luckily, no one was hit or injured.

The cause of the incident was diagnosed as being localized overheating from the way the burner was installed and operated. It's one thing to measure average temperatures. It's another thing to avoid spot overheating when you're applying heat from one place over such a large area. It's also likely that weep holes were covered or obscured and that there was too much moisture in the refractory when it was installed.

Lessons Learned Water can be very powerful when it flashes to steam and expands in volume. Never underestimate what even a little water can do, whether it's in scrap being melted, in refractory being cured, or water that has leaked into an area where it should not be, such as a foundry vessel. Refractory dryout and curing can be a very complex process. It must include considerable planning not only for process results but also for safety considerations. A separate distinct meeting and agenda just for safety issues, including a HAZOP meeting, should be part of any large refractory dryout process.

A number of high-risk activities are associated with fuel systems and combustion equipment. These include setting air/fuel ratios, installing and commissioning new equipment and burners, venting and purging of gases, and doing large refractory cures. None of these are spectator sports, and in all of these cases, personnel in the area should be limited only to those most necessary. Some of these activities should take place in off-hours.

3. *Control panels and components need special attention.* Flood waters usually contain contamination that is corrosive and may be conductive. Do not power panels that have been wet until they are properly dried out, cleaned, and thoroughly tested. You can end up shorting out equipment that was not damaged previously. Replace all relays and critical components such as burner management systems that have been submerged. The replacement cost is not worth endangering someone's life.

   The same applies if you had a fire and a chemical fire extinguisher was sprayed anywhere near a panel. The dry chemical powder has an affinity for anything hot and tends to stick to electrical contact, making a mess of things. Review whether or not you have dry chemical or CO_2 extinguishers in areas that have considerable potential for damage. Consult fire professionals for a review of types of extinguishers in various areas, the number of extinguishers, their capacities, and other safety considerations. Personnel who would be expected to use these extinguishers should be trained on their differences and their use.
4. *Fuel oil tanks need water removed, biocides, and cleaning.* Remember, water can and does get into fuel oil tanks. When it does, the water in the fuel is one problem. However, biological fouling also occurs since some bacteria end up growing rapidly in an oily water environment. In some cases oil can be pumped out and reconditioned with mobile equipment. If you don't clean the oil and treat it with biocides, you will end up fouling all of your oil distribution piping, strainers, pumping systems, control valves, switches, and burners.
5. *All safety interlocks need to be tested.* When coming back from water damage you will need to test all the safety interlocks, including flame

detectors, low-water cutoffs, high-temperature limits, and many others dependent on the type of equipment. Remember, set points need to be correct, and these devices need to work properly the first time.

6. *All fuel valves need to be checked for seat leakage.* All manual and automatic fuel train valves need to be tightness-tested for seat leakage when you are coming back from water damage. Some valve designs are more susceptible than others to water damage. Water could also damage actuators so make sure that where there are covers and housings protecting actuator components and wiring they are opened up and the insides verified to be dry.
7. *Burners need to be removed and checked for corrosion.* Some burner styles with small gas holes are susceptible to water damage and corrosion. As corrosion can occur quickly within days of an incident, such burners need to be checked and cleaned prior to firing.
8. *Fuel/air ratios need to be reset and checked.* Fuel/air ratio control systems come in many types and styles. In some cases sensitive valves and controls could have been compromised. You will need to have burner flue gases checked using a flue gas analyzer over the complete firing range of the burner. A burner that is not operating at its proper fuel/air ratio can be unstable, produce unacceptable levels of carbon monoxide, and be an explosion risk.
9. *Combustion air fans need to be checked and cleaned.* Combustion air fans are the heart of any combustion system. You will need to make sure that if filters exist, they are not saturated and/or clogged. Fan blades will need to be clear. Fan bearings and rotation will need to be checked. Fans will also need to be reviewed for vibration and balance. Fan motors may need to be replaced.
10. *Have procedures in place for emergency response when an event may be imminent.* Think through issues and abatement measures prior to events. What valves will be secured? In what condition will equipment be left? In some cases, sensitive components such as expensive burner management systems and PLC modules can be popped out of a cabinet in less than 5 minutes and removed from the site. Do you have backup PLC programs stored off site?

IMPORTANT

There is likely to be great management pressure to get things started up and running quickly after a flooding incident. *Resist this until you are ready*! Communicate up front to set the right expectations that include all refractory will be dried and all control systems and safety devices and fuel valves will be either replaced or deemed to function properly and not be moisture damaged. This can require considerable time but it's the right thing to do.

12.4 WHEN THERE IS AN INCIDENT

If you have an incident, there will be many issues to consider, including emergency response, business continuity, crisis communication and public information, incident management, and emergency operations centers. The 2010 edition of NFPA 1600, the Standard on Disaster/Emergency Response and Business Continuity Programs, is an excellent reference for understanding how to do this in your organization. There is a lot written on this subject, especially in light of the many terrorist incidents that have occurred worldwide. It is my intent here to address issues that might only be specific to fuels and combustion systems.

12.4.1 Investigation Issues

It seems that in today's world a fire or explosion of any significance will be treated as a crime scene until someone in law enforcement determines that it is not. This means that the facilities and scene are likely to be locked down for some time and then released.

In one such case the facility was tied up over a week by agents from the ATF (Bureau of Alcohol, Tobacco, and Firearms). They set up a perimeter to keep people isolated from the area. Then they were assisted by local and state police. In one case a state had a statute for criminal negligence and the authorities wanted all evidence preserved (Exhibit 12.1). Remember, all it takes is a prosecutor to think that he or she can get a grand jury to indict you or someone on the staff for manslaughter or criminal negligence and suddenly you're possibly a criminal and facing jail time. Each group will have to feel that its own investigation is complete before they release the scene to another group, which may investigate further. In the meantime your customers are without product and you are probably paying employees their full wages just to keep the workforce preserved.

Once all law enforcement personnel are off the site, you have to consider all issues related to evidence handling and proper chains of custody since you will still have to consider civil litigation from insurers and employees who may have claims stemming from injuries. You should have a policy for this and discuss the issue far in advance of when you have to deal with the matter. Rules of evidence involve what may later be admissible at trial. In one state, if a holder of evidence damages or destroys any of it through a conscious act or even through negligence, the jury is instructed that it must assume this to have been done intentionally because the evidence was not favorable to that party. This means you have to be careful collecting and storing things.

There are rules of evidence for every jurisdiction (federal and state). There are also different types of evidence (demonstrative, documentary, and testimonial). It's important that you seek legal advice immediately to keep things preserved. Special rooms might have to be put aside or even warehouses rented for evidence. Then there will have to be protocols and control and possible videotaping when evidence is even touched or moved from the site to a warehouse so that repairs can be made and production resumed.

EXHIBIT 12.1 Scraps of metal GPS located at the site, documented, bagged, and tagged as evidence.

In the case of civil litigation there will no doubt be armies of people wanting access to the site to try to see things, remove things, take pictures, and make evaluations. This will, of course, include news media, both local and national. The first concern has to be the safety of the people at the scene. Safety issues such as structural stability, egress paths, remaining fuels, or explosion hazards all need to be evaluated before letting anyone on site. There will also need to be plans for escorts and PPE for those entering the site.

As an investigation proceeds, interviews will become important. Professional investigators have told me that the best chance of getting unbiased information from personal interviews occurs within the first 24 hours. This is the best time to conduct these if possible. At some point people involved in the incident start speaking with each other and/or with attorneys. Once this happens, accurate accounts of what happen sometimes get contaminated.

Everyone's busy with day-to-day activities, and often there are not enough hours in the day to worry about what is occurring, let alone to think about something that might never happen. However, when it comes to fuels and combustion equipment contingency planning, you will find that the payoff is multiples of your time and money invested now. Share the load and create a team to meet periodically to address these contingency issues. There is no doubt that many of the issues considered and countermeasures taken now will lead to immediate payoffs in ways that not considered previously. At the end of the day, it is my sincere hope that all of this is just an exercise for you that never needs to be actionable.

12.5 CONCLUSION

Be prepared—this motto works for the boy scouts and it can always serve you as well. My sincere hope is that you will never find yourself in a situation similar to one of those described in the book's real-life stories. Hopefully, you have gleaned enough information in these pages to avert tragedy and maintain safe fuel and combustion systems. However, accidents do happen. That's why it's important to ask yourself the big "What if?." What if our plant, building, or facility faces a fuel or combustion equipment emergency? What should we do first? What should we check? How can we ensure the safety of ourselves and others?

12.5.1 Key Items to Implement Immediately

As you complete reading this book, you may be feeling overwhelmed. There's probably likely a lot to do in your organization. Remember that it's a journey, kind of like eating an elephant one bite at a time. I am hoping that as a start you can at least act on the following things right away. Any one of these items can mean the difference between life and death in a time of crisis.

1. Locate natural gas main incoming shutoff valve locations and key distribution system valves within the facility. Have these valves checked for operability in a planned and carefully controlled manner so as to avoid business interruptions. In many cases, natural gas main shutoff valves have not been serviced or checked for function in years. Lubricated plug valves need to be serviced regularly to remain operable. This means that special sealants and equipment have to be on hand and/or available to keep these valves working. Make sure that the valves are well marked and accessible.
2. Review your site's fuel systems and combustion equipment for robustness or vulnerability to water damage. Things to consider are low-lying areas with few drains, fuel trains low to the ground, and even critical control cabinets with water lines directly overhead.
3. Consider steps that you can take today to reduce risks by reviewing sparing philosophies and adding rental boiler connections.
4. Review your site's emergency disaster plans to make sure that proper consideration has been given to fuel systems and combustion equipment. Review the conditions that would trigger a boiler house or process area

evacuation. Make sure that for processes that require orderly shutdown, personnel understand what procedures need to be followed. Review the procedure in a formal meeting with all operators and relevant staff. In some cases, equipment has to be specially prepared for a safe shutdown. Consider the impact and importance that electrical systems and the sudden loss of power can have on combustion equipment. In some cases, a loss in electrical power can cause control systems to restart equipment automatically in an unsafe manner. Review your operations with an eye toward which combustion-related control systems and operations need to be on emergency power or battery backup.

5. Ensure that emergency contact information is up to date. This includes emergency and management personnel phone numbers. Consider that many gas utilities have merged and/or have changed names, which can mean they may also have changed emergency contact information. Consider supplementing your list with disaster resource number information, such as the names and phone numbers of sources for rental boilers, generators, pumps, and/or even fuel suppliers (propane or oil).

NOTES AND REFERENCES

1. Excerpt from the testimony of Jerry N. Ward, consultant, in the following case: State GS Technologies Operating Co. Inc. GST v. Public Service Commission of State of Missouri, September 16, 2003, www.FindLaw.com.
2. Michael Mansur, Christine Vendel and Matthew Schofield, Investigators Hope to Find Answers to Massive Blast at Hawthorn Power Plant, Kansas City Star, February 17, 1999, www.ibew1613.org/library/hawthorn.html.
3. Honeywell, www.honeywell.com.
4. Fireye combustion controls, www.fireye.com.
5. In a February 1985 bulletin, Honeywell provided the following description of NaK: "An alloy of sodium and potassium metals which is liquid above 12°F (−11°C. NaK alloy is a silvery mobile liquid and has no odor. NaK alloy may ignite spontaneously if exposed to water or even humid air." www.uww.edu/adminaffairs/riskmanagement/msds/files/sodium-potassium_alloy-nak_alloy-honeywell_inc._2.8.85.pdf.

Appendix

Analysis of Real-Life Stories

STORY TITLE, STORY NUMBER, CHAPTER AND PAGE NUMBERS

Page	Chapter	Story #	Story Title
1	1	1	Innocent lives lost from a hot water heater explosion
6	1	2	The bulge in the boiler firebox?
19	1	3	But it was inspected recently!
21	1	4	Personal criminal liability at work?
23	2	5	The power of propane
28	2	6	Propane and natural gas systems don't mix well
30	2	7	No one expected a hydrogen explosion
32	2	8	Boiler operations, headaches, and being left alone don't mix well!
40	2	9	A school system in a fog

Fuel and Combustion Systems Safety: What You Don't Know Can Kill You!, First Edition. John R. Puskar.

STORY TITLE, STORY NUMBER, CHAPTER AND PAGE NUMBERS (*Continued*)

Page	Chapter	Story #	Story Title
45	2	10	Flame impingement can be deadly
50	2	11	Black smoke from a stack is never a good thing
53	3	12	Wrong valve pressure rating could have led to many deaths
65	4	13	Never disrespect a gas leak
76	4	14	A small gas smell uncovers big problems
81	5	15	My friend's changed life
83	5	16	Nonidentical twins
86	5	17	Fixing a leak becomes a big problem
96	5	18	Nitrogen-filled vessel work ends in death
100	5	19	Only one whiff of pure nitrogen can kill you
105	5	20	Seemingly simple gas line cleaning costs lives
129	6	21	BMS evolution
131	6	22	Failed light-off attempts result in death
136	6	23	A failed tuning job destroyed a boiler
147	7	24	Ten sailors die in a boiler steam accident
169	7	25	Water hammer causes extreme damage
173	8	26	Culture and design cost a person's life
176	8	27	A death caused by human error when troubleshooting a boiler
180	8	28	Explosion caused by a leaking valve during repair

STORY TITLE, STORY NUMBER, CHAPTER AND PAGE NUMBERS (*Continued*)

Page	Chapter	Story #	Story Title
181	8	29	Another explosion from a leaking valve during a repair
181	8	30	Yet another explosion from a leaking valve during a repair
186	8	31	Poor communications and human factor issues lead to a major explosion
189	8	32	A tricky switch
191	8	33	A fruit cup saves the day
192	8	34	Know who you are dealing with
194	8	35	Selecting the right contractor and defining the rules saves money
197	9	36	High-fire light-off destroys two boilers
214	9	37	Failed instrument tubing could have led to catastrophic results
216	9	38	The inexactness of a detailed procedure
218	9	39	Near-misses screaming for attention
221	9	40	Troubling tuning job
227	10	41	Missing death by moments
233	10	42	Poor commissioning creates problems later
243	10	43	Proper flame detector commissioning is vital
246	10	44	Paint oven explosion shrink-wraps trucks
250	10	45	Small pilot gas valve leaks cause big problems
252	10	46	Where do we start digging up the gas line?

STORY TITLE, STORY NUMBER, CHAPTER AND PAGE NUMBERS (*Continued*)

Page	Chapter	Story #	Story Title
255	11	47	Sometimes the obvious is not so obvious
261	11	48	Quality craftsmanship?
269	12	49	When is a flood a flood?
273	12	50	Failing furnace gave plenty of warning signs
280	12	51	Getting hot and flipping your lid

REAL-LIFE STORIES BY KEYWORDS

Story #	Story Title	Keyword1	Keyword2
1	Innocent lives lost from a hot water heater explosion	Hot water heater	Hazard Recognition
2	The bulge in the boiler firebox?	FA ratio	boiler
3	But it was inspected recently!	Inspections	boiler
4	Personal criminal liability at work?	Due diligence	Personal Liability
5	The power of propane	Training	Emergency Response
6	Propane and natural gas systems don't mix well	Fuel switching	System design
7	No one expected a hydrogen explosion	Purging	Boiler cleaning
8	Boiler operations, headaches, and being left alone don't mix well!	Carbon Monoxide	Hazard Recognition
9	A school system in a fog	Carbon Monoxide	System design

REAL-LIFE STORIES BY KEYWORDS (*Continued*)

Story #	Story Title	Keyword1	Keyword2
10	Flame impingement can be deadly	Flame impingement	Hazard Recognition
11	Black smoke from a stack is never a good thing	FA ratio	System design
12	Wrong valve pressure rating could have led to many deaths	System design	Valve selection
13	Never disrespect a gas leak	Gas leak	Emergency Response
14	A small gas smell uncovers big problems	Gas leak	Hazard Recognition
15	My friend's changed life	Purging	Hazard Recognition
16	Nonidentical twins	Purging	Procedures
17	Fixing a leak becomes a big problem	pressure test	System design
18	Nitrogen-filled vessel work ends in death	Purging	Nitrogen asphyxiation
19	Only one whiff of pure nitrogen can kill you	Purging	Nitrogen asphyxiation
20	Seemingly simple gas line cleaning costs lives	Purging	Procedures
21	BMS evolution	BMS	System design
22	Failed light-off attempts result in death	Procedures	Purging of furnace
23	A failed tuning job destroyed a boiler	Boiler	FA ratio

REAL-LIFE STORIES BY KEYWORDS (*Continued*)

Story #	Story Title	Keyword1	Keyword2
24	Ten sailors die in a boiler steam accident	Steam	Fasteners
25	Water hammer causes extreme damage	Piping	steam water hammer
26	Culture and design cost a person's life	System design	Hazard Recognition
27	A death caused by human error when troubleshooting a boiler	Troubleshooting	Boiler
28	Explosion caused by a leaking valve during repair	Valve leaking	Blinds/Isolation
29	Another explosion from a leaking valve during a repair	Valve leaking	Blinds/Isolation
30	Yet another explosion from a leaking valve during a repair	Valve leaking	Blinds/Isolation
31	Poor communications and human factor issues lead to a major explosion	FA ratio	Procedures
32	A tricky switch	Training	Procedures
33	A fruit cup saves the day	System design	Emergency Response
34	Know who you are dealing with	Procedures	Training
35	Selecting the right contractor and defining the rules saves money	FA ratio	boiler
36	High-fire light-off destroys two boilers	Contractor qualifications	boiler
37	Failed instrument tubing could have led to catastrophic results	Procedures	Hazard Recognition

REAL-LIFE STORIES BY KEYWORDS (*Continued*)

Story #	Story Title	Keyword1	Keyword2
38	The inexactness of a detailed procedure	Procedures	Hazard Recognition
39	Near-misses screaming for attention	Contractor qualifications	FA ratio
40	Troubling tuning job	Contractor qualifications	FA ratio
41	Missing death by moments	Interlock testing	Thermal oxidizer
42	Poor commissioning creates problems later	Commissioning	oven
43	Proper flame detector commissioning is vital	Commissioning	boiler
44	Paint oven explosion shrink-wraps trucks	Valve leakage testing	Hazard Recognition
45	Small pilot gas valve leaks cause big problems	Valve leakage testing	Hazard Recognition
46	Where do we start digging up the gas line?	Valve maintenance	Hazard Recognition
47	Sometimes the obvious is not so obvious	Training	Hazard Recognition
48	Quality craftsmanship?	FA ratio	Hazard Recognition
49	When is a flood a flood?	Water damage	Hazard Recognition
50	Failing furnace gave plenty of warning signs	Training	Hazard Recognition
51	Getting hot and flipping your lid	Refractory	Hazard Recognition

Index

www.ingramcontent.com/pod-product-compliance
Lightning Source LLC
Chambersburg PA
CBHW072111250125
20788CB00003B/32

* 9 7 8 0 4 7 0 5 3 3 6 0 4 *